About Island Press

Since 1984, the nonprofit organization Island Press has been stimulating, shaping, and communicating ideas that are essential for solving environmental problems worldwide. With more than 1,000 titles in print and some 30 new releases each year, we are the nation's leading publisher on environmental issues. We identify innovative thinkers and emerging trends in the environmental field. We work with world-renowned experts and authors to develop cross-disciplinary solutions to environmental challenges.

Island Press designs and executes educational campaigns in conjunction with our authors to communicate their critical messages in print, in person, and online using the latest technologies, innovative programs, and the media. Our goal is to reach targeted audiences—scientists, policymakers, environmental advocates, urban planners, the media, and concerned citizens—with information that can be used to create the framework for long-term ecological health and human well-being.

Island Press gratefully acknowledges major support of our work by The Agua Fund, The Andrew W. Mellon Foundation, The Bobolink Foundation, The Curtis and Edith Munson Foundation, Forrest C. and Frances H. Lattner Foundation, The JPB Foundation, The Kresge Foundation, The Oram Foundation, Inc., The Overbrook Foundation, The S.D. Bechtel, Jr. Foundation, The Summit Charitable Foundation, Inc., and many other generous supporters.

The opinions expressed in this book are those of the author(s) and do not necessarily reflect the views of our supporters.

Can a City Be Sustainable?

Other Worldwatch Books

State of the World 1984 through *2015*
(an annual report on progress toward a sustainable society)

Vital Signs 1992 through *2003* and *2005* through *2015*
(a report on the trends that are shaping our future)

Saving the Planet
Lester R. Brown
Christopher Flavin
Sandra Postel

How Much Is Enough?
Alan Thein Durning

Last Oasis
Sandra Postel

Full House
Lester R. Brown
Hal Kane

Power Surge
Christopher Flavin
Nicholas Lenssen

Who Will Feed China?
Lester R. Brown

Tough Choices
Lester R. Brown

Fighting for Survival
Michael Renner

The Natural Wealth of Nations
David Malin Roodman

Life Out of Bounds
Chris Bright

Beyond Malthus
Lester R. Brown
Gary Gardner
Brian Halweil

Pillar of Sand
Sandra Postel

Vanishing Borders
Hilary French

Eat Here
Brian Halweil

Inspiring Progress
Gary Gardner

Can a City Be Sustainable?

Gary Gardner, Tom Prugh, and Michael Renner, *Project Directors*

Betsy Agar
Perinaz Bhada-Tata
Peter Calthorpe
Alexander Carius
Andrew Cumbers
Geoffrey Davison
Robert Doyle
Richard Friend
Sudhir Gota
Pablo Knobel Guelar
Richard Heinberg

Brian Holland
Daniel Hoornweg
Cornie Huizenga
Jim Jarvie
Madhavi Joshi
Martí Boada Juncà
Gregory H. Kats
Anna Larsson
Haibing Ma
Sanskriti Menon
Sean O'Donoghue

Karl Peet
Simone Ariane Pflaum
Ang Wei Ping
Debra Roberts
Gregor Robertson
Kartikeya Sarabhai
Franziska Schreiber
Kristina Solheim
Juan Wei
Peter Wrenfelt
Roser Maneja Zaragoza

Lisa Mastny, *Editor*

ISLANDPRESS

Washington | Covelo | London

For additional related and updated content, videos, infographics, data, blog posts, and links, visit www.canacitybesustainable.org.

Worldwatch Institute Board of Directors

Worldwatch Institute Staff

Worldwatch Institute Fellows, Advisers, and Consultants

Contents

POLITICS, EQUITY, AND LIVABILITY

List of Boxes, Tables, and Figures

BOXES

TABLES

FIGURES

Units of measure throughout this book are metric unless common usage dictates otherwise.

Foreword

This is a book about hope. A story about possibilities. The sort of story we might entertain while floating in warm water at the edge of the ocean—where life feels more clear, and big ideas rise to the surface. A moment to contemplate what really matters, and to reflect, and dream, and plot. That's what this book is for me.

Clearly, we have a lot to contemplate. Our recent narrative is full of troubling themes, connected and compounding: growing inequality; an increasingly dangerous climate; depleting resources and endangered ecosystems; cultural, ideological, and geopolitical battles intrinsically linked to all of these problems. So complex and daunting are our challenges that it is easy to feel disconnected, powerless.

But this is only part of our story. Reflecting on the state of our ever-changing world requires us to consider not just how things are, but how we would like them to be, and an honest assessment of the course we are on.

This book shares and inspires stories about our potential—about how we can create a more livable and sustainable world—just as much as it reviews our past. It paints a clear picture that we are far from powerless. Across the globe, the narrative of a troubled world is being challenged by stories of leadership, change, and achievement. A growing global movement stands eager to work for the world we want for our children. These stories point to the power of people to craft more thoughtful, sustainable approaches to life on Earth, and our opportunity to write a new narrative of which we can all be proud.

The power of inspired leadership was on full display recently at the United Nations Climate Change Conference in Paris in December 2015. With luck (and work), we may look back on the agreement struck there among nearly 200 nations to tackle the threat of climate change as a defining moment in our history—a bold, collaborative step toward addressing one of the central challenges of our time. The collective leadership of many individuals made that moment possible and will determine our path forward.

The leadership of cities, in particular, played an enormous role in motivating the global agreement that was struck in Paris. Hundreds of mayors from around the world brought bold local goals and action commitments to the negotiations, setting the bar for leadership and demonstrating a readiness to help negotiating nations meet more-ambitious targets.

Speaking in Paris, United Nations Secretary-General Ban Ki-moon recognized the role of city leaders in helping to "inspire national governments to act more boldly." He noted that cities "have formed a remarkable number of alliances that are accelerating and scaling up climate commitments" and recognized cities for "taking [their] leadership to a new level of cooperation and innovation. . . . [They] are the ones who will help turn this global agreement into reality on the ground."

Although the heightened attention to cities during the Paris talks was to some extent a new evolution in the process of international climate negotiations, leaders at the local level have been working to tackle climate change for 25 years. Toronto, Canada, was among the first cities globally to adopt a greenhouse gas reduction target in 1990. The international network of local governments, ICLEI, was founded that same year and launched the Urban CO_2 Reduction Project, which 14 international cities joined in 1991.

From Portland to Ankara, Minneapolis to Helsinki, this was the original vanguard of climate action planning—individuals working at the local level who recognized the need and opportunity to tackle a major global challenge through cumulative work across continents. More and more urban leaders joined their thinking, recognizing that all emissions are generated locally, and that many opportunities to reduce them are under local control. They worked together to develop the methods that cities have used ever since to measure greenhouse gas emissions, establish targets, and craft action plans. Their small ideological movement produced a community of practice.

For the first 20 years that cities began to focus on climate, their numbers grew, but the practice evolved slowly. Targets were comfortably set 15–20 years out, with aims to reduce climate pollution by similar percentages. The exercise was largely internal and independent, focusing on cost-effective actions. Political pressure to achieve adopted goals was generally low. City leaders seldom spoke of the need to adapt to climate change, out of fear that this would create an excuse for inaction.

The local movement grew substantially in 2005 when Seattle Mayor Greg Nickels launched the Mayors Climate Protection Agreement, challenging other U.S. mayors to lead the way in reducing emissions in the face of federal inaction. More than 1,000 mayors, representing nearly 89 million Americans, joined in making the pledge to take climate action.

Today, momentum is building again, spurred in part by the growing pressures of climate change that cities already are feeling. A new wave of collaboration, innovation, and learning is taking place among leaders at the local level. Cities are listening to the science and establishing more-ambitious goals for reducing emissions. Carbon neutrality. Fossil fuel-free. 100 percent renewable energy. These goals change the equation about prioritizing local action.

Now we recognize that solving the climate challenge will require sweeping societal change. Climate leaders are learning that we must work hard on enabling more-equitable access to resources, security, and the power to make and advocate for lower-carbon choices—as hard as we work on the technical nature of the problem. Rather than avoid talk of climate adaptation, we should seek to enhance local preparedness and resilience while reducing emissions. Local action takes place in the local context, and real progress happens when we solve multiple challenges at once. Cities understand this.

I recently spent five years leading sustainability efforts for the City of Oakland, California. Oakland is a case study of the transformations and leadership emerging in cities on these issues. It is a city eager to take bold action, where the most vocal champions of action on urban sustainability and climate represent diverse communities of color whose work is framed primarily through a social justice lens. Climate action is deeply intertwined with other opportunities to enhance quality of life.

Early in my first year on the job, a colleague from the City of Seattle suggested that we get together with other "sustainability" professionals at the government level to see what we could learn from one other. Sixty-five of us met in Chicago in 2009, sharing stories about issues like energy retrofits, zero waste, and green infrastructure, as well as how we were each making the case for and coordinating this work within our respective bureaucracies. We agreed to talk regularly as a group, and to meet again in a year. The Urban Sustainability Directors Network (USDN) was born—the next wave of collaborative local leadership.

Seven years later, nearly 600 local government professionals participate in USDN. They share common traits as change agents tasked with helping their jurisdictions identify and act on opportunities to enhance sustainability. They are trying new things, learning, sharing, inspiring, pushing, celebrating, and evolving. Peer learning and collaboration has emerged as a powerful force in the network, built on trusted relationships. The dialogue has broadened and deepened across 25 streams of more-nuanced group discussions on topics such as climate preparedness, building energy strategies, electric-vehicle-infrastructure planning, and addressing equity in sustainability initiatives.

In cities, the urban sustainability movement is much more than concurrent. It embodies a degree of collaboration that is distinct from many social movements—and that creates space for new leadership and big ideas.

One such idea spurred the 2014 launch of the Carbon Neutral Cities Alliance, a collaboration of international cities committed to achieving ambitious carbon reduction goals of at least 80 percent by 2050 or sooner. These cities are working together to explore how to enact the transformative change necessary to achieve those goals. Some have already proven the value of bold action. From Portland's adoption of an urban growth boundary to concentrate development, to Copenhagen's decision to convert core downtown arteries to bicycle and pedestrian travel, cities in the Alliance are taking some of the boldest action on the planet to address climate change and enhance sustainability.

The stories of this book are inspiring. They offer a narrative of progress and hope. They illustrate that we are not powerless to affect the state of our world. And amid complex topics, they point to simple truths: if we are willing to dream and be bold, if we are willing to work and to work together, if we are willing to embrace and build upon the narrative of leadership that is emerging throughout the world, then a more sustainable future is in our reach.

Garrett Fitzgerald
Strategic Partnerships Advisor, Urban Sustainability Directors Network

Foreword

Cities are synonymous with civilization—in fact, they are the foundation of it. They have always been the major arenas within which high human culture has evolved and flourished, and, since the beginning of the scientific age, they also have been the engines of our expanding knowledge of the planet, its ecosystem, and our place within them. Cities rightly stand as beacons of hope and inspiration to millions and will continue to grow in the coming decades as people on every continent migrate to urban environments seeking better lives.

So it is fitting that cities—where more than half of humanity now lives—are poised on the cutting edge of our attempts to face and master the multiple crises of sustainability that threaten civilization itself. Cities are at a crossroads, confronting historic challenges posed by rising populations, accelerating climate change, increasing inequity, and—all too often—faltering livability.

Fortunately, as this report—the 33rd volume in the Worldwatch Institute's *State of the World* series—abundantly illustrates, cities around the planet are stepping forward to lead their citizenry and to support each other in addressing these challenges and in building the sustainable societies of the future. These are not isolated, solitary efforts; vigorous undertakings are plentiful on every continent and within every category of city, from small to vast, from relatively poor to wealthy, and from ancient to sparkling new. Everywhere you look, cities are striving to achieve smaller greenhouse gas footprints, healthier and less alienating communities, more inclusive governance systems, and greater equity and fairness for all their inhabitants.

Cities also have joined forces to share experiences and solutions via peer-to-peer networks and to help shape policy at the level of their host nations, as well as internationally. The global organization that I am privileged to chair, the C40 Cities Climate Leadership Group, includes more than 80 of the world's major urban settlements, accounting for over 550 million people and one-quarter of global economic activity. Another organization, ICLEI–Local Governments for Sustainability, is more than 25 years old now and boasts 1,200 member cities. The Compact of Mayors, a coalition of city leaders focused on climate change and its impacts, includes the leadership of nearly 400 cities representing almost 350 million people. Organizations such as the Urban Sustainability Directors Network and STAR Communities have sprung up with continental-scale portfolios to promote and support sustainability progress at a more local level.

These and other organizations are both creating and deploying a suite of tools and policy options that is rich, adaptable, plentiful, and designed to address the thorniest sustainability problems. Their determined experimentation with these tools clearly shows that sustainability not only is achievable, but, in many cases, also can save money in the long term.

Cities are where most people now live, and will live in the coming decades. Moreover, because they offer access to all the best that human civilization has achieved, cities are where most people *want* to live. Mayors know that better than anyone. They also know that cities are policy laboratories and have more freedom to innovate than national governments, and that cities are more directly in touch with their citizens and the impacts that sustainability problems—and successes—have on all of us. Building on the new hope created by the breakthrough agreement on climate action achieved in Paris last December, cities stand ready to engage their citizens in building a sustainable future.

State of the World: Can a City Be Sustainable? deeply understands these fundamental facts and assembles an inspiring collection of analyses, stories, examples, and policy options into a vision of a sustainable future that is within our grasp. I urge my fellow mayors and all urbanites to commit themselves to the actions necessary to achieve it.

Eduardo da Costa Paes
Mayor, Rio de Janeiro
Chair, C40 Cities Climate Leadership Group

Acknowledgments

If this book were a city, it would be described as diverse and dynamic, maybe even sprawling. But above all, it would be known for its passionate people. Here we offer our deepest thanks to the many people whose hard work helped to bring this book to life.

We are grateful to our dedicated Board of Directors for their tremendous support and leadership: Ed Groark, Robert Charles Friese, John Robbins, Mike Biddle, Tom Crain, James Dehlsen, Edith Eddy, Ping He, Stefan Mueller, David Orr, Scott Schotter, and Richard Swanson, in addition to our Emeritus Directors, Øystein Dahle and Abderrahman Khene.

We also acknowledge, with deep gratitude, the many institutional funders whose support made the Institute's work possible over the past year. La Caixa Banking Foundation deserves early mention for its double role in supporting this volume: as a generous funder and as the matchmaker that introduced us to the excellent researchers at the Autonomous University of Barcelona, whose work appears in these pages. We are grateful for La Caixa's enthusiastic support of our efforts.

In addition, a host of institutions makes work across the entire Institute possible. We are grateful for the support and confidence of the following: 1772 Foundation; The Aiyer Family Fund of Vanguard Charitable; Ray C. Anderson Foundation; Asian Development Bank (ADB); Aspen Business Center Foundation; Caribbean Community (CARICOM); Collins Educational Foundation; Del Mar Global Trust; Ecoworks Foundation; Folk Works Fund of Fidelity Charitable; The Friese Family Fund; Garfield Foundation, Brian and Bina Garfield, Trustees; German Federal Ministry for the Environment, Nature Conservation and Nuclear Safety (BMU) and the International Climate Initiative; German Society for International Cooperation (GIZ) with Meister Consultants Group, Inc.; Goldman Environmental Prize; J. W. Harper Charitable Fund of Schwab Charitable; Hitz Charitable Fund of Schwab Charitable; Steven Leuthold Family Foundation; MOM's Organic Market; National

Renewable Energy Laboratory (NREL), U.S. Department of Energy; Network for Good; New Horizon Foundation; Paul and Antje Newhagen Foundation of the Silicon Valley Community Foundation; V. Kann Rasmussen Foundation; Robert Rauschenberg Foundation; Renewable Energy Policy Network for the 21st Century (REN21); Serendipity Foundation; Shenandoah Foundation; Sudanshu, Lori & Anand Family Fund of the Silicon Valley Community Foundation; The Laney Thornton Foundation; Turner Foundation, Inc.; United Nations Foundation; U.S. Agency for International Development (USAID) with Deloitte Consulting LLP; Wallace Global Fund; Johanette Wallerstein Institute; Weeden Foundation Davies Fund; and White Pine Fund of Fidelity Charitable.

Friends of Worldwatch offer vital support of the Institute and provide budget stability that assists our financial planning. Dedicated to creating a sustainable civilization, this core group of readers is critical to achieving the Institute's mission.

For this urban edition of *State of the World*, the Institute welcomes submissions from a wide range of authors, all of whom contribute atop the many pressures of their own work. We are grateful for insightful contributions from Betsy Agar, Perinaz Bhada-Tata, Peter Calthorpe, Alexander Carius, Andrew Cumbers, Geoffrey Davison, Richard Friend, Sudhir Gota, Pablo Knobel Guelar, Richard Heinberg, Brian Holland, Daniel Hoornweg, Cornie Huizenga, Jim Jarvie, Madhavi Joshi, Martí Boada Juncà, Gregory H. Kats, Anna Larsson, Haibing Ma, Sanskriti Menon, Sean O'Donoghue, Karl Peet, Simone Ariane Pflaum, Ang Wei Ping, Debra Roberts, Kartikeya Sarabhai, Franziska Schreiber, Kristina Solheim, Juan Wei, Peter Wrenfelt, and Roser Maneja Zaragoza. Their expertise and insights add depth and special value to the book. The Honorable Eduardo Paes, Mayor of Rio de Janeiro, Brazil, and Garrett Fitzgerald, Strategic Partnerships Advisor at the Urban Sustainability Directors Network, were kind enough to cap the book with Forewords. Two other big-city mayors and their sustainability staffs—Gregor Robertson of Vancouver, Canada and Robert Doyle of Melbourne, Australia—graciously agreed to contribute perspectives on the challenges and achievements of their respective cities.

We are also fortunate to belong to a generous community of scholars and activists who shared their knowledge with us and introduced us to other experts. Garrett Fitzgerald along with Hilari Varnadore, Executive Director of STAR Communities, deserve special thanks for orienting us to the world of urban research and advocacy. We also benefited from the expertise of many people who either provided critical insights in discussions with us or who connected us to the expert authors contributing to this book. They include Adam Beck, Maruxa Cardama, Brenden Carriker, Lena Chan, Felix Creutzig, Ruth DeFries, John Fernandez, Tomasz Filipczuk, Marina Fischer-Kowalski, Colin Hughes, Sadhu Johnston, Christopher Kennedy, Cecile Legrand, Matthew Lynch, Jacob Mason, Laurie Mazur, Amanda McCuaig, Leanne Mitchell, Ranjan Nambiar, Danielle Nierenberg, Alexander Ochs, Cathy Oke, Johanna Partin, Andrea Reimer, Kartikeya Sarabhai, David Sedlak, Karen Seto, Chris Smaje, Michael Small, Doug Smith, Sean Sweeney, Jason Vogel, Kristi Wamstad-Evans, and Sandy Wiggins.

Monika Zimmermann, Deputy Secretary General at ICLEI–Local Governments for Sustainability, connected the project team with a range of "City View" authors and provided critical insights from her work. Preeti Shroff-Mehta, Worldwatch India Program Senior Fellow, was an energetic intermediary, strengthening our budding collaboration with the Centre for Environment Education (CEE) in Ahmedabad. We benefited from early discussions with CEE Director Kartikeya Sarabhai, who also contributed to the book.

We are particularly indebted to Worldwatch interns Shashank Gouri and Kristina Solheim for their diligence in digging out obscure information and checking facts for the book. Their thorough and cheerful approach to research made working with them a pleasure.

State of the World is ably edited by Lisa Mastny, who quickly and skillfully sharpens authors' writing and harmonizes their diverse styles. Lisa also manages the production process as a key point person between authors and designer. We greatly value her skill in ensuring that the book is in good shape and delivered on time. Independent designer Lyle Rosbotham showcases the written word through exceptional design, creating elegant graphics and a beautiful layout. And Kate Mertes faithfully and quickly creates an accurate index that makes the book highly accessible.

Once the book is produced, Worldwatch Marketing and Communications Director Gaelle Gourmelon disseminates its messages far beyond our Washington offices and advises staff on the effective use of new communication tools. Director of Finance and Administration Barbara Fallin manages the many details of Institute operations with great efficiency and dispatch. And Director of Institutional Relations Mary Redfern keeps our staff apprised of funding opportunities and manages our relationships with foundations, ever on the lookout for new opportunities for Worldwatch.

We continue to benefit from a fruitful partnership with our publisher, Island Press, which is globally recognized as a first-rate sustainability publishing house. We appreciate the professional and collegial efforts of Emily Turner Davis, Maureen Gately, Jaime Jennings, Julie Marshall, David Miller, Sharis Simonian, and the rest of the IP team.

Worldwatch's publishing partners extend our global reach through their work in translation, outreach, and distribution of the book. We give special thanks to Worldwatch Brasil; Paper Tiger Publishing House (Bulgaria); China Social Sciences Press; Worldwatch Institute Europe (Denmark); Organization Earth (Greece); Earth Day Foundation (Hungary); Centre for Environment Education (India); WWF-Italia and Edizioni Ambiente (Italy); Worldwatch Japan; Korea Green Foundation Doyosae (South Korea); FUHEM Ecosocial and Icaria Editorial (Spain); Taiwan Watch Institute; and Turkiye Erozyonla Mucadele, Agaclandima ve Dogal Varliklari Koruma Vakfi (TEMA), and Kultur Yayinlari IsTurk Limited Sirketi (Turkey).

We are particularly appreciative of the special efforts made by individuals to advance our work overseas, typically on a volunteer basis. Gianfranco Bologna is the force behind the Italian edition of *State of the World*. We have been lucky to enjoy his gracious hospitality on our visits there for more than two decades. Eduardo Athayde is an indefatigable promoter of Worldwatch in Brazil. Meanwhile, Soki Oda labors tirelessly over Japanese translations of the volume. We are grateful for his careful review of our work.

Finally, we tip our hats to the many cities and urban-interest organizations that have shown courageous leadership on climate issues over the past decade and more. In an era when many national governments would not embrace their climate responsibilities, cities have pointed the way forward. Their example and their advocacy helped make possible the 2015 climate agreement in Paris, which keeps alive the hope of a stabilized climate for our world's people. For this gift, we are deeply grateful.

Gary Gardner, Tom Prugh, and Michael Renner
Project Directors
Worldwatch Institute
1400 16th Street, NW, Suite 430
Washington, DC 20036
worldwatch@worldwatch.org
www.worldwatch.org
www.canacitybesustainable.org

World's Cities at a Glance

Gary Gardner

Cities have emerged as the dominant form of human settlement, and they are major economic and environmental actors. The data that follow give a sense of cities as a global phenomenon and of their place in human civilization in the twenty-first century.

People

Since 1950, the global urban population has increased by roughly a factor of five, from 0.7 billion in 1950 to 3.9 billion in 2014. It is expected to increase by another 60 percent by 2050, when 6.3 billion people are projected to live in urban settlements.[1]

As of 2009, more than one-half of the world's people live in cities (see Figure, page xxviii), and the urbanization trend is continuing. More than 90 percent of urban growth is happening in developing countries, although not all developing regions are majority-urban yet. By 2040, all world regions, including Africa, will be majority-urban.[2]

Urban growth rates are stable or slow in highly urbanized regions such as Europe, Latin America, and Oceania, but Asia and Africa are urbanizing quickly. The fastest urban growth is in Africa, where growth rates in some countries exceed 5 percent per year. Europe has the world's lowest urban growth, and in some Eastern European countries, rates are actually negative.[3]

Over the past 65 years, the number of "megacities"—cities with 10 million or more inhabitants—has grown more than 14-fold, from 2 in 1950 to 29 in 2015. (See Map, pages xxx–xxxi.) By 2030, the world is projected to have 41

Gary Gardner is Director of Publications at the Worldwatch Institute.

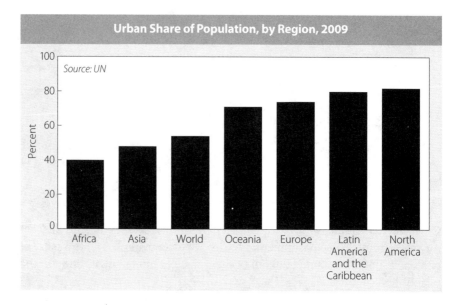

Urban Share of Population, by Region, 2009

Source: UN

megacities. But nearly half of all urbanites live in cities of fewer than 500,000 people. The number of cities with more than 500,000 people has grown nearly sixfold since 1950, from 304 to 1,729.[4]

The Built Environment

The built-up land of cities covers 1–3 percent of global land area, but this could grow to 4–5 percent by 2050 as urban areas expand outward, primarily into prime agricultural land.[5]

Cities are becoming less dense: for decades, across all world regions, the urban land area has expanded faster than the population. If average densities continue to decline, the built-up areas of developing-country cities will increase threefold by 2030 while their populations double. (See Table.) Industrialized-country cities are projected to expand 150 percent while their populations increase by 20 percent. An estimated 60 percent of the built environment needed to accommodate the earth's urban population by 2050 is not yet built.[6]

Household sizes are falling in many countries, which is contributing to an increase in the number of dwellings and the resources required to build them. By 2025, the growth in the number of households is projected to be 2.3 times the population growth rate in the world's top cities. The construction industry

Expected Increase in Area and Population of Cities by 2030		
	Projected Increase by 2030 in	
	Built-up Area	Population
	percent	
Developing-country cities	200	100
Industrialized-country cities	150	20

Source: See endnote 6.

is a major consumer of resources, including 40 percent of all water, 70 percent of timber products, and 45 percent of energy.[7]

By one estimate, cities will need to double their annual investment in physical capital to $20 trillion annually by 2025, most of this in emerging economies.[8]

Urban Economies

Cities are economic engines: some 80 percent of the global gross domestic product (GDP) is produced in cities, and 60 percent is produced in the 600 most-productive cities, where one-fifth of the world's population now lives. Urban economic activity accounts for up to 55 percent of gross national product (GNP) in low-income countries, 73 percent in middle-income countries, and 85 percent in high-income countries. Cities generate a disproportionate amount of revenue for governments.[9]

Urban areas account for a large share of global consumption, including 60–80 percent of energy consumption and more than 75 percent of natural resource consumption. They account for 75 percent of the world's carbon emissions.[10]

Economic power is increasing in cities in emerging economies. By 2025, many of the cities that currently are in the world's wealthiest countries will not even make the list of the 600 richest cities (in terms of GDP) as new cities—in China, the Democratic Republic of the Congo, Nigeria, Indonesia, Pakistan, and India, among other countries—displace them.[11]

An estimated 1 billion people will become part of the global "consuming class" by 2025. They are expected to inject $20 trillion of additional spending annually into the global economy.[12]

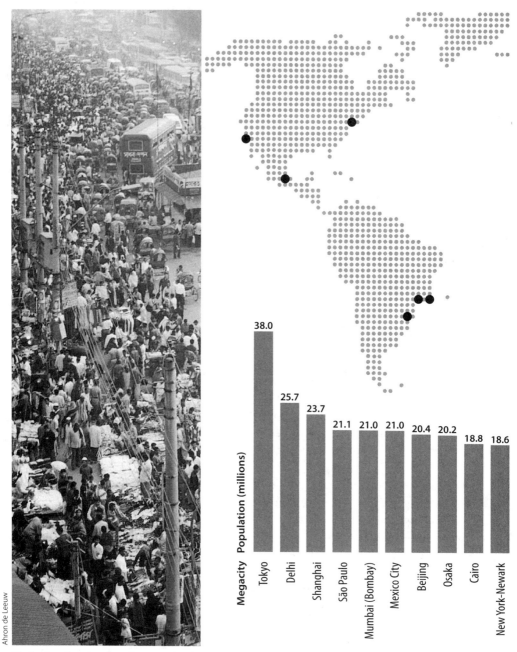

Ahron de Leeuw

Dhaka, Bangladesh, 17.6 million people.

Megacity Population (millions)

Megacity	Population (millions)
Tokyo	38.0
Delhi	25.7
Shanghai	23.7
São Paulo	21.1
Mumbai (Bombay)	21.0
Mexico City	21.0
Beijing	20.4
Osaka	20.2
Cairo	18.8
New York-Newark	18.6

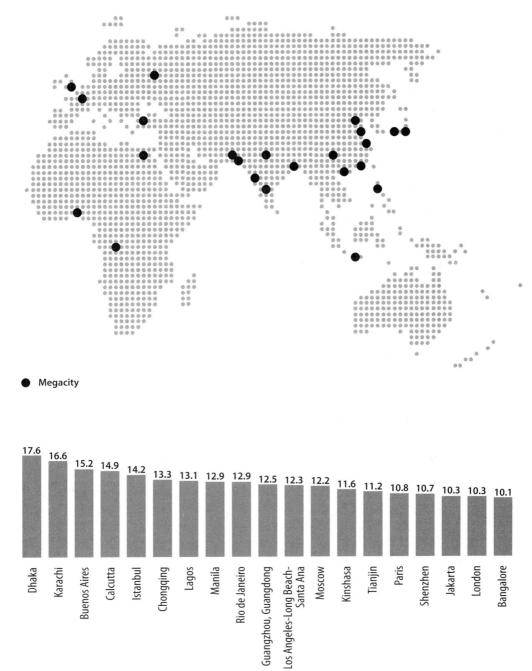

● Megacity

Consumption in the lowest- and highest-consuming megacities differs by a factor of 28 in energy per capita, a factor of 23 in water per capita, a factor of 19 in waste production per capita, a factor of 35 in total steel consumption, and a factor of 6 in total cement consumption. Ten percent of the urban population of developing countries lacks access to electricity, and 18 percent uses wood, dung, or charcoal for cooking. The figures are much higher for urban populations in the least-developed countries.[13]

Poverty, Sanitation, and Health

Although cities are economic engines, they also can be centers of poverty. Approximately 1 in 7 people in urban areas lives in poverty, mostly in informal settlements of the developing world. An estimated 863 million urban residents were living in slum conditions in 2012, up from 650 million in 1990. Yet the overall share of urban populations living in slums fell over this period, from 46 percent to 33 percent.[14]

Less than 35 percent of cities in developing countries treat their wastewater. About 500 million urban dwellers worldwide share sanitation facilities with other households. More than 170 million urban residents lack access to even the simplest latrine and have no choice but to eliminate their waste in the open.[15]

Some 1.5 billion urban dwellers face levels of outdoor air pollution that exceed the maximum recommended limits. In 2012, outdoor air pollution killed an estimated 7 million people, representing 1 in 8 deaths globally and making air pollution the largest single environmental health risk. Meanwhile, indoor air pollution (from the burning of solid fuels for cooking) killed an estimated 4 million people in 2014.[16]

Cities as Human Constructs

CHAPTER 1

Imagining a Sustainable City

Gary Gardner

Describing a sustainable city is no easy task. Cities differ in geography, climate, culture, history, wealth, and a host of other dimensions, each of which precludes any possibility of a one-size-fits-all approach to urban sustainability. A sustainable Riyadh will look and operate differently from a sustainable Reykjavik, because of their disparate climates, among other distinctions. In addition, no mature models of urban sustainability are available today, anywhere on the planet. And even at the definitional level, there is little agreement about what constitutes a sustainable city. Although many of the necessary technologies and policies are well known, recipes for creating a fully sustainable city have not been developed, much less implemented.[1]

Because of these uncertainties, describing a sustainable city is, to some extent, an exercise in imagination. The paragraphs that follow are one possible product of such a visioning exercise.

Imagine a city 20 years in the future, perhaps in Europe, Japan, or North America, that is well on track to becoming the first sustainable city in the world. When it launched its strategic plan for sustainability in 2016, it unfurled the most ambitious sustainability effort ever seen. In this imagined future, you are a 40-year-old accountant and mother of two.

The bedside alarm beeps insistently, nudging you into Monday morning. You surrender to it, emerging from bed into a short shower. Becoming resource-aware was a challenge for you and your neighbors after citizens approved the Our City, Remade strategic plan. But over time, you and your fellow citizens have matured into a world of resource limits, having shed your parents' no-tomorrow approach

Gary Gardner is Director of Publications at the Worldwatch Institute.

to resource use and their misplaced attachment to consumption. Your internalized ethic of restraint gives you the bearing of, well, an adult. You wear it well.

Teeth brushed and fully dressed, you head to the kitchen through your living room, lights illuminating the way automatically as sensors detect your presence. The apartment is snug, with two bedrooms, a small office, a kitchen, a living room, and a balcony. But for you and your spouse, it works well now that "stuff" is kept to a stress-free minimum, and given the common space you share with neighbors: your two kids spend the bulk of their play time downstairs with neighbor children on the nearly traffic-free street, where the occasional car must inch its way through an obstacle course of benches and planters.

The apartment complies with standards set by the city's 100% Renewable Energy initiative, which promotes high levels of efficiency and conservation and is supported by an annual increase in fossil energy prices. The city's energy conservation program helped your landlord swap out inefficient windows and install solar panels and solar water heaters—he had little choice, really, given the large increase in fossil fuel prices. Today, the city has nearly eliminated fossil fuel use, and your energy consumption, at about half its previous level, can now be accommodated by the city's stock of renewable energy.

You walk the little ones to the school three blocks away, engaged in their chatter about today's field trip to the nearby greenway, 1 of 17 large wildlife corridors that radiate from the city's center to its periphery. Rich in habitat and feeding spots for birds, butterflies, frogs, squirrels, and other wildlife, the corridors are an integral part of the city's infrastructure. As extensions of local classrooms, the corridors host field labs for the kids' nature course (they will observe tadpoles today!). The corridors are also recreational havens, featuring trails for hiking and biking, fitness courses, picnic areas, and wildlife education placards. The lush, park-like radials are crisscrossed by green chains of vegetated roofs, community gardens, ponds, street landscaping, and other hubs of natural activity, creating a network of nature that is deeply integrated into city functions. The 17 radials serve as natural flood channels and recharge areas for city aquifers, absorbing the now-torrential rains generated by a changed climate and saving the municipality millions of dollars in construction costs for wastewater conduits and ever-deeper wells.

Arriving at the school, you kiss the kids goodbye and hop on the streetcar to continue on to work, nose in your tablet. Three kilometers down the line, you get off, pull a city bike from the rack, and pedal the last kilometer to the office. Home to office is just 25 minutes, even with the school stop—15 minutes faster than the same trip made by car years ago. New taxes on gasoline and parking had made driving unviable, yet now you rarely miss the car. Between the streetcar, biking, *walking, and car sharing, you have transportation options for every need. And given the city's new emphasis on mixing businesses and residences, core goods and services are often just steps away. Your waistline is smaller and your wallet is fatter without the car, insurance, gas, and maintenance expenses. Above all, your new commute is a calming experience, not a stressful one, as it puts you in touch with the people, sights, and smells of your neighborhood.*

Yours is a full life, with family, work, civic activities, and volunteer work crowding your calendar. Yet most of your daily activities happen within two kilometers of home. The Dense Community, Vibrant Community land-use initiative has brought together more people in neighborhoods across the city, stimulating economic transactions and stronger community ties. Neighborhood outlets meet all of your food needs, most of your recreational and social needs, and a great many of your repair and supply needs. You can easily go one month without traveling more than five kilometers from your home, yet you hardly feel trapped—the wide variety of offerings and extensive social connections within that circle keep you stimulated and alive.

After a six-hour day at work (your hours are reduced through job sharing, giving you more family time while increasing employment), you reverse the morning commute: bike, streetcar, walk. But at the streetcar station, you pause to peruse the offerings at the farm stand, grabbing some fresh vegetables, pasta, and a loaf of bread for dinner and tucking them into the canvas bag that accompanies you everywhere. (No meat today—that once-a-week pleasure is applauded by your doctor, who likes your cholesterol numbers, and by the city's Pollution Control Board, which celebrates lowered greenhouse gas emissions from its Meatless Weekdays program.) Your bounty today is nearly free because you've racked up credit from trading in your homemade compost. The farmer, a local who tends vegetables on three formerly abandoned city lots, values the compost for its structure and organic matter. You value the organic vegetables.

The rhythm of home and work life continues throughout the week, with changes each day to your post-work routine. On Tuesday, you take your toaster in to have its frayed cord fixed. Gone are the days when you would toss out an appliance in favor of a new one, repair now being more affordable than purchasing following the enactment of the citizen-approved Materials Tax, which made metal, plastic, wood, and other materials more expensive relative to labor. Many downtown retailers have evolved into repair shops. The modern culture of repair has renewed an old tradition: handing down household goods to one's children, often over multiple generations. Widely admired are the householders whose goods are old and fully functional—sturdy iron can openers and hand egg beaters from the 1920s, for example, or solid oak tables and chairs kept in good repair. Prized as expressions of resource stewardship, these goods are daily reminders of the new materials ethic at the core of your sustainable city today.

On Wednesday, you remind the kids to take out the discard can. Tomorrow is discard pick-up day for the spring quarter. The city's No Fill for Landfills initiative has cut landfill waste by 93 percent in two decades. Discarded packaging and other waste has been largely eliminated, thanks to a Producer Take-Back initiative that holds companies responsible for any waste associated with their products, giving firms a strong incentive to reduce packaging. It helps that you have developed a new sensitivity to throwing things away: the thought of using a paper towel or paper bag (remember them?) once and tossing it in the trash— your unthinking daily habit years ago—now prompts recoil. "Waste" generation has been reduced so greatly that the city has sold off its fleet of garbage trucks, instead renting small pickups from the car sharing company every three months to collect residual materials. Nearly all of this is recycled.

On Thursday night, you send the kids to a neighbor's apartment to work on homework as you and your spouse head out to a meeting at the kids' school. The facility is bustling with community and civic initiatives. An adult basketball league has games under way in the gym, young and old pump iron in the weight room, and meetings of the historical society, the district music club, and cooking classes are in progress in the classrooms. You and your spouse head to the auditorium for the Budget Consultation meeting—the chance for your district to provide comments about the proposed city budget. You are particularly excited to float some ideas for your district's Community Grant, funds that you and your neighbors can spend as you determine.

Late Sunday afternoon, the family takes its weekly promenade, strolling 10 minutes to the plaza at the district center, a favorite gathering place for people from nearby neighborhoods. You treat everyone to ice cream, but the evening is focused on people more than purchases. The kids soon are surrounded by friends, laughing as they play jacks or hopscotch on grids defined by the plaza paving stones. Parents discuss politics and sports with friends. A music ensemble plays in a corner of the plaza, the notes floating across the square on the warm summer evening. Couples dance to the tunes.

Heading home, your week coming to a close, you lag a few steps behind the family, lost in thought. How much has changed since the Our City, Remade strategic plan was launched 20 years ago! How impossible it all had seemed when the new sustainability goals were approved, with great trepidation, after a contentious campaign. Yet how much richer your life is today! You ponder the irony: less has led to more, living leaner is living richer. Sure, the city still has its challenges, but the great restraint that governs city life somehow has made it more prosperous for more people than ever before. Indeed:

- *Gone is the excess, the wasteful use of so much. In its place is resource stewardship and a deep appreciation for civic resources of all kinds.*
- *Gone is frivolous and thoughtless purchasing. In its place is a restraining ethic characterized by the question, "Will this make my life better?"*
- *Gone is pollution, a noxious sort of waste. In its place is an ethic of cleanliness that extends from the family to industry and the city as a whole.*
- *Gone is homelessness, hunger, and most material poverty. In its place is an ethic of equality and dignity—that every person has value and a place in the community.*
- *Gone is the anonymity of the big city, even as the city has grown through in-migration. In its place are strong and diverse district communities.*

You catch up with the family and turn the corner to your apartment building, energized for a new week.

This imaginary city clearly has made a strong effort in the direction of sustainability. But is it enough? Without a defined set of yardsticks, the answer is unclear. Some analyses—such as the study that Mistra Urban Futures undertook to calculate the lifestyle changes required in Gothenburg, Sweden, to reduce annual greenhouse gas emissions to two tons per person—give results that look much like the lifestyle of our protagonist. But an analysis such as that of Vancouver, Canada, which uses an "ecological footprint" methodology, would restrict our protagonist still further: no meat, and no travel by plane. Other analyses, such as that of the Deep Decarbonization Pathways Project, suggest that keeping global temperature rise to 2 degrees Celsius or less this century is possible but will require aggressive actions immediately. Thus, much work remains to develop a toolkit that allows cities to measure and chart a path to sustainability.[2]

The situation is complicated further by the different sustainability requirements for wealthy and developing countries. Wealthy countries, with the infrastructure and prerequisites for a dignified life already in place, need to shrink their use of fossil energy and materials enough to allow developing countries to expand theirs. If our protagonist's city were in a developing country, her week would be filled with expansion: first of infrastructure, including schools, clinics, transport, parks, and sports facilities, and second of income-generating opportunities that, in turn, would boost consumption to levels required for a dignified life. In sum, while sustainability in our protagonist's imagined city required a degree of scaling back and slowing down, her cousin's poor city across the ocean requires faster economic growth and consumption to lift all citizens to stable lives, even as it also pursues greater efficiency. Thus, the path to sustainability is context-dependent.

Given the variability of approaches to sustainability, this volume does not attempt to prescribe a single path to a sustainable city. Instead, it lays out ideas for moving in the direction of urban sustainability, toward cities that, in their broad outlines, look like our imagined city, with renewables supplying nearly all energy, waste nearly eliminated as a circular economy takes hold, prominent attention to the "people side" of sustainable cities—health, education, jobs, and equity—and a repurposing of modern life away from consumerism. The details will be different in cities worldwide, but most of the prescriptions in this volume head in these general directions.

The book is divided into three main sections. The first, "Cities as Human Constructs," is meant to help readers understand what cities are and how they function. It reviews the historical evolution of cities, examines important urban systems, and elaborates principles for a sustainable city, presenting in a more formal way the ideas lived out by our protagonist above. The section closes with a reality check by Richard Heinberg, who postulates circumstances that could force sustainability on cities by shrinking them to a manageable size in an energy-constrained world.

The second section describes a range of efforts to meet the climate challenge in cities, identifying energy, buildings, waste, transportation, and deforestation as areas that offer the potential for large reductions in greenhouse gas emissions. Although actual gains in these sectors are not yet in the range that would make a city fully sustainable, the section offers numerous policy ideas that could help us move, live, and work much more cleanly. Many of these ideas already are being implemented in cities worldwide, and more cities can be encouraged to replicate the most ambitious and successful of these initiatives.

The third section broadens the lens to consider a number of other issues that are important for urban sustainability, including social justice, biodiversity, and the "remunicipalization" of select urban functions, such as power and water. These are important urban dimensions that are not always at the forefront of the discussion of sustainable cities. The remunicipalization chapter, for example, makes the point that a greater degree of public control of utilities increases the possibility that the public interest is reflected in the provision of some of the most basic city services.

Scattered throughout the volume are a set of "City View" profiles highlighting the sustainability efforts of diverse cities worldwide. All are inspiring and contain measures that could be adopted or adapted for use in other cities seeking a sustainable path. Freiburg, Germany, for example, has taken a wide range of steps to reduce its footprint, while providing a high quality of life to residents. And Jerusalem, Israel, has made considerable effort to maintain its green space and to protect biological diversity within city limits.

But lurking behind each success story is a nagging question: Are these cities doing enough? Have their efforts delivered them to the doorstep of true sustainability? In a world that requires huge reductions in carbon emissions, waste, and materials use, and equally large increases in renewable energy use and in material and energy efficiency, the answer would seem to be "No, not yet." This is not to be discouraging: new initiatives can build on the gains described in these profiles and multiply their benefits. But it is sobering to note that no city can be content with current achievements, no matter how impressive. The successes described in the City View profiles are launching points for a new round of efforts, rather than crowning achievements.

Cities today are in an exciting position to take leadership on the preeminent challenge of our era: the effort to build sustainable economies. Cities are, after all, where many sustainability issues are lived out—where most commerce, energy use, production, and other modern drivers of unsustainability (and one day, of sustainability) take place. Just as important, cities are where people are most likely to understand and engage sustainability concerns, where discussion is no longer abstract but becomes grounded and real. People care about their cities and often are motivated to protect and improve their urban homes. Cities can harness that passion to help advance a sustainability agenda, perhaps more easily than national governments or corporations can. Indeed, cities may be our best hope for shifting economies in a sustainable direction. This volume is offered in the hope that it will motivate citizens and policy makers to create sustainable cities worldwide.

CHAPTER 2

Cities in the Arc of Human History: A Materials Perspective

Gary Gardner

In the first decade of this new century, humans passed a historic threshold when half of us were estimated to be living in cities. We became, for the first time, a predominantly urban species. Our journey toward *homo urbanis* over the past 12,000 years or so was driven by a series of social, environmental, and technological innovations that expanded the materials and energy available to humans to make city living possible and attractive.[1]

Some of these innovations, however, also make today's urban centers unsustainable. Based largely on virgin, non-renewable resources and typically offering only limited opportunities for the poor, no city today stands out as a sustainability success. If cities are the preferred human habitat into the distant future, urban forms and human practices will need to change markedly. Some scholars argue that the sustainable cities of the future will constitute a new urban form, as different from today's metropolises as today's are from pre-industrial cities.[2]

History is an important teacher for cities seeking to become sustainable. It offers a long-range framework that highlights the non-viability of modern economies built on throwaway fossil fuels and wasteful materials use. Yet a large share of the world's people aspires to the development status made possible by modern industrial economies, suggesting that shifting to a sustainable urban development path could be particularly difficult. How cities might offer a dignified life for all, in harmony with nature, remains an open question. But the historical record suggests that materials and energy will have a large role in shaping whether and how such a goal is achieved.

Gary Gardner is Director of Publications at the Worldwatch Institute.

The Urban Base: Energy and Materials

Cities are engines of material transformation. Using tremendous quantities of energy, they digest materials for use in the diverse set of human activities that build cultures, from business and education to medicine and recreation. The fundamental role of energy and materials in shaping society and culture was clear to anthropologist Leslie White, who observed in the 1940s that "culture evolves as the amount of energy harnessed per capita per year is increased," or as societies become more efficient at putting energy to work. In other words, the available quantities and quality of energy, materials, and technology are elemental in shaping a city and setting boundaries for its growth.[3]

For this reason, it is helpful to look at urban flows of materials and energy—a city's metabolism—to understand its environmental footprint and its potential for future development. (See Box 2–1.) Particularly revealing is a historical frame in which metabolism is examined over three great periods: hunter-gatherers (who predate cities by hundreds of thousands of years but whose lifestyle creates a benchmark against which to judge urban materials use); agrarian societies dating back roughly 10,000 years; and industrial societies that have emerged since 1750. Each is a "metabolic regime" with a distinct energy and materials signature that carries important economic and social consequences. (See Figure 2–1.)[4]

The extended time frame throws into relief the enormous energy and

---

Box 2–1. Socio-metabolism as an Analytical Tool

Scholars have long sought to divide human history into a set of overarching narratives that explain the changing way that humans organize themselves, such as Karl Marx's modes of production and Herbert Spencer's stages of civilization. In the 1990s, German historian Rolf Peter Sieferle introduced a new analytical frame, socio-metabolism, which uses flows of energy and materials to understand societal structure and functioning. Like the metabolism of an organism, in which energy is used to digest food for the organism's development (with waste as a byproduct), societies tap energy to transform materials for societal advance, generating waste in the process.

The value of socio-metabolic analysis is its capacity to link societal development with environmental resources, an especially useful service in an era of growing resource scarcity. By highlighting a society's natural boundaries for development, socio-metabolism offers insights into the sustainability prospects of societies.

Source: See endnote 4.

materials appetite of modern cities. It also demonstrates that, at 250 years of age, today's industrial regime is quite young and, while spreading rapidly, is not yet the dominant societal form today. A large share of people today does not live in the industrial mode and instead leads lives that, materially speaking, are closer to those of pre-industrial peoples.

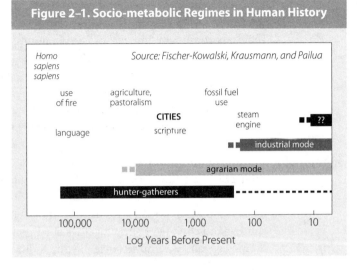

Figure 2–1. Socio-metabolic Regimes in Human History

Source: Fischer-Kowalski, Krausmann, and Pailua

The still-limited breadth of industrial society sets up a critical question for sustainability: What are the resource and environmental implications as people who live at relatively simple levels move into lives of greater comfort and convenience that are characteristic of many urbanites in industrial societies?

The focus here on energy and materials does not lessen the importance of non-material dimensions of city life. The cultural, social, and spiritual dimensions of human life find strong expression in cities and often are a key reason people move there. But for the purposes of considering a city's sustainability, energy and materials are fundamental, as they help to set the broad resource and environmental parameters that determine whether a city can endure for the long term.

Seeds of Urban Living: Hunter-Gatherers

Hunter-gatherers, our early ancestors, did not live in cities, but their footprint sets a materials and energy baseline against which to compare the eventual emergence of urban lifestyles. Hunter-gatherers also displayed early human preferences and tendencies that gained fuller expression much later in cities, revealing, perhaps, some of the most fundamental motivations for living together in cities.

Acquiring all of their energy from biomass and making little effort to cultivate plants or raise animals to augment what nature provided, hunter-gatherers' energy demand per person—largely from securing food and firewood—amounted to only two to four times the basic metabolic rate of their bodies. Their environmental impact was tiny, amounting to less than 0.01 percent of the annual output of biomass (net primary production) of their habitat. Such a low-impact way of living has been called a passive solar existence, meaning that early humans relied on the unmodified byproducts of solar energy—plants and animals—to subsist. But a passive solar existence is also low-yield, and hunter-gatherers needed very large areas of land to support them, around 25 square kilometers per person.[5]

However materially simple their lives were, our ancestors stood apart from animals in their adoption of practices that pointed to their distant future as city dwellers. They lived together in groups of 20–50 persons, invariably fostering a social dimension to their lives as they undertook communal activities such as hunting and cooking. Eating together at a fireplace furthered the development of language and a complex social life. In addition, early humans gathered with others beyond their own clans for cultural and other purposes. Historian Lewis Mumford writes that ceremonial meeting places such as sacred groves, great stones and trees, and caves featuring Paleolithic paintings foreshadow religious and economic functions that would be fulfilled thousands of years later in cities.[6]

Still far from being urbanites, these prehistoric humans nevertheless were transitioning from a passive solar existence to a lifestyle governed increasingly by a controlled solar energy system—agriculture—in which they intervened in the production of plants and the reproduction of animals. Sophistication in these practices eventually would give them, after thousands of years, the material conditions needed for the creation of city life. (See Box 2–2.)[7]

Emerging Urban Life: Agrarian-based Societies

Conditions for cultivation became more favorable as the climate warmed with the retreat of the last Ice Age and as humans became more clever at using the materials and energy sources around them. Seeds from grasses began to be collected, and herd animals—oxen, sheep, donkeys, and horses—began to be employed to work the land, allowing ever-larger plots to be used. The larger population of animals, in turn, meant a growing supply of manure, which was used to fertilize plants, a recycling advance that helped to make villages self-sustaining.[8]

The important effect of these practices, which eventually became the Agricultural Revolution, was to make more food energy available per year and per hectare. This allowed population density to rise—a small but significant

Box 2–2. Density: The Law of Human Attraction

Even before humans had the energy sources and technologies to build cities, they demonstrated a propensity to live more closely to each other. A 2007 study of data for 339 hunter-gatherer societies showed that as the population of hunter-gatherer groups grew, they also became more densely settled. For every 100 percent increase in population, the home ranges of hunter-gatherer groups increased only 70 percent. Science writer Tim de Chant summarizes the finding this way:

> Every additional person requires less land than the previous one. That's an important statement. Not only does it say we're hardwired for density, it also says a group becomes . . . more efficient at extracting resources from the land every time their population doubles. Each successive doubling in turn frees up . . . more resources to be directed towards something other than hunting and gathering. In other words, complex societies didn't just evolve as a way to cope with high-density—they evolved in part because of high density.

Driving this adaptive capacity is the capacity of social networks to facilitate exchanges of materials and information, which leads to economies of scale. These findings may help to explain humans' growing tendency over several millennia to live and work increasingly closely together.

Source: See endnote 7.

---

step toward urban life. Put differently, a new socio-metabolic order was being born, in which the relationship between humans and their environment was changed fundamentally. Cultivated land, grazing land, and meadows replaced forests and other natural landscapes as humans learned to "colonize" nature, appropriating more energy and materials in the process.[9]

The result was a huge jump in the energy and materials use of agrarian societies compared to their hunter-gatherer ancestors. Animal raising played a big role in this jump: when agricultural families owned several large animals per person, the animals' biomass consumption was at least an order of magnitude larger than the biomass used by their human keepers. In energy terms, families with animals used 20–80 gigajoules per person of biomass each year, compared with the 10–20 gigajoules per person for hunter-gatherers. And the share of plant biomass that agriculturalists consumed from the land that they worked rose dramatically as well, to as much as 75 percent, compared with less than 1 percent for hunter-gatherers. (See Table 2–1.) By any measure, the Agricultural Revolution ushered in a huge increase in the human demand for materials and energy.[10]

Table 2–1. Metabolic Profiles of Hunter-Gatherers and Agrarian Society		
Dimension	Hunter-Gatherers	Agrarian Society
Energy use per person (gigajoules per person per year)	10–20	40–70
Materials use per person (tons per person per year)	0.5–1	3–6
Population density (people per square kilometer)	0.025–0.115	Up to 40
Agricultural share of population (percent)	—	More than 80
Biomass share of energy use (percent)	More than 99	More than 95

Source: See endnote 10.

In the agrarian regime, increased energy use was land-intensive. Providing heat requires firewood from woodlands, food comes from cropland, and draft power depends on grassland. (Water and wind power were useful for transport, but they typically account for only a small share of the primary energy supply.) Thus, the quality of each land type and the level of intensification, through the addition of labor, helps to determine agrarian productivity. Overall, the efficiency of converting primary energy (from the sun) into final energy—the energy used by agrarian people—is low. It is high enough to generate the material surpluses needed to create city life, but low enough to limit the size and growth potential of cities.[11]

The Emergence of Cities

The development of cities required human control over the supply of food; labor power from a large, fertile territory; as well as convenient waterways for transport and trade. These conditions took time to develop—in many places, thousands of years would pass between the advent of plant cultivation and the flourishing of cities. But with agriculture in place, the urban share of human populations grew steadily, starting roughly 10,000 years ago. (See Figure 2–2.)[12]

When they did emerge, cities were not simply villages on a large scale—they generated new ways of organizing human life. This is clear in the ruins of ancient cities, which often feature three large buildings—the palace, the temple, and the granary—representing political, religious, and economic power and located within the citadel, the seat of military power. These functions are institutionalized in ways that did not occur under the hunter-gatherer regime.

Whereas hunter-gathers answered to family and clan, the urbanite carried additional, weighty obligations to political, religious, and military authorities. Crop production was no longer merely a source of food— its surplus represented a source of economic power, facilitating the construction of buildings, the raising of armies, and the employment of priests, artisans, merchants, and traders. Thus, the agrarian regime's deep penetration of nature, through the appropriation

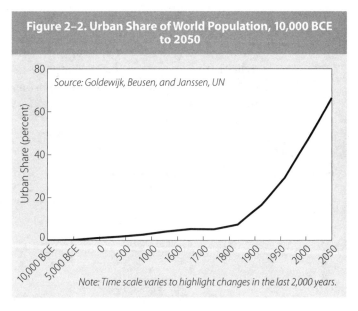

Figure 2–2. Urban Share of World Population, 10,000 BCE to 2050

Source: Goldewijk, Beusen, and Janssen, UN

Note: Time scale varies to highlight changes in the last 2,000 years.

of energy and materials in quantities greater than hunter-gatherers ever had achieved, help to spawn a new social order.[13]

Many cities claim to be the oldest in the world, but the claims are difficult to adjudicate. It is clear, however, that the Fertile Crescent in the Middle East is rich in archaeological ruins of ancient cities. Modern-day Egypt, Iraq, Israel, and Syria are all home to cities dating back a few thousand years before the common era (BCE). Farther east, India and China also host ancient cities.

Cities grew in size over time, with a strong correlation between energy production and city size, according to archaeologist Ian Morris. Measuring energy in calories per person per day (which makes the values comparable to a human energy intake of roughly 2,000 calories per day) and studying cities in the global east as well as the west, Morris establishes energy thresholds for various levels of city size. (See Table 2–2.) He finds that larger cities tend to correlate with more-complex societies: the largest cities, for example, tend to be the administrative centers in the ancient world. Examples include Memphis in ancient Egypt and Anyang in China.[14]

A host of new technologies helped to create, and were made possible by, the new human settlement form that was cities. Mumford observes that the cultivation of grain on a large scale, as well as the development of the plow, the potter's wheel, the sailboat, the draw loom, copper metallurgy, abstract mathematics, astronomical observation, the calendar, and writing and record keeping all came about during the agrarian regime. This was, he writes, "a singular technological expansion of human power whose only parallel is the change that has taken place in our own time. In both cases, men, suddenly exalted, behaved like gods: but with little sense of their latent human limitations."[15]

Table 2–2. Energy Levels Associated with City Size		
Calorie Production per Person per Day	**Status of Human Habitat or City Population Size**	**Time Period**
7,000–8,000	Settlements begin to grow noticeably	3500–3000 BCE (west) 2000–1500 BCE (east)
11,000–12000	Widespread urbanization begins	3500–3000 BCE (west) 2000–1500 BCE (east)
20,000	100,000	1000–0 BCE
27,000	500,000 to 1 million	1000–0 BCE
26,000–29,500	200,000 to 1 million	500–1000 CE (east)
45,000	Many millions	Since 1800 CE

Source: See endnote 14.

The difference between then and now, Mumford observes, is that technology in ancient times was applied to projects of a social nature: great irrigation systems, temples, palaces, and the like. In contrast, innovation today is spread widely across many projects, often originating in the private sector and frequently with no central guiding purpose. The result is city centers that are more diverse and whose power is more diffuse than in ancient times.[16]

The increase in energy and materials flows in agrarian societies also brought a surge in environmental and social challenges to early societies. Soil degradation, such as the salinization of agricultural fields, became a critical problem in Mesopotamia, where it reduced grain production by almost two-thirds between 2400 BCE and 1700 BCE, leading once-great Sumerian cities to shrink to villages or to be abandoned altogether. Organizing agricultural labor and setting rules for the use of common resources, such as water, were other challenges. Although greater dominance of nature gave agrarian societies new capabilities and new opportunities for urban development, it also spelled trouble if resource use was not managed carefully.[17]

Obstacles to Agrarian-era City Growth

Although urban living was an impressive leap over village living and the hunter-gatherer lifestyle that preceded it, agrarian-era cities faced fundamental obstacles to growth in the form of limited energy and labor. Biomass is characterized by low energy density, and this, combined with limited technologies for converting the resource into usable energy, constrained the surplus available in a given area for developing urban centers.[18]

Labor was another constraint that affected the cost of transport. Waterways were cost-efficient for long-distance transport, but overland transport involved prohibitive labor and energy costs after just a few kilometers of hauling. Marina Fischer-Kowalski and her colleagues have calculated that a village of 100 persons would need to devote about 7 percent of its population to the transport of biomass to cities, but that this share increases with growth in the radius of a city and surrounding farmland: a collection of 7,300 villages supporting a large city would need to allocate 15 percent of its labor to transport. The growing demand for human and animal energy to cover larger transport distances—and the additional land needed to support these energy sources—makes the production of surpluses for urban use unviable.[19]

Archaeologist Ian Morris also finds energy to be a constraining factor in city size and has developed estimates of population maximums by subsistence regime (with an additional agrarian regime level added). (See Table 2–3.) Only the development of off-farm energy resources in the industrial era could

Table 2–3. Subsistence Regime and Maximum City Size	
Mode of Subsistence	**Maximum City Population**
Pre-state agrarian	10,000
Agrarian states	100,000
Agrarian empires	1 million
Industrial societies	25 million or more?

Source: See endnote 20.

provide the level of surplus output needed to help cities expand to the sizes and densities we know today.[20]

Cities in Full Bloom: The Emerging Industrial Regime

Around 1500, an important shift in the energy base of societies occurred when coal began to be employed to heat homes in England. Coal mines located close to urban centers facilitated the expansion of cloth manufacturing, and an English trade empire emerged. The invention of the steam engine in the early eighteenth century and its application in mining and railroads brought about the Industrial Revolution and, with it, a new socio-metabolic regime.[21]

Technological and institutional developments helped to shape the way materials were used. The invention of gunpowder early in the fourteenth century made walled cities more vulnerable, no longer defensible with a moat and a simple wall. Walls were replaced with complex fortifications, and the citizen-soldier was replaced by hired troops, with consequences for city design. Unlike the simpler walls that previously had protected towns and could be moved outward to accommodate additional growth, the complex fortification walls made infill and vertical growth the preferred response to population increase. Buildings attained heights of five to six stories in Geneva and Paris, and eight to ten stories in Edinburgh.[22]

In the seventeenth century, military needs became important shapers of cities, at least in Europe. Wide boulevards were plowed through neighborhoods, making possible rapid troop movements as well as smooth transit for the aristocracy. Streets, rather than the neighborhood or quarter, became the central focus of planners. The main lesson of the period, in the view of historian Lewis Mumford, remains relevant today: "Once wheeled traffic is treated as the chief concern of planning, there will never be enough space to keep it from becoming congested, or a high enough residential density to provide enough taxes sufficient to cover its exorbitant demands." Mumford concludes that, "The assumed right of the private motor car to go to any place in the city and park anywhere is nothing less than a license to destroy the city."[23]

The versatility of liquid fuels—oil and natural gas—accelerated the development of the industrial regime, revolutionizing transportation with the invention of motor vehicles and air transport. Technological advances also directly shaped cities—from the subway, which first appeared in London in 1863, to the skyscraper, which first appeared in 1885, both of which made urban density more viable. The standard of a flush toilet connected to a public sewer system, a major public health advance, was established by the end of the nineteenth century, although it took time to spread. In the twentieth century, agriculture became highly productive as well, as tractors and industrially produced fertilizers and pesticides increased productivity, which, in turn, released rural populations to migrate to cities (and, in many cases, drove them off their lands), further spurring urban growth.[24]

The industrial regime also spelled huge increases in the production of consumer goods such as cars, furniture, and household appliances, as well as increases in services such as tourism, all of which stimulated demand for resources. Although new technologies also brought greater efficiency in resource use, efficiency gains tended to be more than offset by the increased consumption made possible by lower resource prices, itself a consequence of efficiency.

The importance of this revolution for urban development cannot be overstated, because it opened up new flows of energy and materials. Fridolin Krausmann and his colleagues explain that, in the industrial era, "All of the socio-metabolic constraints stemming from the controlled solar energy system are abolished: Energy turns from a scarce to an abundant resource, labor productivity in agriculture and industry can be increased by orders of magnitude, the energy cost of long-distance transport declines, and the number of people who can be nourished from one unit of land multiplies, allowing for an unprecedented growth of urban agglomerations."[25]

Released from the energy shackles of the agrarian regime, the industrial era constituted a huge jump in societal metabolism, as the use of materials and energy increased by a factor of three to five. Large increases in agricultural output allowed population density to increase more than 10-fold compared with agrarian societies. (See Table 2–4.) Whereas the global harvest of biomass increased 3.6 times between 1900 and 2005, fossil fuels grew by a factor of 12, industrial minerals and ores by a factor of 27, and construction materials by a factor of 34. Overall, global material resource use increased eightfold, nearly twice the rate of global population growth.[26]

The increased use of resources generated high levels of air and water pollution and other degradation, some of which was global in scope, including a warming climate, species losses, and acidification of oceans. These and other environmental challenges have been documented in many ways. Assessment tools such as the Ecological Footprint; the Human Appropriation of Net Primary Productivity; Planetary Boundaries; the Red List of endangered species;

Table 2–4. Metabolic Profiles of Hunter-Gatherers, Agrarian Society, and Industrial Society

Dimension	Hunter-Gatherers	Agrarian Society	Industrial Society
Energy use per capita (gigajoules per person per year)	10–20	40–70	150–400
Materials use per capita (tons per person per year)	0.5–1	3–6	15–25
Population density (people per square kilometer)	0.025–0.115	Up to 40	Up to 400
Agricultural share of population (percent)	—	More than 80	Less than 10
Biomass share of energy use (percent)	More than 99	More than 95	10–30

Source: See endnote 26.

reports from the Intergovernmental Panel on Climate Change; and the Millennium Ecosystem Assessment all point to economic overreach that is linked to industrial society's aggressive use of fossil fuels and materials of all kinds. This overextension suggests that the industrial metabolic regime is only a transitory stage in human economic history. It is ripe for replacement by a different organization of human affairs based on a different relationship between society and nature.[27]

The growing adoption of this new energy and materials regime resulted in very rapid urban growth. In 1800, only five cities had populations greater than 500,000 people: London, Beijing, Guangzhou (Canton), Istanbul (Constantinople), and Paris. A century later, the number was 46, and, today, more than 1,000 cities worldwide have populations exceeding half a million people. Over the past 2,000 years, as energy, materials, and the innovations to exploit them all expanded, cities have grown in size, as indicated by the populations of the largest cities at various historical points. (See Figure 2–3.)[28]

Collision Course: Cities and Planet Earth

Humanity's massive appetite for materials and energy cannot continue indefinitely. Yet consider the staggering fact that *a large share of the global population still lives at the agrarian level of societal metabolism,* the level of pre-industrial urbanites. It is safe to suppose that virtually everyone in this global majority would prefer the lifestyle offered by the industrial regime. But, if developing countries achieve a level of affluence similar to that of modern wealthy

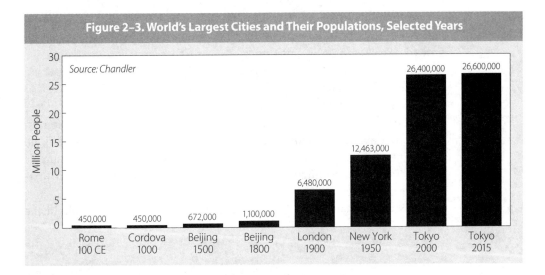

Figure 2-3. World's Largest Cities and Their Populations, Selected Years

Source: Chandler

Million People

Rome 100 CE	Cordova 1000	Beijing 1500	Beijing 1800	London 1900	New York 1950	Tokyo 2000	Tokyo 2015
450,000	450,000	672,000	1,100,000	6,480,000	12,463,000	26,400,000	26,600,000

countries, scholars estimate that *global material use will grow to three to five times its current level*. Herein lies a conundrum: the train that urbanites ride is now too heavy for the rails beneath. Additional passengers—or additional baggage for current passengers—could buckle the tracks and derail the cars. Yet the masses on the station platform clamoring for a seat are our family. What are we to do? [29]

Many look to technology to resolve the dilemma. But socio-metabolic analysis suggests that, unless technology is built and guided by parameters of sustainability, it actually could worsen our plight: over the past 500 years, technological advances are estimated to have increased environmental impact by 1.5 times. Technology often is powerful enough to overwhelm local environments: dragline excavators now remove entire mountaintops to extract coal, and factory trawlers can rapidly reduce regional fish populations. In addition, efficiency gains achieved by technological advance often translate to *increases* in consumption, as new and efficient extraction and production methods make commodities cheaper. Economist Peter Victor observes that while steam engine efficiency increased an impressive 36-fold between 1760 and 1910, this advance was dwarfed by a 2,000-fold increase in the use of steam power over the period. [30]

The same historical analysis of environmental impact shows that of the three widely cited drivers of impact—population, affluence (consumption), and technology—affluence has become the greatest driver in the industrial era, accounting for three times as much environmental impact as population growth. (See Table 2–5.) This finding suggests that gains in redesigning cities to accommodate all people sustainably may need to be found in reducing resource consumption. Many ideas exist today for creating economies that meet human needs with fewer resources—substituting services such as car sharing in place of goods like cars, among other strategies—but reductions in the resource appetites of citizens in high-income cities will be required as well. Returning to the train metaphor, the most direct way to create a sustainable journey may be to reconfigure the train for greater efficiency and to make first-class travel more expensive.[31]

In sum, the challenge is huge. To accommodate all of the people in the

Period	Increase in Environmental Impact	Distribution
1 BCE through 1500 BCE	5-fold	Population and affluence were roughly equally responsible.
From 1500 BCE to present	10-fold	Affluence is responsible for about three times more impact than population growth. Technology increased impact by a factor of 1.5.

Table 2–5. Relative Contribution of Population, Affluence, and Technology to Environmental Impact Over History

Source: See endnote 31.

world who seek an industrial-level urban life will require large reductions in materials use compared to the business-as-usual path. And if technology is to contribute to the solution, strict parameters around its use will be needed to ensure that it does, in fact, help to reduce humanity's footprint overall. Industrial ecology scientists have calculated that 4- to 10-fold reductions in the material and energy footprint of industrial nations are technically possible, which would go a long way toward meeting the reductions required. But no society is gearing up to achieve such reductions, and more will need to transition to reductions in consumption.[32]

Creating sustainable cities for all will require great creativity as well as decidedly lower levels of consumption. Technology undoubtedly will play an important role, but only if it is bounded by sustainability values and is used in service of, in the words of Krausmann and colleagues, a "novel industrial transformation, a transformation that does not build human communication, creativity, and happiness upon gigatonnes and megajoules." Only to the extent that a fourth socio-metabolic regime is created—one that reconfigures the relationship between humans and their energy and material base and respects natural boundaries—will sustainable cities become possible.[33]

CHAPTER 3

The City: A System of Systems

Gary Gardner

Cities are places of human convergence, where people live, work, and play. But beneath the bustle of any city are systems that make these hubs of humanity function. Cities are akin to living things that take in energy, metabolize material, and spit out waste. They consume and grow, using digestive, respiratory, and circulatory systems. And, like living things, cities can, with a nudge from citizens and their leaders, evolve in directions that increase their prospects for survival.

Informed citizens and city leaders understand these systems and how they can be designed to advance human well-being. In a world of ever-rising consumption of energy, materials, food, and water, modern urban systems are ripe for critical review, and for exploration of more-sustainable solutions.

Energy

Cities have voracious appetites for energy, accounting for about three-quarters of the world's direct final energy use in 2005—far more than their 49 percent share of global population that year. They used 82 percent of the world's natural gas, 76 percent of its coal, 63 percent of the oil, and 72 percent of non-biomass renewable energy in 2005. Only the share of biomass energy consumed in cities worldwide, at 24 percent, was less than cities' share of population.[1]

Wealth, urbanization, and energy use tend to rise together, with the wealthiest regions generally being the most urbanized and also using the most energy per person, as shown in Table 3–1. This suggests a conundrum for those interested in building sustainable cities. On one hand, cities are centers of wealth

Gary Gardner is Director of Publications at the Worldwatch Institute.

Region	GDP per Capita	Urban Share of Population	Estimated Urban Energy Use per Person	Urban Share of Total Final Energy Use
	U.S. dollars	percent	gigajoules	percent
North America	42,893	80	235	86
Pacific OECD	35,480	86	107	78
Western Europe	31,217	74	114	81
Central and Eastern Europe	7,401	60	79	72
Latin America	4,973	77	40	85
North Africa and Middle East	4,384	60	72	84
Former Soviet Union	3,566	64	112	78
Other Pacific Asia	3,442	48	61	75
China and Centrally Planned Asia	1,738	42	52	65
Sub-Saharan Africa	907	33	34	54
South Asia	703	29	23	51
World	—	47	—	76

Table 3–1. Urban Direct Final Energy Use by Region, 2005

Source: See endnote 2.

generation and, in principle, centers of opportunity. But, if they also tend to drive up energy use per person, cities become victims of their own success: high levels of per capita energy use have tended to mean greater pollution and adverse health and climate impacts.[2]

The energy appetite of cities is also seen in comparing column 3, the urban share of population, with column 5, the urban share of energy use. In almost every region, the urban share of energy use is higher, suggesting that city living stimulates energy consumption per person. And, moving down the table toward poorer regions, the gap between columns 3 and 5 grows larger: urbanites in poorer cities use much more energy than their energy-impoverished rural cousins. These poorer regions are expected to urbanize fastest in the decades to come and, therefore, could see large increases in energy use.

Cities can break the tight coupling of prosperity and polluting energy in two ways. First, they can replace fossil fuels with clean and renewable sources of energy. Some cities already have declared 100 percent renewable energy to be a goal. (See Chapter 10.) Second, cities can increase their energy efficiency and better conserve energy. Calculations of data in Table 3–1 reveal that North America's gross domestic product (GDP) per capita is 21 percent higher than that of Pacific countries that are members of the Organisation for Economic Co-operation and Development (the OECD, a grouping of wealthy nations), but North America's urban energy use per person is 120 percent higher. The superior performance of the Pacific OECD nations (and Western Europe) should be encouraging to cities that wish to provide a high quality of life without a correspondingly high energy bill.[3]

Energy in cities is used to power the residential, commercial, industrial, and transportation sectors, but the relative appetite of each sector varies with the level of development, among other factors. In Figure 3–1, national and regional data are used as a rough proxy for city data (because most energy is used in cities) to illustrate the relationship between development level and the breakdown in energy use. The Figure shows the clear distinction between OECD and non-OECD patterns of energy use, with developing countries using more than half of their energy in industry as they endeavor to build their economies. Wealthier countries typically add commercial services such as banking and insurance to their national economic portfolio, which raises the commercial share of energy use; the commercial share of OECD energy usage is roughly

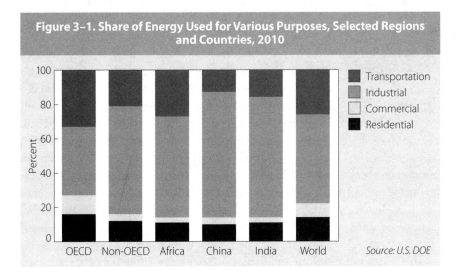

Figure 3–1. Share of Energy Used for Various Purposes, Selected Regions and Countries, 2010

Source: U.S. DOE

triple the level of poorer countries. Also noteworthy is the relatively high share of transportation energy in OECD countries, where cars, an energy-intensive transportation option, are often a large part of the transportation mix.[4]

Drivers of Urban Energy Use

Energy use in cities is driven by several factors. Citizens and leaders can do little to change two of these—their city's geography and climate—but they can respond to them. Developers, for example, can choose "cool" roofs, which reflect sunlight and minimize heat absorption in a building, or "hot" roofs,

Michael Renner

High-density, high-rise condominium building in São Paulo, Brazil.

which absorb sunlight and warm a building, depending on a city's climate profile. Other drivers of urban energy use are more directly manipulable by urban citizens and leaders. These fall into two categories: the built environment and economic and lifestyle factors.[5]

The built environment. Improving the energy efficiency of buildings, through better lighting, heating, and cooling systems, offers tremendous potential for energy savings. (See Chapter 9.) Buildings consume more than one-third of the world's final (end-use) energy; the share can be as high as 80 percent in regions that are highly dependent on traditional biomass. Within buildings, space and water heating account for 60 percent of energy use in cold climates and about 43 percent in mild and warm climates. Electricity use is a smaller share, but is growing quickly: between 1990 and 2010, electricity use increased by 66 percent in wealthier countries and by 320 percent in

poorer countries. Thus, heating, cooling, and electricity are areas ripe for performance upgrades as a way to moderate continued high demand for energy in buildings in the decades ahead. Residential surveys and energy audits as well as updated building codes and efficiency standards are high-payoff tools for efficiency gains. (See Chapter 8.)[6]

The pattern of building and the habitation intensity of a city are also key to energy savings because they greatly influence activity patterns and energy use. Density is important: a minimum of 5,000–15,000 people per square kilometer is cited as a threshold below which energy efficiency declines and energy use increases, especially for transport. Transport systems can be revamped to emphasize public transport and non-motorized transport options for energy savings. Offering alternatives to automobile-based individual transport in urban life could greatly reduce fossil fuel use, and it also reduces pressures for sprawling development and offers a multitude of health benefits. (See Chapter 7.)[7]

Finally, urban energy systems can be better integrated, for example through the use of cogeneration or waste-heat recycling technologies. Centralized electricity generation and separate on-site heat generation, the traditional configuration, have a combined efficiency of about 45 percent, whereas cogeneration can be as much as 80 percent efficient. In the United States, in 2008, cogeneration accounted for only 9 percent of the nation's electricity generating capacity; raising that share to 20 percent by 2030 could lower carbon emissions by the equivalent of removing 109 million cars from U.S. roads, according to Oak Ridge National Laboratory.[8]

Economic and lifestyle factors. A city's economic structure—whether it is dominated by manufacturing or services, for example—has a large bearing on its energy profile. Income and household size can be other predictors of energy use. Income and energy use traditionally have tended to rise together, as people who earn more tend to spend more in ways that use more energy. For example, urbanites who use a car daily, live in large apartments or houses, and consume high levels of livestock products tend to drive up their personal energy use. The pattern is not universally true, however: Beijing and Shanghai have higher rates of energy use than Tokyo, despite lower incomes per person. Meanwhile, when household size is greater than two people, economies of scale set in to reduce energy use per capita.[9]

Finally, the way energy is organized and delivered also affects the level of energy use and greenhouse gas emissions. Energy utilities may not align easily with a city's efficiency goals. Vienna, Austria, for example, owns its electricity, gas, and district heating utilities and thus has greater influence compared to cities with completely privatized and deregulated utilities. (See Chapter 16.) Many industrialized cities have put in place Climate Action Plans, which are expected to reduce or dampen energy use or to promote shifts to renewables in the coming decades, but their success will depend on the links between city government and their energy providers.

Looking Ahead

A 2014 analysis of energy trends in 274 cities projected that, under business-as-usual conditions, urban energy use would balloon more than threefold between 2005 and 2050, from 240 exajoules (EJ) to 730 EJ. However, the study also concluded that cities could limit the increase to 540 EJ in 2050, shaving more than 25 percent off of the projected increase. Key to such success will be for cities to focus on efficiency, conservation, and management of energy demand, rather than on ever greater increases in energy supply.[10]

Materials

Cities today must deal with growing stress on raw material supplies. Extraction of metals, minerals, and fuels is increasingly complex now that the easiest sources have been tapped. As a consequence, researchers and activists regularly call for materials efficiency and conservation, even advocating for "circular economies" in which goods and materials of all kinds are designed for comprehensive reuse or recycling. Increasingly, city leaders will be pressed to build urban economies that do more with less.

City-level data on materials consumption are sketchy and incomplete, but data at the national level give a sense of consumption in rich and poor countries and may be roughly indicative of urban consumption, given that cities consume some 75 percent of all natural resources. Materials flow analysis reveals interesting patterns that city leaders may want to be aware of. On average, each human being consumed 10 tons of materials in 2009, a 25 percent increase over 1980. (See Table 3–2; note that the high rates of contraction for Europe and North America are largely a function of the end year of the analysis, 2009, the first full year of the Great Recession.)[11]

But averages obscure: per capita consumption is 60 times higher in the highest-consuming country than in the lowest-consuming one. This differential suggests that, as poor countries continue to prosper, consumption levels are likely to increase greatly, and global materials use—and the environmental burden it brings—could surge. In an illustrative example, the people of Taipei, Taiwan, consume 30 kilograms of copper per person, with consumption growing at 26 percent per year, whereas residents of Vienna, Austria, use 180 kilograms per person, at a growth rate of just 2 percent per year. As cities in poor countries prosper, the challenge is for wealthy countries to create the environmental and resource space needed for poor cities

Table 3–2. Domestic Material Consumption per Person, by Region, 1980 and 2009			
	Domestic Material Consumption per Person		
Region	1980	2009	Change
	tons per person		percent
Africa	5.0	4.8	- 4.5
Asia	4.9	9.2	87.2
Europe	16.3	13.0	-19.8
Latin America	10.6	13.1	23.4
North America	24.8	20.1	-18.9
Oceania	34.6	35.6	- 2.9
World	7.9	9.9	25.4

Source: See endnote 11.

to prosper, and for poor cities to provide dignified lives to residents on a moderate materials budget.[12]

As city leaders consider how to dampen the appetite for materials, scientific insights suggest that city development may be influenced by a set of predictable, although bendable, relationships. A team of researchers interested in applying scaling laws to urban development has used data covering hundreds of cities from all continents to identify three sets of relationships that they say apply across myriad cities, cultures, and historical periods:

- For *infrastructure*, cities tend toward efficiency as they grow. A doubling of population size tends to produce only an 85 percent increase in sewer lines, power lines, roads, and other infrastructure.[13]
- For *human needs*—measured, for example, by employment, water consumption, electricity consumption, and housing—a doubling of population leads to a doubling of these indicators, a 1:1 relationship.[14]
- For *socioeconomic measures*—including information, innovation, and wealth, but also serious crime and disease—a doubling of population produces roughly a 115 percent increase in these measures, or a 15 percent increase per capita.[15]

This analysis suggests that many urban development variables unfold within roughly a 15 percent variance (above or below) of population growth. Deviations from the 15 percent rule can be viewed as measures of over- or under-performance relative to the expectations for a city's size. Researchers Luis Bettencourt and Geoffrey West note that relatively large deviations (as much as 30 percent) are found for city phenomena with small values, such as number of patents or number of murders, whereas economic variables often have much smaller deviations—less than 10 percent. These insights offer rough benchmarks to city leaders seeking to evaluate their performance relative to cities of similar sizes. But the benchmarks are not sustainability metrics; a city may perform better than similar-sized cities yet still be far from sustainable.[16]

Some scaling-laws analysts hypothesize that these urban dynamics emerge from the networks of connections found in cities, and that these connections are a function of density. Density drives connectedness, which drives innovation, which, in turn, drives the dynamics of urban development. But they worry that innovations in cities must occur at an increasing rate to support continued growth, and that, without increasingly rapid innovations, the indicators of urban advance could slow or stop. Their thinking has emerged within the past decade, and it remains to be seen how critical the role of continuous innovation is in urban prosperity.[17]

Waste

One of the many disadvantages of the linear, use-and-discard pattern of materials use—the model for most industrial economies—is their large streams of waste, much of which ends up in cities. Waste comes in many forms, including municipal solid waste (MSW, or garbage), construction and demolition waste, hazardous waste, and other streams. Data on waste volumes are often scarce or unreliable, but a World Bank study on MSW shows that the volume is huge and growing—and at a faster rate than urbanization. The study also shows that waste generation correlates with affluence, with a nearly fivefold difference in average waste generation per person between the world's richest and poorest regions. (See Table 3–3 and Chapter 13.)[18]

The World Bank study projects that waste levels will increase 69 percent by 2025 over 2012 levels. Other scholars, using growth in income and population projections for various world regions, have projected that the peak in global waste production, under business-as-usual conditions, will not occur before 2100. With more aggressive sustainability policies, the peak could occur around 2075 and could be reduced in intensity by some 30 percent. The

Table 3–3. Municipal Solid Waste Generation per Person, Selected Regions, 2005	
Region	**Waste Generation per Capita**
	kilograms per person per day
South Asia	0.45
Africa	0.65
East Asia and Pacific	0.95
Europe and Central Asia	1.10
Latin America and Caribbean	1.10
Middle East and North Africa	1.10
OECD	2.20

Source: See endnote 18.

projections depend heavily on how waste generation unfolds in sub-Saharan Africa, a region with high population growth rates.[19]

Policies can make a big difference. New Yorkers produce some 1.49 tons of MSW on average, while a Londoner generates only 0.32 tons per capita, about one-fifth as much, in part because of a landfill tax in the United Kingdom. The tax reduced the share of waste landfilled in the country from 80 percent in 2001 to 49 percent in 2010. City leaders will need to consider a wide range of policies to re-engineer their economies away from waste and toward recycling and reuse. (See Chapter 13.)[20]

Despite greater awareness of the need for materials to circulate more broadly, no economy is close to being circular yet. A 2011 study of 60 metals found that, at the global level, only 18 metals had an end-of-life recycling rate (the share of discarded metal that is recycled) exceeding 50 percent. And the United Nations Environment Programme reports that, of the metals found in the 50 million tons of electronic waste produced annually around the world, only 15–20 million tons is recycled.[21]

Food

As growing centers of human population, cities have a robust appetite for food, most of which come from well beyond a city's own limits. A city's food system—the production, processing, distribution, consumption, and waste of its food—has impacts that extend to a city's host region and country, and often to other countries as well. Citizens and their leaders who are concerned about a city's food security and its "foodprint"—its environmental impact related to food—pay attention to the food supply chain, as well as to the accessibility of food for all residents.

Emerging Food Systems

Food systems today are increasingly centralized. A supply chain that once was short, involving small farmers, local distributors, a minimum of processing, and neighborhood grocery stores, has been replaced over a period of decades with large farms, a few global-level distributors, and expansive mega super-markets, even in emerging economies. In Argentina, Brazil, Chile, South Korea, and Taiwan, supermarkets' shares of food sales have grown rapidly, from 10–20 percent in 1990 to 50–60 percent in the early 2000s. And a food item's travel to those supermarkets can be far: some 70 percent of Chilean grapes are produced for export, many to grocery stores in Europe, the United States, and China.[22]

City leaders will want to be aware of the impacts of their city in this increasingly centralized global system. In the United States, for example, concern has arisen about what have controversially been termed "food deserts": areas that lack ready access to fresh, healthy, and affordable food but that often feature convenience stores and fast food outlets. These areas can have clear health impacts. A 2008 study in California found a 20 percent higher prevalence of obesity and a 23 percent higher prevalence of diabetes in populations living near convenience stores, compared with people living near supermarkets and produce vendors.[23]

Status of Nutrition in Cities

Urbanization tends to produce more-diversified diets, including increases in consumption of vegetables and fruits; meat, fish, eggs, and milk; and pulses/oilseeds. It also increases consumption of processed products that are pre-pared at home and of prepared foods from restaurants and other outlets. In sum, urbanization offers opportunities for better and more-diverse nutrition compared with rural diets, along with risks for unhealthful eating. In this con-text, two paradoxical challenges have emerged in many cities.[24]

The first is hunger and undernourishment, especially in rapidly expand-ing cities whose infrastructure and social systems cannot keep up with the influx of migrants. Rapid urban expansion can eliminate farmland along the outskirts of a city that once supplied food, and can overwhelm transportation and logistics systems that bring food into urban markets. These problems can produce volatile food prices and hardships for the poorest. The second challenge is obesity. No longer primarily a wealthy-country problem, obesity is caused in part by the food environment created in cities, which often includes "cheap calories," animal foods, refined grains, sugary drinks, and fast food. City leaders are challenged to develop strategies that address the chief nutritional concerns in their communities.[25]

Food Systems in Different Countries

Cities get their food in different ways, often related to their income level. The cases of Lusaka (Zambia), Bogotá (Colombia), and Manchester (United Kingdom) give a sense of the different food concerns faced by leaders of low-, middle-, and high-income cities. (See Table 3–4.) Hunger is very much a concern in Lusaka, but not in Manchester, whereas Bogotá is working to maintain traditional food supply chains even as supermarket access expands. In each case, food insecurity, poverty, the extent of access to affordable food, and the origins of food carry different weights for policy makers.[26]

City Foodprints

A city's foodprint merits the attention of leaders and citizens because food production is a resource-intensive activity. Nearly 40 percent of our planet's ice-free land surface, and roughly 70 percent of water consumption, is used for agriculture. Moreover, the supply chain for food accounts for 19–29 percent of global carbon emissions. Significantly, the share of emissions associated with postharvest activities is larger in high-income countries, where these activities are more likely to be associated with urban life than in lower-income countries.[27]

A city's foodprint can rise or fall based on several factors, including the kind of food eaten (grain-fed versus grass-fed meat, meat versus vegetables, water-intensive versus less-thirsty crops, etc.), the amount of food wastage, the distance food travels to the city, and other factors.

Consider food type. Urbanization is a major driving force influencing global demand for livestock products. Animal raising is the largest land-use system on Earth, with wide-ranging environmental impacts. The livestock sector occupies some 30 percent of the world's ice-free surface and accounts for one-third of freshwater consumption while using one-third of global cropland to produce feed. The city of Oxford in the United Kingdom has found that it could reduce these dimensions of its foodprint by 30–40 percent.[28]

Table 3–4. Comparison of Food Parameters in Lusaka, Bogotá, and Manchester			
Parameter	Lusaka	Bogotá	Manchester
Share of Income Spent on Food	46 percent	No data, but 28 percent of residents are below the poverty line	12 percent
Level of Food Insecurity	69 percent severely food insecure	33 percent of households face food insecurity	Little food insecurity, but diets often are too heavy in fats and sugars and lacking in fruits and vegetables
Supermarket Share of Sales	10 percent for staples, with most shoppers from high-income households	25 percent	In the U.K. overall, 95 percent of the grocery sector is controlled by supermarkets
Non-supermarket Access to Food	Most meat bought from small shops; eggs, milk, and vegetables often bought from informal sector street sellers	135,000–140,000 small shops, markets, and the informal sector supply most of the food	Traditional groceries are at just a quarter of the level of the 1950s; direct sales, such as through farmers' markets, account for about 1 percent of sales
Where Food Comes From	Most vegetables are produced in the peri-urban area, and most food comes from the greater regional area; 80 percent of processed food is from South Africa	33 percent comes from within or around Bogotá; 44 percent comes from the greater region; 10 percent is imported; in total, 80 percent of staple food is produced within a 300-kilometer radius, and over 60 percent is produced by small-scale farmers	One-third of food is imported from Europe and 20 percent from the rest of the world; food produced in the Greater Manchester area is purchased mostly by centralized supply chains and is distributed nationally

Source: See endnote 26.

Meanwhile, the United Nations Food and Agriculture Organization (FAO) estimates that about one-third of all food produced in the world is wasted, and that this loss occurs increasingly in cities. Urbanization contributes to wastage for a variety of reasons: city dwellers often earn more than rural workers, making waste less economically painful; they buy more food from supermarkets that have high standards for food appearance, leading produce to be rejected; and they live farther than rural people typically do from the source of food.[29]

Of the world's food wastage, 46 percent occurs in downstream activities such as processing, distribution, and consumption, which often occur in and around cities. (The other 54 percent of waste occurs at the farm level or in post-harvest storage.) Overall, consumption waste in middle- and high-income countries—very likely an urban phenomenon because most people in these countries live and consume food in cities—accounts for 31–39 percent of waste in those countries.[30]

Food waste represents a tremendous waste of agricultural resources. Excluding greenhouse gas emissions from land-use change, the carbon footprint of wasted food amounts to about 3.3 gigatons of carbon dioxide (CO_2) equivalent. If food waste were a country, it would rank third in carbon emissions after the United States and China. In addition, according to the FAO, the consumption of surface water and groundwater associated with food wastage totals some 250 cubic kilometers, about the annual discharge of the Volga River. Wasted food is produced on the equivalent of nearly 1.4 billion hectares of land—about 30 percent of the world's agricultural land area. In addition, wastage represents a loss of nitrogen, an energy-intensive input to farming, and phosphorus, a finite element of fertilizer that is critical to highly productive agriculture.[31]

Steven Depolo

Food waste headed to a municipal compost facility.

Because a growing share of food waste occurs in city-based stages of the food system, especially during consumption, cities can be key actors, through waste reduction efforts, in greatly lowering the environmental impact of food production.

Finally, one can consider food kilometers (food miles), or the distance that food travels to a city. Typically, the greater the distance that food travels, the greater the carbon emissions that are associated with transport. In the United States, a 2001 study of the transport of 28 fruits and vegetables found that they traveled an average of about 2,440 kilometers from farm to consumer, whereas locally sourced food traveled only 72 kilometers.[32]

But too much can be made of food kilometers. First, they are a relatively small part of emissions associated with a person's foodprint. In the United

States, transport of food from producer to consumer accounts for only a small share of food-related carbon emissions—about 4 percent, compared with 83 percent that is associated with food production. Second, food sourced at distances can supplement local supplies and provide stability to food markets. In Zambia, for example, crops are largely rain-fed, making production levels variable from year to year. And surveys of households in Lusaka found that only 20 percent reported having enough food to eat during the low season of April to July each year. Thus, food sourced from a distance can play a role in food security in many cities.[33]

Water

City leaders work to ensure that citizens get clean water daily, that sewage is removed, and that stormwater is managed to prevent flooding. But this work is more difficult as cities add new residents and as the context for water and sewer provision shifts away from nineteenth- and twentieth-century priorities, such as ever-increasing supply and the construction of centralized systems. The challenge is particularly great in developing-country cities, which frequently lack the financing or political will to provide systems for purifying the public supply of water, for treating sewage, or both.

A growing number of cities are grappling with inadequate supplies of water. A study of the world's 70 largest cities whose water comes from rivers and reservoirs found that 36 percent do not meet the full demand posed by human, environmental, and agricultural users, and that the share would rise to 44 percent by 2040. Meanwhile, the World Health Organization reports that between 1990 and 2012, the number of urban dwellers without access to an improved drinking water source increased by 34 percent, to 149 million, as water provision in many cities failed to keep up with the rapid pace of urbanization.[34]

A similar trend is found for sanitation, where the number of urban residents without access to improved sanitation increased by 40 percent, to 754 million, over the 1990–2012 period. The United Nations estimates that up to 90 percent of all wastewater in developing countries is untreated and is released directly into rivers, lakes, or the oceans. As cities experience rapid growth, city leaders are challenged to keep up with the growing demand for water and sanitation services.[35]

Yet the context for water and sanitation provision is different today than when most systems were designed. A changing climate, increasing population densities, growing recognition of nature's claims on water, and financing challenges are changing the way that city leaders think about infrastructure and policy for water and sanitation.[36]

Most of today's urban water systems were developed in the twentieth century, when water and energy were cheap and abundant. Water utilities could afford to focus on expanding supply to meet a growing demand

The R. C. Harris Water Treatment Plant in Toronto opened in 1941.

City of Toronto

spurred by population advance and, in many regions, rising prosperity. Planners preferred large, centralized systems because of the economies of scale they provided. With an emphasis on expanding supply, little attention was given to efficiency and conservation. Today, water and energy are increasingly difficult to find, which pressures city leaders to find new sources of water, increase efficiency and conservation, or do both.[37]

But in many cities, "additional" water can be found within city limits, simply through better management. Most cities deliver more water than is billed for, the difference being known as "non-revenue water (NRW)." NRW consists of water lost to leakage, theft, inaccurate metering, or use in city functions, and its share of a city's water can be considerable: in 2010, the average NRW share across more than 1,800 utilities serving half a billion people was about 27 percent. NRW tends to be higher in developing- than in developed-country cities, although many cities in wealthy countries have NRW rates of greater than 20 percent. Reducing the level of NRW would lead to greater available supply (in the case of fixed leaks) or greater revenues (in the case of stymied thievery and accurate metering). Active city attention can make a meaningful difference: Phnom Penh, Cambodia, reduced unaccounted-for water in its system from 72 percent in 1993 to just over 6 percent in 2008.[38]

Cities also might more actively collect rainwater from rooftops or urban catchments as a supplementary source of water. Rainwater harvesting can reduce pressures on rivers, lakes, and other water sources, and it can help to prevent urban flooding by reducing storm flows. In Singapore, nearly all of the urban runoff is harvested for deposit in drinking water reservoirs. (See City View: Singapore, page 211.) This is possible in part because of monsoon rains that deliver heavy quantities in concentrated events. But it also is made possible by reducing the level of contaminants in runoff through regulations governing land use, automobile maintenance, and the use of chemicals on buildings and land.[39]

Another often-untapped source of water for cities is recycled wastewater. Treated wastewater can be used (after additional filtration and disinfection) in limited applications, or treated to a higher level of quality. Only in high-income countries is most wastewater treated. (See Table 3–5.) But even when treated, most wastewater is not reused. In the United States, 75 percent of 85 cubic kilometers of wastewater is treated each year, but only 3.8 percent of the treated wastewater is reused. The National Research Council reports that water discharged to oceans and estuaries in 2005 amounted to about 6 percent of total U.S. water use and about 27 percent of municipal use. Capturing this discharge water would constitute a meaningful increase in the nation's water supply.[40]

Table 3–5. Share of Wastewater Treated, by Country Income Level	
Country Income Level	Share of Wastewater Treated
	percent
High Income	70
Upper-middle Income	38
Lower-middle Income	28
Low Income	8

Source: See endnote 40.

Perhaps the easiest applications of wastewater reuse are at large industrial sites and landscaping of expanses such as parks. These areas can demand large quantities of water, pose little chance of accidental ingestion of water by humans, and often are located close enough to treatment plants to avoid unacceptably high transportation costs. Recycling also is possible at the household level. Household graywater systems capture wastewater that has not come into contact with food or sewage—typically from showers/bathtubs and washing machines—and use this in toilets and on landscaping. But these systems tend to be expensive.[41]

Some cities, typically in areas of water scarcity, exploit the potential of wastewater. In St. Petersburg, Florida, nonpotable wastewater accounted for 40 percent of the city's water use in 2009, largely for irrigation purposes. But in other cities, wastewater becomes part of the drinking water supply. Reclaimed water has supplied almost one-quarter of the water supply in Windhoek, Namibia, home to the oldest direct potable reuse plant in the world. The water agency in Orange County, California, supplies about 227 million liters per day of highly treated sewage water to recharge a drinking water aquifer.[42]

Wastewater also can be used in agriculture, although care is required to ensure that water is fit for such use. In Israel, about 75 percent of effluent goes to agriculture. However, the use of urban wastewater on farms is limited by the cost of transport. Wastewater-fed agriculture is most feasible on farms near a city, perhaps another argument for cities to build up, rather than out, in order to preserve farmland on a city's perimeter.[43]

Desalination—making fresh water from sea water—is another supply option for coastal cities that can afford expensive water and can manage the technology's high requirement of energy. Technological advances have lowered costs enough to make desalination an option to consider for coastal cities facing serious water scarcity. Desalination plants in Israel supply about 17 percent of the country's water supply, and this share is expected to reach 30 percent by 2020. Australia's cities generate about 15 percent of their water from desalination, and, in Perth, the share is about one-half. Perth built a solar energy plant to reduce greenhouse gas emissions that otherwise would have been generated at the energy-intensive desalination plant, and the city used technology to reduce fish kills at the desalination plant's intake pipes. But the desalted water is expensive, the result of some $9 billion in investments over a five-year period. In sum, desalination may boost water supply in select cities, but it must be undertaken with attention to environmental issues, and with residents willing to pay a premium for the resource.[44]

Finally, city leaders must safely remove stormwater. Stormwater infrastructure can be expensive, but many cities are looking at using "green infrastructure"—such as parks, waterways, and wooded areas—to manage water and control flooding. (See Chapter 17.) Aquifers also can be used to store stormwater. These options can obviate the need for constructed infrastructure, at a savings to the city.[45]

Lessons for Cities

This review of key city functions yields some general lessons for city residents and their leaders. First, urban density matters: supplying energy and water to a city generally is cheaper and more efficient, and requires a lighter materials load, when a city is more compact. Second, in contrast to most of the twentieth century, when city leaders focused on increasing the supply of resources, leaders today are challenged to manage demand by delivering more energy, water, or material per unit of resource input. Third, urban form matters for energy and materials, because of the way it affects density and transportation options. Fourth, consumption levels, primarily in wealthy-country cities, generally require moderation, perhaps by substituting services such as car sharing for goods such as cars, or by providing more opportunities for public consumption (in parks, plazas, and municipal swimming pools, for example). Finally, the extent of waste and pollution in each sector can be regarded as measures of failure and as opportunities for increased efficiency. To the extent that cities re-engineer key systems around these lessons, they increase the prospects for the creation of sustainable cities in the decades ahead.

CHAPTER 4

Toward a Vision of Sustainable Cities

Gary Gardner

The path to a sustainable city starts with a vision, a description of a city's future that articulates its aspirations for sustainability. A well-crafted vision can rally public support and mobilize civic energy for a long-term urban makeover that touches virtually every sector of a city. Many cities have produced and published their own sustainability visions. The Aalborg Charter, the Leipzig Charter, Melbourne 2030, and Sustainable Singapore are among the city visions that spell out, in broad, overarching strokes, the general features of their envisioned future. Each is unique, reflecting the particular characteristics and context of their cities.[1]

Taken together, the global collection of visions contains what could be seen as a common set of principles that provide helpful guidance to almost any city, and to urban practitioners within those cities. Seven key principles—covering physical structures, the natural environment, and human needs—can be used to summarize the broad spectrum of areas relevant to sustainability:

1. Reduced, Circulating, and Clean Flows of Materials
2. A Prominent Place for Nature
3. Compact and Connected Patterns of Development
4. Creative Placemaking
5. Centers of Well-being
6. People-centered Development
7. Participatory Governance

These principles might be encapsulated in a concise vision of urban sustainability: *A sustainable city is a vibrant human settlement that provides ample opportunities, in harmony with the natural environment, to create dignified lives*

Gary Gardner is Director of Publications at the Worldwatch Institute.

for all citizens. Although simple in concept, this overarching vision is a challenge to realize.

1. Reduced, Circulating, and Clean Flows of Materials

Perhaps the biggest single step that cities can take toward a sustainable future is to create economies that greatly reduce materials use, (re)circulate most materials, and rely largely on renewable energy. (See Table 4–1.) The challenge is different in wealthy countries—which largely have materials-intensive infrastructure in place—than it is in poorer nations that need additional schools, hospitals, and transportation networks. Scientists have suggested that the needed increase in resource productivity could be huge in wealthy countries—on the order of 80 percent. Developing countries can focus on designing their additional future infrastructure to be as efficient as possible. This challenge is mammoth and will require all the ingenuity and moral strength that citizens and their leaders can muster.[2]

Table 4–1. Reduced, Circulating, and Clean Flows of Materials: A Checklist for Urban Practitioners	
Traditional Practice	**Policies That Point Toward a Sustainable City** (first in wealthy-country cities, eventually in all cities)
Economies based on a linear flow of materials	Circular economy, with essentially zero waste to landfills
Resource efficiency is sufficient to solve environmental problems	Absolute reductions in the use of materials and energy as an essential materials policy metric
Acceptance of high volumes of materials and energy use	Commitment to major reductions—on the order of 4- to 10-fold—in materials use
Encouragement of mass consumption	Emphasis on enhanced well-being, which may involve a reduction in consumption levels
Consumption is overwhelmingly private and material	Consumption is increasingly public, often in civic places, and consisting of services

Reduced Urban Material Footprint

City leaders will need to seek absolute reductions in virgin material and fossil energy use, not just efficiency gains that merely slow the rate of increase in consumption. In recent decades, global resource use has grown 1–2 percent more slowly each year than economic growth, largely in developed-country cities where most infrastructure is already built. But this decoupling has not ended growth in resource use in developed countries, nor does it open up the "ecological space" needed to allow poorer cities to grow. Efficiency gains will be only one part of a much broader materials reduction strategy in cities.[3]

That comprehensive strategy will need to include the creation of "circular economies" that substantially increase the rates of recycling and reuse of metals, plastic, and other materials. In a circular economy, products are designed for durability, disassembly, and refurbishment. Recycling is greatly enhanced. Production is designed to minimize waste, through co-location of factories that can feed off of each others' wastes. (See Chapters 13 and 14.) Stated conceptually, sustainable cities "close nutrient loops," whether those nutrients are technical (the metal and mineral inputs to factories) or biological (the food and yard wastes that are composted for plant growth) in character. Such cities also increase energy recovery.[4]

To meet the goal of an 80 percent increase in resource productivity, much more is needed than higher rates of post-consumption measures such as recycling. Clever strategies, many centered around providing people with services rather than goods, can accelerate reductions in materials use. A good example is car sharing, which offers people private transportation without multiple private cars per family, reducing a person's materials footprint. Tool libraries, such as the one in Berkeley, California, which is a branch of the city's library system, are another example. City residents can choose from hundreds of tools, from power drills to ladders to carpet cutters, providing a less-expensive, and materials saving, alternative to tool ownership. Sustainable cities will feature many more service providers and repairers than are found in cities today.[5]

Cities on the path to sustainability can establish a rational hierarchy of effort for materials reductions. A product study by the Joint Research Centre of the European Commission found that food and drink, transport, and housing were consistently the consumer items with the greatest environmental impact. These economic areas, taken together, accounted for 70–80 percent of the lifecycle impact of the products studied, suggesting that these areas might be prioritized to achieve materials reductions.[6]

Cities also can jumpstart the market for green products by instituting a green procurement policy that reduces waste, conserves natural resources, eliminates the use of toxic materials or pollutants, and promotes the use of recycled content. The policy can apply to virtually all city purchases, from paper and cleaning products to cars. In Santa Monica, California, green purchasing priorities have resulted in a municipal vehicle fleet consisting of more than 80 percent alternative-fuel and advanced reduced-emission technologies. The European Union, meanwhile, has set standards for 18 products for green procurement, as an assist to meeting its goal of 50 percent green-product purchases.[7]

Although efficiency alone is insufficient for creating sustainable cities, it is still critical, and opportunities abound. The experience of Los Angeles, California, is instructive. In 2013, the city completed a huge street lighting retrofit, installing 140,000 bright but efficient LED (light-emitting diode) fixtures across the city. The $57 million program has yielded energy savings of 63 percent (equivalent to removing 9,500 cars from Los Angeles streets) and a reduction of 47,000 tons of carbon emissions. Crime in the better-lighted streets is down by 10.5 percent. And program financing was manageable: loans are paid back over just seven years, using energy and maintenance savings. Once the city's loans are paid off, it expects to realize $10 million per year in savings.[8]

Finally, large reductions in materials and energy use may require changes in consumption, including more emphasis on public rather than private consumption. Attention to placemaking (see Principle 4 on page 55) can help create venues for "environmentally light" consumption. People who enjoy an evening stroll to a public plaza are consuming city assets—streets, streetlights, and public space—in a way that involves a minimum of materials and energy. Thus, in a sustainable city, people may spend much more time at concerts, sporting events, and festivals, and much less time at shopping malls.

Reduced Urban Energy Footprint

Cities also will need to shrink their energy footprints by creating more energy-efficient infrastructure and converting their energy supplies to renewable sources. Cities can target the largest users of energy, especially buildings and transport, for efficiency upgrades. Setting a passive-house standard for new buildings and renovations—a globally recognized benchmark of building energy efficiency—can dramatically reduce energy use. (See Chapter 9.) Carbon-free transport can be prioritized, with heavy emphasis on biking and walking, then on public transport. (See Chapter 11.) In addition, cities can promote district heating (and cooling), and combined heat and power to capture wasted energy. Efficient appliances can be promoted through regulation, standard setting, and green procurement policies.[9]

Greater efficiency facilitates the transition to renewable sources of energy, which some cities now see as the eventual source of 100 percent of energy use. (See Chapter 10.) Many sources of renewable energy are found in cities themselves, including power from solar photovoltaics (PV), small-scale wind power, heat pumps and geothermal systems, biomass, and methane capture from sewage. Cities also can connect to regional and national renewable energy grids.[10]

2. A Prominent Place for Nature

Nature is the very ground on which cities and urban activities are built. The natural environment supplies resources such as clean air, water, trees, and species, as well as services, from water filtration to pollination, that are vital for city functioning and that make cities beautiful and livable. (See Chapter 17.) A sustainable city operates in harmony with nature, respecting and implementing the ecological principles of diversity, adaptiveness, interconnectedness, resilience, regenerative capacity, and symbiosis in its development and planning activities. (See Table 4–2.) In a sustainable city, nature is no longer an urban afterthought.[11]

Table 4–2. A Prominent Place for Nature: A Checklist for Urban Practitioners	
Traditional Practice	**Policies That Point Toward a Sustainable City**
Nature is tamed and segregated	City initiatives work in harmony with nature
Over-reliance on constructed facilities	"Green infrastructure" is used where possible to meet urban development needs
Development without regard for the needs of wildlife	Preserve natural habitat by providing wildlife corridors and avoiding excessive fragmentation of land
Neglect of nature as teacher	Integrate nature into education for all ages

Benefits of Nature in Cities

The extensive presence of nature in cities carries a range of benefits, starting with a more livable environment. Robust greening of cities can purify air and water, reduce artificial warming from buildings and streets (known as the heat-island effect), and increase biological diversity. Toronto, Canada, has calculated that vegetating 5,000 hectares of roofs would result in a reduction in ambient air temperature of 0.5–2 degrees Celsius, decreasing energy demand for cooling. As a result, since 2010, any building in the city with more than

Michael Renner

A portion of The High Line, New York City.

2,000 square meters of floor space must devote up to 60 percent of its roof area to vegetation.[12]

A greened city provides clear economic benefits. Property values often increase in beautified areas, and studies show that homes with trees fetch a higher market price than similar homes without trees. The High Line, a former elevated freight line in New York City that has been converted to a park, has attracted $4 billion in private investment to the area. Additionally, flood control from well-designed green areas makes houses and other buildings more secure and more desirable: preserved coastal wetlands provide an estimated $23 billion in hurricane protection alone.[13]

A healthy environment can improve human health as well. Green spaces in Denmark correlate with lower stress levels and lower levels of obesity. Purified air from trees and other plantings reduces exposure to pollution and contaminants. Green space is shown to attract people outdoors, as in Stockholm, where green urban features often entice people to walk or bike to work, incorporating exercise that they otherwise might not get. (See Chapter 18.)[14]

Managing Green Assets

"Green infrastructure"—the use of wooded areas, creeks or rivers, and other natural areas to provide economic services—can help cities avoid construction of costly new water management facilities. Aquifers can be recharged for water storage, sidestepping the need for new holding tanks. Parks and fields can be designed to provide flood protection, reducing outlays for concrete channels. Even at the scale of the neighborhood, green infrastructure can soak up and store water. As the U.S. Environmental Protection Agency observes, "bioswales, rain gardens, permeable pavement, green roofs, and other innovations help to channel, store, and filter water that would otherwise flow out of the city, sweeping up pollutants from streets in the process."[15]

Management of a city's natural assets extends beyond city limits. The Nature Conservancy notes, for example, that watersheds supplying the world's 100 largest cities cover an area 12 times greater than the cities themselves. At no charge to cities, these watersheds collect, filter, and deliver water to nearly 1 billion people "before it ever enters a pipe." Careful conservation of watersheds is therefore a smart way to ensure clean water for urban areas. New York City understands this, having opted to invest in conserving its upstate watershed rather than spend billions on an expensive plant to treat what otherwise would be impure water from a contaminated watershed.[16]

Cities also can collaborate with regional neighbors to secure water supplies. Consider San Diego, California, which pays farmers in the nearby Imperial Valley to conserve water and pump the savings westward to the city. Farmers line irrigation canals to prevent water loss, invest in drip irrigation and other efficiency enhancements, and fallow some fields every few years. By 2020, conservation in the Imperial Valley is expected to supply 37 percent of San Diego's water supply.[17]

Indicators of a Green City

Timothy Beatley's book *Biophilic Cities* paints a vivid and detailed picture of what an environmentally grounded city might look like, from urban infrastructures to municipal activities and city governance. Beatley has developed a set of indicators of "urban biophilia" and offers sample values that might serve as guidelines for cities. His values for infrastructure, for example, might vary by city, but they give a sense of the extent to which nature can be integrated into urban life. (See Table 4–3.)[18]

3. Compact and Connected Patterns of Development

Sustainable cities bring people together in close but livable quarters, abandoning the low-density model that has prevailed in many cities. Compact cities offer two overarching advantages. First, they generally require fewer resources per person: the land, pipes, and communications and transport infrastructure needed to serve each person in a community decreases as people live closer together. In compact cities, structures and spaces also tend to get greater use, with fewer or shorter periods of idleness each day or season than is characteristic in low-density cities.[19]

Second, compact cities tend to enhance connectedness of all kinds—physical, social, and economic—generating innovation, economic activity, and

Table 4–3. Beatley's Indicators for Biophilic City Infrastructure	
Indicator	Beatley's Guideline Values
Share of population within 100 meters of a park or green space	100 percent
Existence of an integrated, connected, ecological network; "green urbanism from rooftop to region"	Ideally, unbroken green corridors from the center of the city to the edges
Share of city land area in wild or semi-wild nature	10 percent
Share of forest cover in the city	40 percent (less in the core, more near the periphery)
Extent and number of green urban features (green rooftops, green walls, trees)	1 green rooftop per 1,000 inhabitants (minimum 1 per block)
Kilometers per capita of walking trails	1.61 kilometers (1 mile) per 1,000 population would be a high level
Number of community gardens and garden plots	1 community garden per 2,500 city residents

Source: See endnote 18.

social and cultural capital. Some scholars see cities fundamentally as social networks whose primary role is to expand connectivity per person and to increase social inclusion—prerequisites for realizing cities' full socioeconomic potential. (See Chapters 18 and 19.) Human connectedness is central to civic life, and a compact city is best able to enhance connectedness of all kinds. Connectedness has many dimensions, including accessible and affordable transportation, robust digital infrastructure, ample indoor and outdoor public meeting spaces, and unobstructed corridors for wildlife. When these networks of connection are dense and functioning properly, they correlate with vibrant and prosperous cities.[20]

Compact cities are created through interventions in many different facets of city life, including land use, housing, transportation, buildings, and the digital sphere. (See Table 4–4.)

Land Use: Compact cities use land intensively. They tend to feature interconnected streets, mixed-use buildings and spaces, and development that encourages local self-sufficiency of daily life. City layout is in the pattern of a dense web of similar-size streets, rather than the hierarchy of freeways, arterials, and side streets—often featuring cul-de-sacs, dead ends, and other

Table 4–4. Compact and Connected Patterns of Development: A Checklist for Urban Practitioners	
Traditional Practice	**Policies That Point Toward a Sustainable City**
Sprawling development	Compact, mixed-use development, with more-intensive use of buildings and public spaces.
Car-centric land use	Easy access to walking, cycling, and public transport
Blocked flows of people and information	Connectivity that stimulates cultural and economic interchange
Housing distant from work and shopping	Affordable, location-efficient housing
Use of cul-de-sacs and other terminal roadways	Interconnected street systems
Parks and green space as isolated entities	Connected natural areas that promote biodiversity and natural services

disconnected patterns—that characterize low-density cities. (See Chapter 7.) In less-compact development, the hierarchy of streets often leads to choke points as rivers of traffic are channeled into a few major roads, requiring widening over time, typically making biking and walking dangerous. These arterials also tend to attract "big box" outlets, whereas smaller streets with a mix of uses tend to support small retail establishments. More-convenient land-use patterns reduce car journeys, congestion, and the energy needed for transport, and increase air quality, cycling, and walking.[21]

Housing: Many cities feature a hub-and-spoke transportation system that brings workers from the city's periphery to its center. This development pattern is a function of car dependence and cheap fuel, which supported the logic of putting housing in the suburbs and jobs in the cities. But with congestion and other limitations of car dependence and the likelihood that fuel will not always be cheap (see Chapter 5), cities increasingly understand the need to bring workers closer to employment. In regions where housing is expensive (in the United States, some 12 million households spend half of their income on housing, a share that rises to 60–70 percent when transportation costs are added in), housing located near multi-modal transportation options can help ease the housing and transportation burden on citizens. Incentives to promote this include measures—such as density bonuses, fee waivers, expedited permits, and tax credits—that encourage developers to include affordable housing units within a given project, or to undertake infill redevelopment.[22]

Transportation: Transportation done well enhances connectedness, making flows of people easy, convenient, and affordable, in contrast to car-centered development, which can block connectedness because of congestion-related delays and the presence of major arterials that cut off cross streets. Sustainable cities will need to shift from car-oriented urban patterns (such as cul-de-sacs and expressways) to transit-oriented patterns (such as mobility hubs, intensified corridors, and transit-oriented development). (See Chapter 11.)[23]

Sustainable cities will emphasize walking, biking, and public transit in their transportation mix. Multi-modal approaches to transportation ease the pressure on any single mode and attract high-density development where different modes meet. They also increase access to transportation by helping to meet the needs of diverse sets of people, such as the elderly, children, and low-income residents.[24]

Michael Kodransky

A "shared space" street in Zurich, Switzerland. Eliminating the continuous curb found in traditional streets puts pedestrians, bikes, and cars on an equal footing, resulting in slower car speeds, greater safety, and increased social interaction.

Streets will likely look different in a sustainable, compact city. In residential areas, concepts such as shared space, home zones, and *woonerfs* ("living streets") can be used to re-create residential streets as a social space, with plentiful planters, benches, and other ornamental features that create inviting areas where children play and adults mingle. Only secondarily are such streets a throughway for vehicles; cars appear infrequently and navigate carefully and slowly. In commercial areas and downtown, the "complete streets" concept can create a lively but safe environment in which auto traffic, bus lanes, bike lanes, and parking have designated spaces, each safely segregated from the others.[25]

Buildings: As cities work to create new housing and transportation options, they can adopt green building standards for both new and existing stock, starting with city-owned buildings to create a market for green building materials and practices. (See Chapter 9.) They might take their lead from the City of Los Angeles, which adopted a Green Retrofit and Workforce ordinance that not only promotes energy-efficient retrofits, which saves energy, but creates jobs and improves social equity by doing so in low-income communities.[26]

Digital Sphere: Access to the Internet and other digital tools is increasingly a necessity for robust civic life because it is the entry point to many cultural and economic activities. A study of 275 U.S. metropolitan areas conducted for the social networking site LinkedIn found that regions with the highest levels of connectedness experienced job growth of 8.2 percent between 2010 and 2014, compared with just 3.5 percent in the least-connected regions. Notably, the strong correlation held even when controlling for region size and for low and high levels of technology industry presence. The study also cites research concluding that connected entrepreneurs are more likely to be successful and that connected scientists are likely to have more patents. A sustainable city will offer robust opportunities for accessing the Internet and for other digital connections, perhaps through citywide provision of a public Wi-Fi system.[27]

4. Creative Placemaking

Cities are rich in character, with historical landmarks, parks, plazas, courtyards, civic buildings, rivers, lakes, and parks that give urban spaces personality and serve as gathering spots for the public. By investing in these assets and adding to them—for example, through reclamation of dormant and empty spaces such as the tens of thousands of vacant lots in Philadelphia, Los Angeles, Cape Town, and other municipalities worldwide—cities can create attractive places that advance civic pride and unity and create a strong sense of community. This cultivation of public spaces is known as "urban placemaking." (See Table 4–5.)[28]

Placemaking is linked to the pedestrianization of city life: public spaces are best located within walking distance of a resident's home or workplace. Each should be easily accessible and serve multiple purposes, becoming known and appreciated by large numbers of citizens. In this way, placemaking activity can be a driving force for the creation of a strong civic culture. The city of Pickering, Canada, in its advice for placemaking activity, urges participants continually to ask: Is it beautiful? Is it comfortable? Is it welcoming and accessible to all? Do people want to use the space?[29]

Table 4–5. Creative Placemaking: A Checklist for Urban Practitioners	
Traditional Practice	**Policies That Point Toward a Sustainable City**
Inattention to vacant lots and other poorly utilized spaces	Creation of incentives to minimize the inventory of underutilized spaces
Poor understanding of the value of neighborhood-level gathering spots	Investment in public spaces to make them community-usable
Inadequate pedestrian access to heritage and other assets	Provision of adequate foot, bicycle, and public-transit options near public spaces
Development projects conceived with little attention to public space	Inclusion of opportunities for creating public space in new development projects

Medellín Metrocable

When Medellín, Colombia, built its aerial tram to link poor hillside neighborhoods with the central metro system, it leveraged the project to maximize its development impact. (See Chapter 18.) The city employed a strategy called "social urbanism," part of which refers to compensating the poor for a "historic debt to them" through the construction of high-quality infrastructure and impressive architecture—a sharp departure from development projects that often feature inferior materials and construction. The aspiration was to build, "a new 'social contract' through the provision of spaces of citizenship, places for democracy and environments of conviviality," according to scholars who have studied the project.[30]

As part of the project, the areas around giant pylons that support the tram were made into plazas that feature food vendors, benches, and landscaping, while parks, schools, and libraries were built or upgraded a short walk away. New lighting, pedestrian bridges, and street paths also were built. The libraries included a great many community services in addition to provision of books: information technology, training courses, cultural activities, social programs, and support for the creation of micro-businesses, to name a few.[31]

Not only did the Metrocable give poor hillside residents access to the city's metro, but it brought many social benefits as well. Local labor was used in construction, a stimulus to impoverished neighborhoods in the area. Residents had a voice in the disposition of 5 percent of project funds, an effort to change longstanding patterns of patron-client politics. Area improvements, especially better lighting, also likely helped lower crime rates: homicides in neighborhoods served by the Metrocable dropped by 66 percent more than in nearby neighborhoods that did not receive Metrocable-related investments.[32]

Bryant Park, New York

In the late 1970s, Bryant Park, in the heart of Manhattan, was a dirty, crime-ridden, and drug-infested space. To clean up the area, a nonprofit

Medellín Metrocable aerial trams, with the city in the background.

organization, the Bryant Park Corporation (BPC), contracted with the City of New York to manage the park and invested some $18 million over 10 years in restoration efforts, raising capital from grants, business improvement district assessments, bond funds, city capital funds, and private capital. The investments paid for new landscaping, renovated restrooms, and 2,000 lawn chairs, in addition to revenue-generating initiatives including two restaurants and six kiosks selling specialty items from coffee to ice cream. The BPC operates with a staff of 55 persons in the summer who manage security, landscaping, and special events, including fashion shows, jazz festivals, and Monday-night films. Other pastimes, such as chess and bocce, as well as ice skating in the winter, are also part of life in the park.[33]

Bryant Park now records some 6 million visitors annually and is credited with helping to revitalize midtown Manhattan. In the two years following its rehabilitation, rental activity in the area increased by 60 percent and crime fell: 150 robberies were recorded in the park the year before BPC moved in, compared with just a single robbery since 1980. The turnaround is an example of a highly successful public-private partnership for placemaking. Although its success would be difficult to replicate in a city without the investment funds and disposable income found in Manhattan, the example of Bryant Park gives a sense of what is possible when citizens and city officials pull together to advance the public interest through creative placemaking.[34]

5. Centers of Well-being

Because cities typically generate and accumulate wealth, they are in a strong position to promote good health, security, decent employment, and robust social opportunities—the foundation blocks of well-being. The need is great. Experts convened by the World Health Organization assert that outdoor urban air pollution causes 1.3 million deaths each year, while sedentary lifestyles cause 3.2 million deaths, traffic injuries 1.3 million deaths, and violence some 1.6 million deaths. Sustainable cities will reduce these incidences of mortality dramatically by structuring cities to avoid their causes.[35]

In most cities, well-being is lacking for many people or even for most people. Well-being requires a broad set of policy choices that ensure that everyone has access to the basics for a dignified life. (See Table 4–6.) Fortunately, the pursuit of urban sustainability can help to advance well-being: a 2013 study comparing the Gallup Healthways Well-Being Index with four indices of sustainability covering 50 or more U.S. cities found that those cities that are more sustainable also tend to score highly on well-being indices.[36]

Table 4–6. Cities as Centers of Well-being: A Checklist for Urban Practitioners	
Traditional Practice	**Policies That Point Toward a Sustainable City**
Unwitting encouragement of inactive lifestyles	Transportation policies that promote walking and cycling, and other policies that get people outside
Inadequate enforcement of regulations governing clean air and water	A no-waste ethic that is intolerant of pollution, the most noxious form of economic waste
Tolerance of unemployment and homelessness	Jobs for all who need them and safety-net measures
Insufficient attention to providing access to medical care citywide	Availability of affordable medical and dental care for those lacking it

Health

Cities will be challenged to ensure that all citizens have access to health care. Attention by a doctor or dentist early in the development of a medical condition can catch maladies before they become serious sickness. Cities, too, can minimize or prevent problems by helping residents stay healthy. By enacting and enforcing strict standards for air and water cleanliness, for example, cities can prevent respiratory and digestive ailments that affect many citizens. In other words, cities can be designed for health.

Decisions about land use and transportation have important health effects. By designing cities for walkability and bikability, with short distances to work and shopping, residents can build exercise—a key to good health—into their daily routines. Many cities see the promotion of walking and biking as key not only to a multi-modal transportation system, but to prevention of chronic diseases such as heart disease and diabetes. Parks, in particular, are an attractive way to promote good health, with sports fields, exercise circuits, and bike paths being popular ways of keeping people active. Where parks are not available, exercise can be brought to people: in China, authorities have created some 4,000 free outdoor gyms since 1998 that contain many of the types of exercise equipment found in private gyms, but adapted for outdoor use. The national and municipal governments have promoted these facilities through public events such as the annual National Fitness Day and Beijing Olympic City Sports Culture Festival.[37]

Income

Central to well-being is secure income, typically from a steady source of employment. Employment can be public, private, or in the nonprofit sector, and it can be informal or formal in nature. Although a wide host of policies can be used to promote employment, at an overarching level, cities might follow what the International Labour Organization calls a Decent Work Agenda, which includes four objectives: creating jobs, guaranteeing rights at work, extending social protection, and promoting social dialogue. Commitment to these objectives can help minimize unemployment, improve workplace relations and safety, and provide for secure incomes.[38]

Many cities have a large share of the population working informally, largely outside the protection of municipal rules governing work. But in some cities, innovative schemes exist to improve the conditions of informal workers. In the Indian state of Gujarat, for example, the conditions of homeworkers are monitored by the Self-Employed Women's Association, which also helped establish minimum piecework rates consistent with the minimum wage. In Sudan, women working in the informal sector have formed associations to cover their health needs.[39]

Because the bulk of employment in most cities is in the private sector, city governments can seek to influence wages and working conditions in this sector. Contractors at large institutions, such as airports, can be required to provide standard wages and benefits. Wage and benefit minimums also can be set as a condition for economic development subsidies. In the United States, the city of Santa Fe, New Mexico, requires a Community Workforce Agreement with wage and benefit stipulations on any city-funded construction project valued at more than $500,000.[40]

When workers are out of work or otherwise cannot get employment income, cities can establish social protection floors—safety nets that keep citizens from falling into extreme poverty. These are best designed not just as crisis management tools, but as extensions of development. In Brazil, the government uses its Bolsa Familia federal cash transfer program to provide the poorest families with financial support, with the proviso that children attend school, are vaccinated, and are monitored for growth and weight, and that pregnant women receive pre- and postnatal care. In 2011, Brazil launched an amended version of the program, Bolsa Verde, which gives approximately $150 each trimester to poor families that adopt environmental conservation actions, generating additional revenue for low-income residents. Women are the primary recipients of funds from the programs and account for 93 percent of program debit card holders, increasing family security and women's negotiating power vis-à-vis their spouses.[41]

6. People-centered Development

The fundamental purpose of a city is to serve its people. Yet in many cities, development priorities are set based on the needs of builders, financial brokers, and the city's privileged, while sidelining the needs of the city's majority. (See Chapter 19.) To be more inclusive, city administrations can weave people's interests into the very fiber of city initiatives, involving citizens in city governance (see Principle 7) and reflecting the needs of the majority in daily city administration. (See Table 4–7.)

Applying Maslow's Hierarchy to City Administration

Placing people's needs at the center of city initiatives in a systematic way is a challenge. An innovative approach proposed in the state of Victoria, Australia, imagines incorporating standards for a healthy society into city programs across the state by simplifying and adapting Abraham Maslow's "hierarchy of needs" schema. In this simplified framework, known as Existence, Relatedness, and Growth (ERG), "Existence" refers to meeting basic survival needs, "Relatedness" is about facilitating interactions among people and with nature,

Table 4–7. People-centered Development: A Checklist for Urban Practitioners	
Traditional Practice	**Policies That Point Toward a Sustainable City**
Strategic plans based on scenarios that benefit particular interests	Strategic plans based on citizen needs
People's interests made to fit into needs of economic interests	People's interests drive the planning process; economic interests build around people's interests
Access to basic services governed by market prices	Broad access to basic services, increasing equity across a city

and "Growth" corresponds to promoting equity, justice, beauty, and other higher-level values. Standards for each area can be applied across city departments and programs.[42]

Researchers in Victoria have imagined applying ERG to a city's water provision. In their framework, Existence needs for water are met when a city can supply basic services (meeting survival needs of drinking, cooking, washing, and bathing); when sanitation advances health through the prevention of disease; and when stormwater drainage protects against flooding (meeting an important security need). These basic water services are supplied through conventional and usually centralized water management infrastructure, such as reservoirs, treatment plants, and pipes, canals, and other conveyances.[43]

Smart water policy can help a city meet water-centered Relatedness needs as well. Parks, sports fields, and open spaces are places where people socialize and enjoy nature, but these areas often are neglected during droughts, reducing opportunities for interaction. What if recycled stormwater and wastewater were used during droughts to augment water supply and keep parks green? Those resources typically are wasted: Victoria reused only 2 percent of reusable stormwater in 2010 and only 7 percent of the sewage available for treatment. A city can hardly be called water-scarce, the researchers argue, if it throws away large quantities of water.[44]

Finally, water policy can meet Growth needs of equity and justice by giving greater control to citizens, which often translates to diversity of offerings and decentralized management. At the household level, rainwater harvesting can help families reconfigure their water supply options and cut water costs. Neighborhood-level systems can manage wastewater and stormwater. These systems can be made available to all citizens, rich and poor alike.

Beautification can be created by exposing at the street level once-piped storm-water in garden-like drainage systems, helping to achieve both aesthetic and equity goals.[45]

7. Participatory Governance

Sustainable cities are participatory organisms that encourage self-governance. On questions large and small, citizens in sustainable cities find roles open to them in governing, as individual voters, and as members of stakeholder groups. Power and decision making are shared between city hall and smaller jurisdictions within the city. And members of civic groups of all kinds are consulted regularly and included in major decision-making processes. (See Table 4–8.)

Table 4–8. Participatory Governance: A Checklist for Urban Practitioners	
Traditional Practice	**Policies That Point Toward a Sustainable City**
Chief decision makers are elected officials	Power is distributed and devolved, with districts and neighborhoods having strong voices
Important decisions are made in secret	Decision-making processes are public, including posting of progress on mass communications media, including a website
High bar of qualifications for citizen participation	Citizens of all kinds are encouraged and recruited to participate
Powerful interests have extensive access and influence over decision makers	Meeting calendars of civic officials are made public

Participatory Budgeting

An excellent example of people-centered governance is the participatory budgeting process pioneered in Porto Alegre, Brazil, in 1989 and now used in more than 1,500 localities worldwide. In the Porto Alegre process, citizens are mobilized through 16 regional and 5 thematic plenary assemblies to offer ideas for how to spend a part of their city's municipal budget, including having a voice in deciding how funds will be distributed among districts. Evaluations of the process suggest that participatory budgeting in Porto Alegre has strengthened democracy and civil society in the city and brought new investment to marginalized areas. The process is being adopted in larger cities as well, including

New York, which in 2012–13 had 13,000 people participating to determine the distribution of $10 million in city funds.[46]

Beyond budgeting, several cities in Europe offer models of greater democratization of civic governance. In 2014, Amersfoort in the Netherlands introduced the Year of Change, a shift in administrative practices to emphasize shared responsibility and collective leadership. It borrowed a process from Belgium called G1000, under which the city chose a panel of 1,000 citizens randomly and invited them to a deliberative event. Some 600 people showed up and selected and developed 10 project plans to pursue, in partnership with the city administration. City officials noted that "the process was quicker, less expensive, and achieved a wider consultation than when normally done by the municipality."[47]

Participatory governance can address a number of issues of widespread importance to the poor. These include: property tenure, especially on land that the poor have taken the initiative to occupy; securing an "official address," which often is required to gain access to city services and to vote; and the challenge of regulations that get in the way of making a livelihood or of securing housing or land.[48]

The Vision in a Nutshell

Citizens driven by a sustainability vision are committed to building a city that sets important limits, promotes a healthful society, and creates widespread opportunity and well-being. These are summarized as follows:

- *Setting important limits* refers to respecting boundaries set by nature, avoiding waste by circulating materials to the extent possible, reducing the city's energy and materials footprint, and relying primarily on renewable sources of energy.
- *Promoting a healthful society* means ensuring access to health care and the building blocks for healthy lifestyles; promoting walking, cycling, and transit while de-emphasizing automobiles; and promoting well-being in lieu of endless increases in consumption.
- *Creating widespread opportunity* refers to ensuring full employment, expanding connections among people, businesses, organizations, and city hall, and giving a strong voice to the community in decision making.

A sustainable city brings people together, maximizing the stimuli, innovation, and enrichment that is born of connectedness. It is proudly public—with a strong commitment to parks, transit, festivals, community gardens, and civic spaces, all broadly accessible. It treats squirrels and robins, streams and trees, as neighbors rather than artifacts. And it provides for people's basic needs, the springboard for higher-level necessities such as fulfillment and belonging. In sum, a sustainable city offers the prospect of creating the next great stage in human civilization.

CHAPTER 5

The Energy Wildcard: Possible Energy Constraints to Further Urbanization

Richard Heinberg

Many urban analysts assume that standard projections regarding growth in human population, agricultural output, and energy supply and demand over the next several decades are correct. If current trends continue, all of these factors would increase significantly. However, an alternative scenario also is plausible—one in which the increased costs and reduced availability of energy, especially fossil fuels, sharply constrain further growth and diminish or even reshape some of these trends.

The last few decades have seen dramatic urban expansion as a result of both global population growth and the influx of rural migrants to cities. Most futurists assume that this trend will continue throughout the current century, and many environmentalists welcome the prospect of urbanization because cities seem to impose lower environmental costs, per person, for a given standard of living. In industrialized countries, urban living often correlates with higher energy efficiency.

Residents of Manhattan in New York City, for example, use less energy per capita for transportation than do typical suburban or rural Americans, who tend to drive cars more often and for longer distances. (The latter have little choice, as good U.S. public-transit options are less prevalent outside dense urban centers.) This correlation between energy efficiency and urban density is weaker—or even reversed—in countries that are not highly industrialized, where rural subsistence farmers typically do not own private vehicles and travel little. Thus, there is a loose correlation between the relative energy efficiency of urban living and society's overall energy consumption levels: as total energy consumption rises, so does the relative efficiency of urban living.[1]

Richard Heinberg is Senior Fellow-in-Residence at the Post Carbon Institute.

In any case, ongoing urbanization means that societies need to keep providing more food, more employment, more housing, and more transport to growing ranks of city dwellers. All of these provisions require energy, yet, somehow, they all must be delivered without raising greenhouse gas emissions—and ideally, while *decreasing* emissions globally.

How long can the trend toward urbanization continue in the face of this century's energy and climate constraints?

Energy, Climate Change, and Urbanization

Energy is essential to all human activity. Industrial civilization—the biggest energy extravaganza in human history—arose due to the historically anomalous advent of fossil fuels, with their ability to deliver dense, portable, easily storable, and cheap energy in vast, unprecedented quantities. Fossil fuels replaced most human labor in agricultural production and also provided faster and more-efficient means of bringing distant resources to urban hubs. Coal, oil, and natural gas now account for 81 percent of global energy consumption.[2]

But our current fossil fuel-based energy regime faces two serious challenges: depletion of the "low-hanging fruit" of global petroleum supplies, and the need to reduce carbon emissions to avert catastrophic climate change.

For the past decade, conventional oil production rates have been flat-to-declining, and nearly all growth in oil production has come from unconventional sources such as tar sands, tight oil, and deepwater oil. The higher expenses associated with these options resulted in roughly a 10 percent annual increase in exploration and production costs during the years 2009–14. Oil prices have become more volatile, however, and U.S. tight oil production—one of the main categories of recent oil production growth—is now declining, while exploration in the Arctic and elsewhere is going nowhere fast. There no longer seems to be a "Goldilocks" oil price that is high enough to justify new marginal production but not so high as to hamper general economic expansion.[3]

Many industry analysts point out that average well productivity in the Bakken and Eagle Ford tight oil "plays" in the United States is increasing, and they forecast that, once supplies tighten, oil prices will rise, production will become profitable, drilling rates will increase, and overall production will begin expanding again. However, geologist David Hughes of the Post Carbon Institute, who has published several detailed reports on U.S. tight oil resources, notes that the current increase in per-well productivity is due simply to drillers cherry-picking the very best sites. If drilling rates (which fell more than 60 percent between July 2014 and August 2015) increase again, drillers will be forced to target lower-quality resources, causing average well productivity to decline. Profitable drilling sites are limited in number and dwindling. Petroleum geologist Art Berman observes that, in the current oil price environment, despite drillers homing in on the very best areas, only 1 percent of wells break even financially.[4]

The situation with natural gas is parallel, although it is complicated by the fact that gas is more of a regional, and less of a globally traded, commodity. The United States—the world's top natural gas producer and consumer—now obtains roughly half of its supply from unconventional shale gas resources produced through hydrofracturing ("fracking") and horizontal drilling. However, production from individual shale gas wells tends to decline quickly, requiring high rates of drilling to maintain a constant production rate. Well quality varies dramatically within productive regions, and good drilling locations within each play are limited in number.[5]

Barnett Shale gas drilling rig near Alvarado, Texas.

David R. Tribble

Some of the regions where fracking was first deployed just over a decade ago (including the Barnett and Haynesville plays) have seen their total production rates fall sharply in recent years. Studies conducted by Hughes suggest that the overall U.S. shale gas production rate is likely to begin declining before the end of the present decade (and it is already in decline, as of this writing). Although prospects exist for shale gas production elsewhere in the world (notably in China and the United Kingdom), the regulatory, financial, and political contexts in these countries may not be as favorable as they have been in the United States. Moreover, the North American experience suggests that production elsewhere will be characterized by the same relatively brief boom-and-bust cycle.[6]

Efforts to replace coal and natural gas with wind and solar energy have begun in the electricity sector, largely in response to the climate change dilemma. However, the challenge of a full transition from fossil fuels to renewable energy sources is more daunting when it comes to petroleum, the world's dominant energy source. Crude oil and the products derived from it account for nearly two-thirds of global transport fuels.[7]

Substitutes for oil are problematic in many cases. Biofuels perform poorly from an energy-accounting perspective and often entail unacceptable environmental costs. Batteries, due to their low energy density per unit of weight, work well only with smaller vehicles such as cars and bicycles. Hydrogen, while conceivably viable as a fuel for shipping and trucking, will entail enormous investments for the redesign and retooling of vehicles and supporting infrastructure. The challenge of powering commercial aviation with renewable energy is so daunting that a major downsizing of this sector may be unavoidable. Altogether, a full transition away from oil will require decades and many trillions of dollars in investment. It may result in systems that are more expensive to operate than current ones or that simply fail to deliver all the services that we currently expect.

The two economic sectors most vulnerable to oil supply limits (imposed either by depletion or by efforts to mitigate climate change) are agriculture and transport, and both are pivotal to continued urbanization.

What Might Post-Fossil Agriculture Look Like?

Although agriculture was the primary source of energy for pre-industrial societies, industrial agriculture is a net consumer of energy. The Green Revolution, which more than doubled cereal production in developing countries between the years 1961 and 1985, succeeded by vastly increasing inputs—nitrogen fertilizers, pesticides, and herbicides—all made from fossil fuels. The U.S. food system, which served as a model for the Green Revolution, consumes approximately 12 calories of energy (mostly from fossil fuels) for every food calorie delivered to the final consumer. (See Figure 5–1.) Fossil fuels are used not only in manufacturing nitrogen fertilizers, pesticides, and herbicides, but also in fueling farm equipment and in transporting inputs to, and outputs from, the farm. Further energy use occurs in food storage, packaging, processing, and sales.[8]

Ways to reduce fossil fuel inputs to food systems include the use of farm machinery powered by renewable electricity or farm-produced biofuels; the localization of food systems to reduce transport (perhaps entailing vertical urban agriculture); the adoption of organic and ecological production

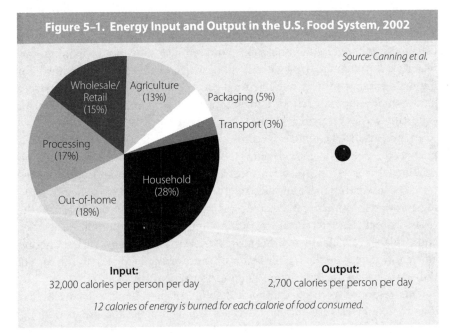

Figure 5–1. Energy Input and Output in the U.S. Food System, 2002

Source: Canning et al.

Wholesale/Retail (15%)
Agriculture (13%)
Packaging (5%)
Transport (3%)
Processing (17%)
Household (28%)
Out-of-home (18%)

Input:
32,000 calories per person per day

Output:
2,700 calories per person per day

12 calories of energy is burned for each calorie of food consumed.

practices to reduce the need for nitrogen fertilizer, pesticides, and herbicides; and an overall reduction in the consumption of highly processed foods.[9]

It is unclear, however, whether such measures could be substituted for fossil inputs while maintaining current levels of agricultural production, as studies have led to sharply conflicting conclusions. If smaller-scale organic production cannot produce high-enough yields, or if the transition is hampered by a failure to train farmers adequately (or quickly) enough or to undertake land reforms required, then the consequences of a significant involuntary reduction in the availability of energy (especially oil) to food systems would be severe. Food likely would become more expensive (high food prices tend to correlate with high oil prices), and less of it would be transported long distances. Crucially, even if small-scale organic practices did prove sufficiently productive, more agricultural labor would be needed, perhaps encouraging (or requiring) many people to move back to the countryside—slowing or even reversing the long-term trend toward urbanization and perhaps entailing social upheaval. It is possible to imagine methods by which this "re-ruralization" could be accomplished that are either progressive in nature (ecological farm co-ops, for instance) or regressive (a new serfdom).[10]

Post-Petroleum Transportation

Cities depend overwhelmingly on powered transport for obtaining raw materials for manufacturing (nearly all of which occurs in or near cities); for importing and exporting manufactured goods; for moving people to and from home, work, shopping, school, and cultural events; for importing food; and for exporting wastes of various kinds.

In industrialized countries, the lion's share of oil consumption in the transport sector occurs in private automobiles. (See Figure 5–2.) Trucking and aviation vie for second place in most countries, with rail and public transportation assuming more-modest roles. However, the fuel efficiency of moving 1 kilogram of material a distance of 1 kilometer runs more or less in the reverse order. Under typical operating circumstances, automobiles, trucks, and airplanes (constantly fighting against friction and gravity) are less energy-efficient, whereas public transit and rail (maximizing load and, in the case of rail, drastically reducing friction) are more energy-thrifty.[11]

One response to the declining availability of petroleum is to discourage low-occupancy car use and air travel while encouraging the use of walking, bicycling, and public transit for short distances and rail for medium and long distances. Making such a shift requires significant investments and changes to land-use and transportation policies, which inevitably are politically contested. But, as many cities around the world have been discovering, the results often are quite popular and beneficial. (See Chapters 11 and 12.)

Another way to tackle the problem is to pursue substitutions in the energy sources for current modes of transport: batteries for cars, sails for ships, electricity for railroads, and biofuels or hydrogen for trucks and airplanes. Some of these substitutions, which are being widely proposed as climate solutions, are likely to be more successful than others. For example, kite sails can make container ships more fuel-efficient, but sailing ships without engines would have vastly lower freight capacity. Electric cars are becoming more common, but hydrogen-powered airliners are not. As suggested earlier, aviation may not have much of a future without oil. Although specialty biofuels are able to power jet engines, they are unlikely ever to be as affordable as petroleum-derived jet fuel; and while helium-filled dirigibles could provide slower long-distance passenger air transport, helium is itself a depleting, non-renewable resource.[12]

Industries that are highly transport-dependent, such as tourism, will be more vulnerable, whereas work that can be done largely at distributed locations and online—such as an ever-growing amount of computerized office

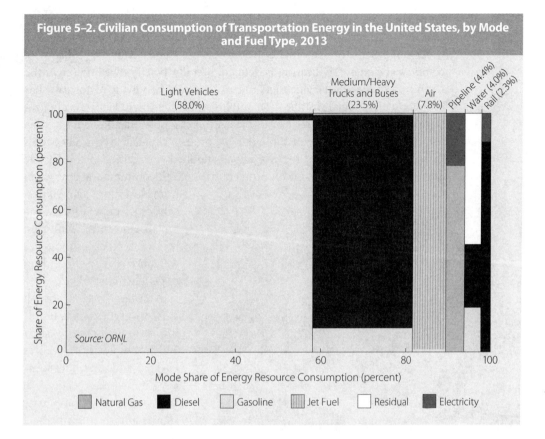

Figure 5–2. Civilian Consumption of Transportation Energy in the United States, by Mode and Fuel Type, 2013

work—will be less vulnerable (although manufacturing and transporting computers still requires fuel). Overall, cities may need to adapt to more-localized raw materials sourcing, manufacturing, and waste disposal.

Near the Tipping Point?

In a world with less liquid transport fuel, it is hard to see how cities could continue growing as they have in recent decades.

Prior to the fossil fuel-driven Industrial Revolution, the share of the global population that lived in cities was small—less than 10 percent. Today, just over half of humanity lives in cities. It is extremely unlikely that we will ever get to the point where all people reside in urban centers. At some stage, the trend toward urbanization will taper off or even reverse itself. It is impossible to know how close we are to that tipping point, but it could well occur during this century, and a decline in available energy is likely to be the key driving factor.[13]

Analogous historic moments have been associated with the collapse of complex societies. Civilization is defined, literally, by city dwelling (*civis,* the Latin root of the word *civilization,* means "city"). Urban living historically has brought with it social stratification and full-time division of labor. As civilizations expand, they urbanize; when they fail, cities empty out and people return to subsistence agriculture or foraging. The process of collapse typically entails massive mortality. When the Mayans abandoned their cities of the Classic period during the eighth and ninth centuries of the common era, and when Rome lost more than 90 percent of its population in the fourth and fifth centuries, literacy waned, technologies were lost, and great numbers of people perished.[14]

Brush Park, one of Detroit's earliest affluent districts of wealth and distinction, is now home to crumbling houses and vacant lots.

Archaeologist Joseph Tainter argues that building and maintaining societal complexity requires energy, and that the strategy of investing energy in sociopolitical complexity as a response to immediate problems tends to reach a point of declining marginal returns, resulting in eventual collapse. A process of protracted, chaotic civilizational collapse similar to what happened in the cases of Imperial Rome and the classic Maya clearly would be the worst-case outcome of energy scarcity in the remainder of the twenty-first century and beyond.[15]

It is possible to imagine a non-catastrophic pathway to de-urbanization. For optimum success, however, it almost certainly would need to be guided by sound policy. Financial and intellectual resources (presumably emanating from government) would be required to train large numbers of farmers in organic methods, and to rebuild rural culture and support infrastructure. Import substitution and land reform would imply a substantial reorganization of the economy. Success would require the creation of economic opportunities throughout large regions that have stagnated for decades. And all of this would have to occur at the same time that unprecedented levels of investment are being directed toward a historic switch in energy sources.

Society's managers would be unlikely to undertake such an immense and counterintuitive project unless they were convinced that re-ruralization is inevitable and necessary and that the attempt to maintain current levels of urbanization is futile. But by the time they were sufficiently convinced of this, it might be too late to initiate an organized program in lieu of a disorganized (and disorganizing) spontaneous process of societal de-complexification.

Surely few, if any, ancient Romans or Mayans understood urbanization or collapse in terms of energy, complexity, and the law of diminishing returns. No doubt they attributed the failure of their societies to immediate challenges such as barbarian invasions and drought (which they might have interpreted further in religious terms). In the same way, managerial elites of this century may see society's greatest challenges in terms of financial crises, geopolitical destabilization, and climate change, while failing to appreciate the underlying erosion of the energetic foundations of industrial cities.

As ancient civilizations crumbled, absent sound leadership from elites, people responded by creating new, simpler social and economic arrangements that required smaller energy investments. We can see the faint beginnings of similar trends today in the "sharing economy," in organizations of idealistic young organic farmers, and in "transition town" initiatives. Although the 1970s "back-to-the-land" movement (which coincided with historic energy crises) largely fizzled, perhaps due to cheap oil flowing from the North Sea and Alaska during the 1980s and 1990s, we may be on the cusp of similar, more widespread, and possibly more desperate trends today.[16]

The Urban Climate Challenge

CHAPTER 6

Cities and Greenhouse Gas Emissions: The Scope of the Challenge

Tom Prugh and Michael Renner

Since at least 2008, cities have hosted half or more of the earth's human beings, a share that continues to grow. Cities also account for more than 80 percent of global gross domestic product (GDP) and about 70 percent of global energy consumption and greenhouse gas emissions. If present trends continue, urban populations are expected to increase to 6 billion by 2045, at which point two-thirds of all people will live in urban environments. These figures suggest that while cities tend to be associated with higher per capita wealth than rural communities, they also account for higher per capita greenhouse gas emissions. In any comprehensive attempt to address climate change, therefore, cities and their inhabitants must play a vigorous and leading role.[1]

It is no surprise that cities collectively account for a large share of greenhouse gas emissions, because they concentrate economic activity. However, cities vary widely in their per capita emissions (see Table 6–1), depending upon a wide array of variables that may or may not be under their control. These include climate (which affects heating and/or cooling requirements); location (which helps determine climate and whether a city is a gateway for people and goods via ports and airports); primary sources of energy consumed (hydroelectric power and/or other renewables, coal, nuclear; these often are not under city control); urban form (transport energy use and greenhouse gas emissions are inversely correlated with settlement density, for example); technology (such as the use of methane capture in landfills); and the age, characteristics, and condition of the building stock. Economic factors, such as the wealth and income of residents and the level of economic activity, also play

Tom Prugh and **Michael Renner** are senior researchers at the Worldwatch Institute and codirectors of the *State of the World* 2016 project.

Country, City	Greenhouse Gas Emissions	Year
Table 6–1. Greenhouse Gas Emission Baselines for Selected Cities and Years		
	Tons of carbon dioxide-equivalent per capita	
Rotterdam, The Netherlands	29.8	2005
Denver, Colorado, USA	21.5	2005
Sydney, Australia	20.3	2006
Washington, D.C., USA	19.7	2005
Minneapolis, Minnesota, USA	18.3	2005
Calgary, Canada	17.7	2003
Stuttgart, Germany	16.0	2005
Austin, Texas, USA	15.6	2005
Dallas, Texas, USA	15.2	
Baltimore, Maryland, USA	14.4	2007
Juneau, Alaska, USA	14.4	2007
Houston, Texas, USA	14.1	
Frankfurt, Germany	13.7	2005
Seattle, Washington, USA	13.7	2005
Boston, Massachusetts, USA	13.3	
Los Angeles, California, USA	13.0	2000
Portland, Oregon, USA	12.4	2005
Chicago, Illinois, USA	12.0	2000
Miami, Florida, USA	11.9	
Shanghai, China	11.7	2006
Cape Town, South Africa	11.6	2005
Toronto (Metropolitan Area), Canada	11.6	2005
San Diego, California, USA	11.4	
Bologna (Province), Italy	11.1	2005
Philadelphia, Pennsylvania, USA	11.1	
Bangkok, Thailand	10.7	2005
New York City, New York, USA	10.5	2005
Athens, Greece	10.4	2005
Beijing, China	10.1	2006
San Francisco, California, USA	10.1	
Hamburg, Germany	9.7	2005
Turin, Italy	9.7	2005
London (Greater London Area), U.K.	9.6	2003

Table 6–1. continued		
Ljubljana, Slovenia	9.5	2005
Toronto (City), Canada	9.5	2004
Prague, Czech Republic	9.4	2005
Glasgow, Scotland, United Kingdom	8.8	2004
Singapore	7.9	1994
Geneva, Switzerland	7.8	2005
Brussels, Belgium	7.5	2005
Porto, Portugal	7.3	2005
Helsinki, Finland	7.0	2005
Madrid, Spain	6.9	2005
Paris, France	5.2	2005
Tokyo, Japan	4.9	2006
Vancouver, Canada	4.9	2006
Mexico City (City), Mexico	4.3	2007
Barcelona, Spain	4.2	2006
Seoul, South Korea	4.1	2006
Naples (Province), Italy	4.0	2005
Buenos Aires, Argentina	3.8	
Stockholm, Sweden	3.6	2005
Oslo, Norway	3.5	2005
Amman, Jordan	3.3	2008
Mexico City (Metropolitan Area), Mexico	2.8	2007
Rio de Janeiro, Brazil	2.1	1998
Colombo, Sri Lanka	1.5	
Delhi, India	1.5	2000
São Paulo, Brazil	1.4	2000
Ahmedabad, India	1.2	
Kolkata, India	1.1	2000
Bangalore, India	0.8	
Dhaka, Bangladesh	0.63	
Thimphu, Bhutan	0.33	
Kathmandu, Nepal	0.12	

Note: Data are not always directly comparable due to differing years and methodologies; they are meant only to give a general sense of relative emissions. Data without a year were provided by ICLEI–Local Governments for Sustainability, with no year specified.
Source: See endnote 2.

a major role, as does economic structure: urban areas with extensive manu-facturing industries have a very different footprint than cities where service activities predominate.[2]

Much is known in broad terms about the major drivers of urban greenhouse gas emissions that can be shaped or influenced by public policy. These include building energy use, transport, the forms of urban development, waste han-dling and disposal, and deforestation. According to one recent analysis of 274 cities with a total population of 775 million and drawn from all regions and city sizes, four factors—economic activity, transport, geography, and urban form—account for 37 percent of the variability in urban direct energy use and 88 percent of the variability in urban transport energy use.[3]

To some extent, the driver categories overlap and influence one another. For instance, modes of transportation and settlement patterns shape each other over time; favoring automobiles and the roads they require tends to encourage sprawl, while more-compact patterns obviate automobile use for many city residents and also enable more-efficient public transport. In addition, gener-ally speaking, all of these factors and the greenhouse gas emissions associated with them can be said to be direct or indirect functions of lifestyle.

The major common thread running through the categories is energy use: how much and of what kind. Many cities have little control over their energy supplies (although there is a recent trend toward remunicipalization; see Chapter 16). However, nearly all cities have numerous options and room to maneuver on the demand side of the energy equation. It is largely because of their ability to influence or control so many decisions—about building effi-ciency, transport modes, development patterns, and even, to some extent, the consumption practices of city dwellers—that cities are among the key actors in the effort to constrain global energy use and greenhouse gas emissions.

Motives for City-Level Action

Urbanites increasingly realize that climate change poses tremendous chal-lenges and, in more than a few cases, may even threaten cities' continued hab-itability. If climate change is allowed to proceed unchecked, urban life will be conditioned by sea-level rise, storms, flooding, droughts, and heat waves. These phenomena will claim growing material and financial resources in response to disasters or due to mitigation efforts, while undermining urban economies, destroying jobs, and imposing rising health costs. Beyond the direct climate consequences for cities are the effects on the inflow of food and other natural resources on which cities depend. Climate impacts are added, and often linked, to many other, longstanding sustainability concerns, such as the threat of air and water pollution or hazardous waste flows. These affect not just the health of the urban population, but also cities' livability and attractiveness to businesses and to visitors.

Cities also are discovering that other drivers and concerns may justify and stimulate action toward sustainability, including economic development and innovation through various "greening" measures. Creating and securing local jobs is a key concern of any city administration. Traffic congestion increases the costs of business, wastes fuel, and pollutes the air, while measures to reduce traffic and to shift from cars and trucks to public transit have multiple benefits. Concerns about the

Elevators to a futuristic-looking RandstadRail light rail station in The Hague, The Netherlands.

security of energy supplies—including worries about volatile prices (and the risk of growing energy poverty)—may lead mayors and city councils to procure locally produced renewable energy supplies and require more energy-efficient buildings. Cities suffering from deindustrialization or dramatic changes in their economic base may seek to revitalize former industrial areas through efforts that can be focused on sustainability measures. Well-designed policies can address both socioeconomic and environmental problems, increasing the cohesion between these goals and reducing conflicts and contradictions.

Spurred by these motivations and by the disappointingly slow progress toward addressing climate change at the national and international levels (notwithstanding the climate agreement reached in late 2015 in Paris), for some years now, cities have increasingly recognized their role in contributing to the global burden of greenhouse gas emissions and accepted their responsibility for reducing it. More and more cities around the world are taking action to address the climate crisis and other environmental challenges, in the hope and expectation that action will be swifter and more meaningful at the local level.

Growing numbers of cities have pledged themselves to climate commitments as well as to broader sustainability goals and are banding together with like-minded counterparts in peer-to-peer networks to facilitate and reinforce movement toward sustainability. The C40 Cities Climate Leadership Group (which, as of 2015, had expanded to over 80 cities) is a prominent network that has pledged to take action to reduce greenhouse gas emissions. Other noteworthy initiatives and groupings are pursuing eco-mobility, renewable energy, green buildings, zero waste, and the like. The Compact of Mayors, launched at the 2014 United Nations Climate Summit, is the largest coalition of city leaders addressing climate change. ICLEI–Local Governments for Sustainability has a long track record of working with more than 1,000 cities around the world as well as with international agencies such as UN-Habitat. Other organizations with narrower geographical focuses, such as STAR Communities and the Urban Sustainability Directors Network (both in North America), work to support and assist cities toward their sustainability goals.

These initiatives for collaboration and mutual support have begun to bear fruit in the form of encouraging steps toward concrete action. C40's *Global Aggregation of City Climate Commitments* details how 228 cities (with a combined population of 439 million people) have set climate reduction goals or targets that would, if achieved, lead to significant reductions in annual emissions compared with a business-as-usual scenario. To date, the reductions are stated in terms of emissions "savings" from business-as-usual scenarios that assume ongoing growth in population and economic activity, rather than in terms of their effect on atmospheric greenhouse gas concentrations. It appears that these reductions alone would not actually reduce the global rate of emissions but only slow the rate of continued increase; moreover, most of the commitments are set for 2020 or 2050. However tentative and conditional, these commitments nevertheless are vital as public acknowledgments of the climate challenge and the urgent need to address it. They constitute crucial underpinnings for countries' Intended Nationally Determined Contributions (INDCs)—the national greenhouse gas reduction pledges embodied in the Paris Agreement.

Cities' Powers to Act

Although cities across the world face many similar challenges, their particular circumstances, needs, and capacity to act—which are typically a product of their historically grown structures and their political cultures—can vary enormously. The shares of cities' greenhouse gas emissions attributable to each sector, for example, differ widely from one city to another, and each of the major drivers of urban greenhouse gas emissions has its own suite of shaping forces and policy options. Plans tailored to each city's circumstances, while sharing certain broad features, therefore will be highly individualistic.[4]

Depending on an urban area's specific economic base and profile, the most effective focus for emissions reductions and other pro-sustainability changes may be in the industrial sector, in transportation, or perhaps in the built environment. Cities with pollution-intensive industries, such as refineries or heavy manufacturing, face a much greater challenge than those that are more service-oriented. (Such factors need to be accounted for in setting fair reduction targets for individual cities and nations, for the simple reason that, in a globalized world, nominally low-pollution cities may account for significant amounts of greenhouse gas emissions via their consumption of products made in higher-polluting cities.) Rich cities may be able to act in ways that poor cities can only dream of. Dense cities are able to build attractive public transportation networks, while sprawling megalopolises struggle to make them work. Cities with growing populations confront a more confounding, rapidly changing situation than cities with more stable populations. This is especially the case in many developing-world cities that have large slum areas and informal settlements.[5]

Similarly, the capacity and freedom to act effectively on sustainability problems are far from uniform across cities. A recent assessment by C40 of its member cities shows that mayoral powers vary considerably from one city or policy area to another. Cities exert different degrees of control in terms of ownership, management or operational authority, regulatory power, and enforcement, as well as with regard to budgetary control, taxation, and financing. Land-use planning is a critical element in many urban decisions and is important for climate adaption. Two-thirds of C40 cities have operational control of relevant actions and have strong powers to set/enforce policy.[6]

In transportation, nearly all C40 cities have strong control over various assets such as roads, cycling lanes, sidewalks, and parking, but powers over mass transit differ greatly from one case to another. More than 80 percent of C40 cities own and/or operate their own municipal bus fleets, while just 44 and 39 percent, respectively, own/operate subways and light rail systems. (State or regional authorities manage these systems in the remaining cities.)[7]

In the building sector, C40 cities have the broadest powers over municipal buildings, with at least 70 percent reporting strong ownership or operational control, policy setting and enforcement, and budgetary control. Influence over privately owned buildings is more indirect and tenuous. Cities tend to have less control over energy policy. Just 42 percent have direct control of their municipal energy supply; about half of them (23 percent of surveyed cities) actually own and operate their own power utilities, while the other half has some influence on utility price setting and energy mix. One-quarter of C40 cities own and operate their own district heating/cooling systems.[8]

The C40 report concludes that mayors exert more power over waste management than in any other sector surveyed. Most of the cities own or operate street sweeping and waste collection functions, and more than half of them set policies for waste collection. Interestingly, poorer cities tend to operate only small-scale recycling programs, relying heavily on landfilling instead. Recycling is found most often in C40 cities with higher GDP per capita.

These particular distributions of power may not hold for other cities around the planet. Consistent with the theme of variance from one place to another, it is important to note that, in many places, the specific authority or capacity to act may not be statutorily provided for or, even if it is, may be subverted or overwhelmed by more fundamental governance problems, such as corruption, incompetence, turf rivalries and conflicts, a high rate of in-migration, lack of funds, and/or other impediments. Policies imposed by regional or national governments may constrain cities' room to maneuver in addressing their own local needs as well. (See Chapter 19 for further discussion of some of these issues.)

Means for Implementing Change

Many cities clearly have grasped the importance of action on climate change and other sustainability problems, and a great deal is known about the sector origins of greenhouse gas emissions and the means to reduce them. The big question is, besides the policy options mentioned above, are the tools that are needed to act on this knowledge close at hand?

One requirement is strong and comprehensive data. Any systematic attack on urban greenhouse gas emissions will falter unless data are developed that accurately characterize current emissions by sector and enable the tracking of changes over time. Otherwise, policy makers will remain blind not only to the progress, or lack of it, being achieved, but also to which investments are yielding the greatest returns in reductions.

Sophisticated methodologies have been developed for cities to use in identifying and assessing their emissions. For example, STAR Communities, a nonprofit organization focused on enabling progress toward sustainability in North American cities, has developed a detailed rating system with seven major sustainability goals, including for climate and energy. The World Resources Institute, C40 Cities, and ICLEI have collaborated on a comprehensive protocol that is accessible, is applicable to all cities, and enables clear and consistent identification, categorization, and measurement or estimation of greenhouse gas emissions from all sources. The protocol offers detailed guidance on reporting and accounting principles, inventory boundary definitions, sourcing data and calculating emissions, and how to use the data to track progress and set goals.[9]

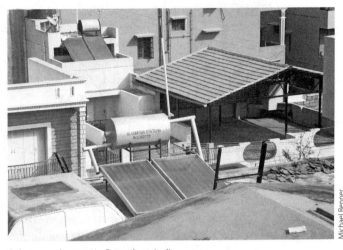

Solar water heaters in Bangalore, India.

Michael Renner

A second key requirement is money. Sustainability aspirations may soar, but financing constraints are the ballast that keeps urban policies closer to the ground, either because cities have limited borrowing and spending authority or because they do not receive adequate support from state and national governments. According to the World Bank, only 4 percent of the 500 largest cities in developing countries are deemed creditworthy in international financial markets, rising to 20 percent in local markets. Among C40 member cities, three-quarters have budgetary control over property/municipal taxes. Just one-half are able to directly retain tax revenues for local investment, while one-quarter receive an allocation of these taxes from higher levels of government. Close to 40 percent of C40 cities are able to issue their own bonds, while another 18 percent can do so only with approval from higher authorities. Unconditional borrowing power lies with just one-third of these cities; one-quarter need authorization. Finally, 25 percent of C40 cities indicate they have their own municipal bank, which gives them additional wherewithal for investment decisions.[10]

The ability of cities to rely on revenue-sharing arrangements, or on cash grants and loans by national authorities, varies considerably. In poorer cities of the developing world, multilateral development banks and a variety of donors may play an important role.

Dilemmas and Daunting Challenges

The challenge over the next several decades is an enormous one, given the need to dramatically reduce greenhouse gas emissions as well as other environmental and resource impacts. This requires not change around the edges, but a fundamental restructuring of how cities operate, how much they consume in resources and how much waste they produce, what they look like, and how they are structured. The remaining chapters in this section examine the functional characteristics that are chiefly responsible for cities' greenhouse gas emissions—buildings, urban form, transport, waste, and lifestyle-related deforestation—from a variety of perspectives and suggest some principles that could guide a transition to urban sustainability in general.

The dilemma facing the world's cities today is that the path to low urban greenhouse gas footprints is strewn with daunting obstacles, yet leads to major payoffs if those obstacles can be negotiated. For example, if current trends in urbanization continue unabated, urban energy use will more than triple, compared to 2005 levels, by 2050. Although hundreds or even thousands of cities worldwide are developing local climate action plans, their collective impact is unpredictable due to uncertainties in baseline data, the level and stability of commitment to implementation, and the suitability of the plans to specific local circumstances.[11]

Because cities consist heavily of buildings and other long-lasting infrastructure, decisions about them tend to have consequences that unfold over years and decades, making this tailoring of action plans to localities critical. As noted earlier, cities often do not control their energy supplies, so action options often are restricted to demand-side policies. The capital costs of the necessary greenhouse gas reduction measures will be substantial, yet securing the financing could be problematic. Despite what is already known about how to reduce urban greenhouse gas footprints, some trends still are running in the wrong direction; for instance, urban population densities in China actually are declining. (See Chapter 7.) Finally, cities also must contend with unhelpful national-level subsidy structures and the challenge of working within complex political/bureaucratic/physical systems, with many actors having different and sometimes conflicting interests and agendas.[12]

Daunting, too, are the depth and extent of the efforts required to achieve significant greenhouse gas reductions. A recent analysis of Toronto, Canada, a progressive and affluent city in an affluent country, estimated that it could reduce its per capita emissions 71 percent by 2031—if it undertook "aggressive" measures, including retrofitting all existing buildings for higher efficiency, using renewable heating and cooling systems, and promoting the proliferation of electric cars. This is a deeply ambitious plan even without factoring in the costs, which the study did not consider. All in all, perhaps it is not surprising that while "commitments" to greenhouse gas reductions and "targets" to aim for are plentiful, concrete and implementable plans for achieving them are rarer.[13]

And yet this inertia is not attributable to a lack of relevant options or tools, which abound. The study mentioned earlier (which predicted a tripling of urban energy use under a business-as-usual scenario) concluded that straightforward modifications in urban form, in conjunction with significantly higher gasoline prices, could reduce projected increases in urban energy use by more than 25 percent. And because the potential of energy efficiencies has only begun to be explored systematically, significant gains remain to be harvested there as well.[14]

A 2007 study by the international consulting firm McKinsey (updated in 2015) analyzed the potential carbon savings from various measures in power generation, manufacturing, transportation, residential and commercial buildings, forestry, and agriculture in North America, Western and Eastern Europe (including Russia), and other developed as well as developing countries, including China. According to the report, power generation and manufacturing account for less than half of the low-cost carbon-avoiding potential. Most of the potential lies in transportation and in residential and commercial buildings via improved efficiency. About one-quarter of the abatement potential comes at no net cost and would pay for itself; these are mostly efficiency measures. Nearly three-quarters of the abatement potential measures are technology-independent or use already-mature technologies. (The report bases these conclusions on an assumed carbon abatement cost of €40 ($43) per ton, which is significantly higher than trading prices for carbon registered in the European Union Emission Trading Scheme to date. If all abatement potential available for up to €40 per ton were captured, the report estimates that global atmospheric concentrations of carbon dioxide would remain at or under 450 parts per million at a cost of 0.6 percent of global GDP in 2030.)[15]

The particular tasks and opportunities facing various cities, especially in

JohnPickenPhoto

Urban wind turbines on a parking garage in Chicago provide power for the building's exterior lighting.

developing versus developed nations, may differ in important ways. A common denominator is supporting compact urban forms and relatively high population densities. Cities in developed countries may find the greatest additional gains in raising carbon prices for individual motorized transportation (i.e., higher gasoline prices), whereas cities in developing countries that are still building infrastructure can aim for compact forms integrated with careful transport planning so as to avoid locking themselves in to carbon-intensive transport systems. Likewise, although strong efficiency standards are necessary to reduce energy use in the built environment, in typical cities, the building stock changes by only about 2 percent per year. Retrofitting existing buildings is more difficult than building them to high standards in the first place, giving developing cities another advantage.[16]

Moving Beyond Technical Fixes

As with so many sustainability challenges, many observers stress the central importance of politics and stakeholder engagement in formulating plans and acting upon them. Technological fixes alone cannot reduce emissions deeply or rapidly enough. Particularly in the United States and other developed nations (U.S. per capita carbon emissions are nearly 10 times as high as the levels thought necessary to stabilize the climate, and European Union per capita emissions are 5 times as high as safe levels), behavior change is seen as indispensable. Creating incentives for that change will require persistent and careful efforts to build political support.[17]

Conversely, a study of patterns of urban sustainability in Asian cities argues that because many or most of the required technologies are already available, failure to engage stakeholders and communities is the principal barrier to progress. In the study, overcoming that failure and finding ways to link projects to job creation were the most important factors in allowing successful projects to be scaled up and spread to other communities. Thus, local government's role lies at the heart of successful implementation. (The value of local control and ownership of projects also was revealed in the general failure of efforts sponsored by international aid agencies to be replicated.) However, a local success might be too expensive to scale up, a national or regional policy might thwart it, or the strong local leadership might not be duplicated at higher administrative levels.[18]

Urban greenhouse gas emissions are reducible, but, like most dimensions of sustainability, this is a moving target and will only pose greater challenges with every year of delay. Rising populations, along with current and prospective climate changes, will increase stresses and demands on energy, agricultural lands, and other resources for cities worldwide. Coastal cities will face additional challenges: 8 of the world's 10 biggest cities lie along coasts, and several hundred million people (13 percent of the world's urban population, by a 2006 estimate) are directly vulnerable to sea-level rise. Rising waters, storm surges, saltwater intrusion, periodic or episodic submergence of buildings and infrastructure, and related problems, such as the cost of public works projects to armor coasts or relocate people and buildings, will complicate those cities' efforts to cope.[19]

Finally, cities might bear in mind that, although renewable energy supplies are increasing at encouraging rates, these energy sources are unlikely to permit substitution of carbon-based energy sources rapidly enough to avoid further climate disruption. As energy analyst Vaclav Smil has noted: in human history, it has always taken many years or decades to transition from one energy source to another, and all such transitions have involved adding new sources of energy to existing ones. With the possible exception of whale oil, no major energy source has ever been largely eliminated from the world's energy mix—yet that is exactly what the renewable revolution must accomplish. Only demand-side policies that succeed in sharply reducing energy consumption in transport, buildings, waste handling, and agriculture can address the urgent need to decarbonize energy. It is cities that must step up to the front lines of that battle.[20]

CHAPTER 7

Urbanism and Global Sprawl

Peter Calthorpe

Cities affect our lives in profound, self-reinforcing ways: they can be a source of economic innovation, a pathway for poverty reduction, a brake on logarithmic demographic growth, and a solution to climate change—or they can reinforce economic isolation, heighten environmental impacts, and engender social strife. They represent 80 percent of global economic output and 70 percent of total energy and greenhouse gas emissions. Cities are the superstructure for the culture, lifestyles, aspirations, and well-being of half of the world's population today and an estimated 70 percent by 2050. If they fail and become matrixes of gridlock, poisonous air, economic segregation, and environmental pollution, the planet will follow. If they succeed in lifting the next generation into sustainable productivity, integrating immigrants and working families into the next economy and living lightly on the land, they will contribute significantly to a civilized and sustainable future.[1]

Although issues and solutions in individual cities are unique, many of the best urban development strategies are universal and simultaneously address social, economic, and environmental challenges. Mixed-use, walkable, economically integrated, and transit-rich places define good urbanism in any city. More often than not, the positive outcomes that result cost less in upfront infrastructure, ongoing maintenance, and the average household cost of living. Cities that persist in low-density development that isolates activities and income groups and has poor transit will heighten economic and social ills as

Peter Calthorpe is Founder and Principal of the sustainable urban design and planning consultancy Calthorpe Associates and the author of many works on urbanism, including *Urbanism in the Age of Climate Change.*

well as emit more carbon. The latter effect has become a global crisis, especially in the developing world.[2]

The developing economies of the world will account for the vast majority of urban growth in the next half century. More than 90 percent of urban growth is occurring in the developing world, adding an estimated 70 million new residents to urban areas each year, much of it in the world's poorest regions. However, it is the emerging middle class within cities that drives carbon emissions, not the poor: 86 percent of energy-based carbon emissions come from upper-income populations. Therefore, it is the upper economic half of the global population that must adjust. Those economies and cities that are transitioning to a higher standard of living in the developing world must lay the groundwork for sustainable, low-carbon futures. In these cities, sustainable urbanism—places that are compact, mixed-use, walkable, and transit-oriented—is essential.[3]

It is important to keep in mind that regions and cities struggling with extreme poverty are not the source of the planet's climate change problem. Urban citizens in the developing world typically account for just one-twentieth to one-hundredth of the per capita greenhouse gas emissions of people in high-income nations, making carbon emissions in developing countries a lower priority. The average person worldwide accounted for 4.9 tons of carbon dioxide (CO_2) emissions in 2011, whereas the bottom quarter of the global population emitted only 0.3 tons per capita and the second quarter emitted 1.5 tons—slightly below the world target of 1.6 tons per capita for 2050 identified by the Deep Decarbonization Pathways Project. These populations do not own cars or air conditioners, live in large homes, or eat steaks. If they succeed in the next 30 years, their carbon emissions will still be reasonable.[4]

Urban sprawl in the developing world has many manifestations and just as many challenges. Clean water, adequate sewage treatment, consistent power, social services, affordable housing, gridlock, health care, economic development, and environmental decay—these are a short list of chronic issues in emerging cities. These challenges are all interconnected in a self-reinforcing cycle that either enhances or destroys opportunity and progress. And, in a systemic way, the form of the city affects each of these challenges. Urbanism at its best reduces per capita environmental demands while it makes services, infrastructure, and economic development more efficient, more cost-effective, more accessible, and more interconnected.[5]

There are three types of sprawl challenging cities around the planet, two of which dominate the developing world: high-density sprawl, which is unique to China, and low-income sprawl, seen across Latin America, Africa, and much

Left: High-density sprawl in Shanghai. *Right*: Low-income sprawl, the Kibera slum in Nairobi.

of Asia. Although such a taxonomy is reductionist and simplistic, it helps identify characteristics that cluster issues and opportunities in ways that are useful. (The third type is the well-known North American version of sprawl, so-called high-income sprawl, which has low densities, isolated uses, and an auto-dominated transportation system. Put in the context of the developing world, its most salient difference is that, since World War II, the wealthy and middle class have abandoned the city for suburbs.)[6]

Low-income sprawl dominates most of the developing world. In this case, relatively low-density housing at the metropolitan edge isolates the poor from access to jobs and services, while the wealthy remain in the urban center along with the concentrations of jobs and economic opportunity. Here, the low-income population—those most in need of economic and social access—are isolated and condemned to debilitating commutes on substandard transit. In contrast, the high-density sprawl typical in China does not isolate the poor at the urban edge, but it builds housing towers in a single-use "superblock" pattern that compromises local connections, walkability, and transit. Even though development is dense, land uses are isolated in the superblocks and surrounded by vast arterial roads, compromising the fundamental fabric of a healthy city: walkable streets and convenient transit.[7]

China's High-Density Sprawl

Of the two types of sprawl infecting growth in the developing world, high-density sprawl, found mainly in China, is unique, ironic, and tragic. One thinks of the high-rise, high-density buildings in many Chinese cities as inherently urban, but they are not. Smart growth and urbanism is more about connections, human scale, walkability, and mixed uses than it is about gross density. China's pattern of gated superblocks (often over 40 acres, or 16 hectares, each) and isolated uses is actually a high-rise version of the American suburb or a literal version of the failed American low-income projects of the 1950s and 1960s.[8]

In China, single-use residential blocks of largely identical units are clustered in superblocks surrounded by major arterial roads. Vast distances separate everyday destinations and create environments hostile to pedestrians. Sidewalks rarely are lined with useful services, and crossing the street is death-defying. Job centers are distant and commutes are long, especially for lower-income groups. In major Chinese cities, the gridlock expands to all hours of the day. The simple truth is that an auto-based city, even at low densities, cannot work. At the scale of China's development and density, it is impossible, no matter how many freeways and ring roads are built.[9]

Gridlock in major Chinese cities is ubiquitous, even though only one-third of households own a car. Here, freeway traffic and smog in Beijing.

In the last five years, China has built more than 30,000 kilometers of expressways, finishing the construction of 12 national highways a whopping 13 years ahead of schedule and at a pace four times faster than the United States built its interstate highway system. Over the last decade, Shanghai alone has built some 2,400 kilometers of road, the equivalent of three Manhattans. China's urban population is projected to grow by 350 million people by 2020, effectively adding today's entire U.S. population to its cities in less than a decade. China already has passed the United States as the world's largest automobile market, and, by 2025, the country will need to pave up to an estimated 5 billion square meters of road just to keep moving. With it all has come gridlock and poisoned air.[10]

Nonetheless, China's love affair with the car has blossomed into a torrid romance. Recently, nearly 1 million people poured into the Beijing

International Automotive Exhibition to coo over the latest Audis, BMWs, and Toyotas. And, like the U.S. cities of the 1950s and 1960s, Chinese cities are working to accommodate the explosive growth of automobile travel by building more highways, ring roads, and parking lots. But China is in danger of making the same mistakes that the United States made on its way to superpower status—mistakes that have left Americans reliant on foreign oil from unstable parts of the world, staggering under the cost of unhealthy patterns of living, and struggling to overcome the urban legacy of decades of inner-city decay. The choices that China makes in the years ahead will have an immense impact not only on the long-term viability, livability, and energy efficiency of its cities, but also on the health of the entire planet.[11]

If anything, due to China's high population density, the Chinese urban reckoning will be even more severe than that in the United States. Already, traffic in Beijing is frequently at a standstill despite the incredible pace of road construction (a "solution" akin to trying to lose weight by loosening your belt). The situation is so dire that Beijing, Guangzhou, and Shanghai are using a lottery to allocate a limited number of vehicle registrations. (See City View: Shanghai, page 109.) In August 2010, a 96-kilometer (60-mile) traffic jam clogged a highway outside Beijing for 11 days. There is a reason that no high-density city has ever been designed around the car: it simply does not work.[12]

The form of China's urban growth also will shape much of the country's environment—and not for the better. As Beijing orders up ever more freeways and parking lots, walking, biking, and public transit are declining. Since 1986, auto use has increased sixfold in Beijing, whereas bike use has dropped from nearly 60 percent of trips to just 17 percent in 2010. The congestion, air quality, and greenhouse gas impacts of this shift have been massive: Beijing remains one of the world's most polluted major cities. Merely to ensure blue skies during the 2008 Olympics, the city spent some $17 billion restricting traffic and shutting down factories. It even employed 50,000 people to fire silver iodide at clouds to release rain. The health damages caused by local air pollution, resulting largely from auto use and local coal and oil combustion, are very large: China's air pollution was linked to 1.2 million premature deaths in 2010—or, put in monetary terms, damages equivalent to 9.7–13.2 percent of the country's gross domestic product (GDP). (See Figure 7–1.) The problem is so severe that curbing local air pollution has become a major item on the government's policy agenda, driving plans to curb China's coal consumption.[13]

Across China, injuries to drivers, pedestrians, and cyclists are on the rise.

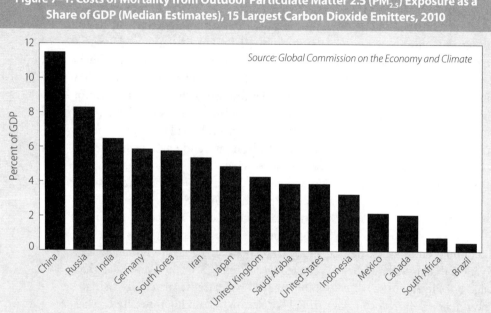

Figure 7–1. Costs of Mortality from Outdoor Particulate Matter 2.5 (PM₂.₅) Exposure as a Share of GDP (Median Estimates), 15 Largest Carbon Dioxide Emitters, 2010

From 1992 to 2004, the bicycle-related mortality rate increased 99 percent in Shanghai. Traffic fatalities in China are a severe problem by any measure, with various estimates ranging from 160 to more than 700 per day, among the highest in the world. The underlying reason for these trends is no mystery: bad urban planning.[14]

At the center of this planning is the superblock: a weapon of mass urban destruction developed in 1935 by the Swiss architect Charles Edouard Jeanneret (better known as Le Corbusier) and embraced wholeheartedly by China's efficiency-minded traffic engineers. Based on a network of wide, arterial streets, China's superblocks feature large, single-use development areas, often more than a quarter mile (0.4 kilometers) per side and designed like barracks, inconveniently located far from workplaces and shopping centers. The goal is to move cars efficiently; people are an afterthought. The ironic result is an alienating landscape that makes walking and biking difficult, which in turn increases congestion on the streets, with all the attendant social and environmental costs. Culturally, it is a tragedy for Chinese cities, which are seeing traditional neighborhoods, where friends and family could easily pop in for tea and conversation, destroyed by misguided development. Now, people have to take a crowded bus or, if they are lucky, a car.[15]

The congestion will only get worse. The international consulting firm McKinsey projects that nearly 64 percent of China's population will live in urban areas by 2025, up from 48 percent in 2010; by then, there will be 221 Chinese cities with more than 1 million people. Can China afford it? Transportation already accounts for 40 percent of China's oil demand, according to the International Energy Agency, and is expected to reach 65 percent by 2035. The Carnegie Endowment for International Peace projects that the country's vehicle fleet could grow from more than 200 million today to as many as 600 million by 2030. By that year, oil consumption is projected to have nearly tripled. Needless to say, finding all those resources is going to be a challenge— that is, if Chinese cities don't choke on pollution and gridlock first.[16]

The figures are daunting. But the engineers who run the Chinese ship of state are nothing if not good at math, and they have committed to making real changes: building mass-transit systems, introducing alternative fuels such as ethanol, and promoting fuel efficiency and electric cars. There are still other things Chinese cities can do at the margins, such as introducing the sorts of "congestion pricing" schemes—taxes on vehicles as they enter certain areas—that have worked wonders in places like London and Singapore. Unfortunately, numerous studies have shown that the numbers don't quite add up, as these technical fixes tend to ignore China's fundamental problem: cities designed around cars, not human beings.[17]

China's superblocks result in auto-oriented environments. Here, at the intersection of two arterials in Kunming, even the crosswalks leave a long and challenging trip for pedestrians.

Peter Calthorpe

The problem is not just the increase in cars in high-density environments; it is the coarse nature of the typical Chinese road network as well. The current unholy alliance of superblocks and oversized arterials not only frustrates pedestrians and cyclists, but it fails for cars as well. An arterial system of wide, "canyon-like" streets creates a hostile environment for pedestrians and bicyclists. Wider streets lead to increased crossing distances, longer distances to intersections for pedestrians, higher traffic concentrations on fewer roads, few alternative routes for emergencies, and complex traffic movement at intersections that threaten pedestrian and bicycle comfort and safety. Often with few entrances, superblocks add to the circuitous access routes for cars as well as pedestrians.

The alternative is a more traditional city grid of streets with higher intersection density and a broader range of street types. In this tried-and-true street network, high volumes of through-traffic are dispersed over parallel and smaller roads or onto pairs of one-way streets. Pedestrian and bike zones are protected and enhanced on all streets. Transit lines and bus rapid transit (BRT) systems gain dedicated lanes, and auto-free streets enhance alternate modes. Such a street network creates a radically different urban landscape, one that replaces China's isolated superblocks with small courtyard blocks. Streets are the DNA of a city; their scale and how they mix public spaces, shops, pedestrians, bikes, and cars is critical to the health of a city.

The transportation and carbon emission problem in China cannot be solved without fundamental changes in urban design and land-use planning. For the past five years, the Energy Foundation has sponsored demonstration projects in six Chinese cities with planning for a combined population of over 10 million to show that the alternative is feasible, efficient, economically strong, and socially advantageous. The projects have worked so well that the national government has adopted design standards for areas within walking distance of transit stations that reinforce walkability, mixed-use, and small-block urban design. These transit-oriented development (TOD) standards, adopted in 2015, ultimately will affect a growing population as the national investment in high-capacity transit expands by 10,000 kilometers in the next 10 years.[18]

The use of small blocks is a radical departure from the superblock pattern prevalent in most of China. As a result of this more-diverse street network, small blocks create a human-scaled environment of shared courtyards; smaller, more walkable, local streets; a fine-grained network of public spaces around the blocks; and a greater land-use mix in a smaller area. (See Figure 7–2.) This structure, along with the presence of only one-quarter of the inhabitants of a superblock, enables a stronger sense of community. The intrinsic courtyard pattern recalls the historic city forms throughout China, from the traditional *hutong* courtyard housing to the form of the Forbidden City. It emerges here at a different scale, but provides the same urban layering, from public street to semi-public courtyard to private home.[19]

Figure 7–2. Small Blocks versus Superblocks

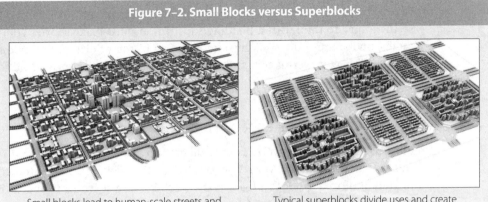

Small blocks lead to human-scale streets and public spaces.

Typical superblocks divide uses and create large arterials.

Note: The two images contain the same land area and quantity of development.

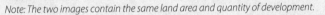

The typical small block should have dimensions of approximately 100–200 meters per side, with a block area of 1–1.5 hectares. This area will result in just 400–700 dwellings, housing at most 1,500 people. This number is small enough for most people to recognize one another and establish strong social connections. In contrast, superblocks contain easily 5,000 people, a scale in which many people become anonymous and children are more frequently exposed to strangers. In addition, small blocks increase the opportunity for the kind of street-side shops and local services that support street life and neighborhood identity. All of this adds up to more walking and less auto use, along with convenient transit.

It turns out that such urban forms have a big impact on travel behavior and therefore energy consumption and carbon emissions. The Energy Foundation conducted studies in Jinan in which people from a variety of neighborhoods recorded their trips and distances. Regardless of income, the average citizen living in a superblock drove four times the distance as others living in more walkable mixed-use areas. This fourfold increase, if expanded by more superblock construction, will cascade into traffic jams, polluted air, more energy imports, and more carbon emissions. With China's cities projected to swell by 350 million additional people, mainly from rural-urban migration, over the next 20 years, this difference alone could represent massive quantities of new CO_2 in the air.[20]

China's leaders have a limited window of opportunity to plan for prosperous, livable, low-carbon cities. They have the resources and the wherewithal to make the sweeping changes required to avert an impending social and environmental disaster of proportions unknown in human history. It might seem strange to think that a budding superpower must make shorter commutes, public transport, walking, and bicycling its top priorities. But unless it does, China's powerful economic engines—its cities—will slowly grind to a halt.[21]

Mexico and the Challenge of Low-Income Sprawl

Most of the developing world is on a different trajectory than China and is suffering from a different type of sprawl. Rather than government-controlled migration of rural poor to urban districts of high-rise apartments, much of the rural poor around the world access cities through slums, *favelas*, *barrios*, and informal (illegal) housing. Instead of oversized streets and new metro lines, the streets are undersized, discontinuous, and uncontrolled—and are overwhelmed by cars, trucks, rickshaws, tricycles, and jitneys. In the place of public bus systems, there are jitneys, *colectivos*, or other types of privately owned buses. These are largely the organic, privately operated minibuses with polluting engines that run on chaotic routes with irregular schedules, causing congestion with their ad hoc stops. Rather than state-of-the-art infrastructure, state-sponsored schools, and health services, there are instead chronic shortfalls in all municipal services.[22]

Much of this difference is the product of low per capita incomes and weak or corrupt government. An estimated 1 billion people live in urban slums in developing countries. This leads to a different set of priorities than those in China: poverty and slum revitalization, economic integration and workforce productivity, basic public services, and environmental cleanup are all urgent needs. Until the urban poor can become productive, and their communities are secure and integrated into the life of the city, their lives and the city's economy will suffer.[23]

Improving the lives of poor urban dwellers is a big part of the global city-building challenge of the coming decades. It is well known now that transforming slums involves multiple challenges: foremost land security, safe and consistent utilities, a range of social services, and efficient transportation. Revitalization initiatives have been particularly effective when they integrate social, economic, and infrastructure programs. For example, programs in

Jamaica and Brazil combine microfinance, land tenure, crime and violence prevention, investments in day care, youth training, and health care along with physical upgrades. Such efforts take money, consistent governance, and good urban design.[24]

Many programs and policies have demonstrated that slums can be transformed into vibrant and well-integrated parts of a city. But one key challenge stands out: many of today's urban poor live in remote locations due to the high cost of housing at the city core and because social housing policies push them toward cheap but remote land. Living in peripheral urban locations, particularly without adequate access to efficient transport services, can mean exclusion from a range of urban facilities, services, and jobs—and very long commutes for those lucky enough to have jobs.[25]

The good news is that solving the cluster of issues surrounding urban poverty—such as air quality, congestion, water pollution, and affordable housing—leads inevitably to a low-carbon city. As is the case in the developed world, smart urban design strategies solve a range of social, economic, and environmental ills. A new regional study for Mexico City by Centro Mario Molino and Calthorpe Associates connects the dots.[26]

Mexico City is not a poor region by global standards; as a whole, Mexico's per capita gross national income is about $10,000, whereas about half the globe subsists on less than $4,000 a year. Even though Mexico City is wealthier than many cities, it still struggles with the all-too-familiar urban challenges: *barrios*, informal housing, disastrous air quality, gridlock, social stratification, and chronic water shortages, to name a few. Perhaps more important, it represents a metropolis in transition to a more middle-class economy and, as such, could be a model for urban forms that are critical to the climate change imperative and its list of social challenges. Although the poorest cities of the world should focus on basic health, well-being, and equity, emerging economies like Mexico must find a path to living well while living lightly on the land and air. As Mexico City's population gains in wealth and consumption, must the global pattern of higher carbon emissions lock in?[27]

During the past half century, the Mexico City region has become less centralized, as the poor have been pushed to sprawling edge communities. More than half of the city's 20 million people live outside of the Federal District, the historic core. The poor generally have moved to unstable and flood-prone areas with limited infrastructure. Meanwhile, the middle class and wealthier residents occupy areas with stable soils and gentle slopes in the southern and western areas of the city, closer to the job centers. Although the region's poor and rich have never lived side by side, recent growth and expansion have magnified the scale of segregation.[28]

As Mexico City spreads outward and trends toward lower densities at the periphery, the global pattern of low-income sprawl is manifest. Physical expansion has outpaced population growth. From 2005 to 2010, the average annual growth rate of the region's population was 0.9 percent, whereas that of the built surface was 1.2 percent as economic activities have remained far more centralized. Moreover, the growth has been at the perimeter of the region; jobs and the wealthy are clustered in the historic center of the region, while the poor are spread to the edge. (See Figure 7–3.)[29]

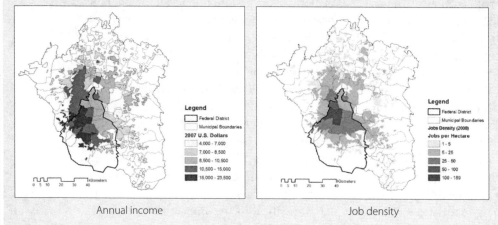

Figure 7–3. Income and Job Segregation in the Mexico City Metropolitan Area, 2008

Source: Centro Mario Molina and Calthorpe Analytics

Annual income

Job density

Mexico's urban geography parallels many economically emerging metropolitan regions. The poor are isolated in informal developments or social housing projects. More often than not, these are remote, creating a debilitating disconnect between the poor and the economic opportunities, social capital, and social services that the central city enjoys. Compounding the physical isolation, the lack of dependable, efficient transit makes commutes an ordeal for the poor. In some cases, those at the urban edge have two-to-three hour one-way commutes. This is the plight of the autoless population that dominates those at the urban edge.[30]

Like many cities in China, Mexico City is yet to fully vest in cars; auto ownership is about 170 vehicles per thousand people (compared to more than 530 vehicles per thousand people in Los Angeles or 320 in London). As in most emerging economies, the urban poor cannot afford cars. But Mexico City is headed in the wrong direction. From 1980 to 2010, the light-duty vehicle fleet grew from 1.8 million to 5.4 million cars; the modal share of automobiles jumped from 20 percent in 1995 to 28 percent in 2005 and continues to climb. Even while accounting for less than one-third of all trips, private cars emitted more than half of all road-based greenhouse gas emissions, including from freight vehicles and public transportation.[31]

Mexico City's congestion and deadly air quality are legendary, even though most residents do not own cars. This is because they depend largely on ad hoc *colectivo* buses and *combi* minivans, rather than on public bus routes or high-quality express transit systems. Their trips are slow, indirect, and, more often than not, in polluting vehicles. In addition, a massive taxi fleet, augmenting the fragmented informal transit routes, produces 1.75 times more greenhouse gas emissions than the notably inefficient transit system. This transportation system drives up congestion, emissions, energy consumption, and air pollution while providing very low average travel speeds and long commutes for the poor. As the city geography has expanded overall, door-to-door transit and auto travel speeds are dropping, and air quality remains toxic. This air pollution disproportionately harms poorer residents, since they are more likely to live in neighborhoods with higher levels of pollution—particularly near highways—and experience more-frequent direct exposure to fumes from low-quality transit. In 2007, the average commute to work by car for someone in the urban center was less than two-thirds of the average public-transit commute by someone in the suburbs (47 minutes compared to 73 minutes).[32]

Like those of so many emerging cities, Mexico City's ills are driven by urban form and transportation systems, the DNA of the city. Where development happens, what form it takes, and what kind of transit is available is fundamental to meeting each challenge. In 2015, Centro Mario Molina helped conduct a scenario process for Mexico City with the goal of revealing the impacts and tradeoffs of differing urban growth strategies through the year 2050. The process developed differing future-growth scenarios and analyzed those scenarios across a range of metrics. One key metric was the quality of transit and travel times. Another was equity: the remoteness of most low-income and affordable housing. The urban design imperative was to connect low-income neighborhoods to jobs and regional assets.[33]

To better understand the social, economic, and urban form of the region, the scenario team developed a new type of mapping that combined accessibility, urban form, and income to produce 48 distinct neighborhoods or "Place Types." Four types of accessibility were defined by proximity to the major job centers of the region and access to formal transit. Urban form was defined by combining two factors: housing density and walkability. Finally, three levels of income were combined with the 16 proximity and urban types. This mapping of income, regional location, transportation facilities, density, and urban form created a unique and revealing picture of the city.[34]

Each Place Type told a story. Only 26 percent of the city's populations live within 5 kilometers of a regional center, the study's definition of good proximity. Only 29 percent of households live close to transit, defined as high-capacity buses or trains running on fixed schedules with high-quality service—effectively the city's metro and BRT lines. Sadly, more than two-thirds of the population is close to neither a job center or structured, high-quality transit. They live far from the economic and cultural assets of the city.[35]

Things look better in terms of urban form. Three-quarters of the population lives in areas of more than 50 houses per hectare (not unlike the densities of London's townhouse neighborhoods), and 68 percent live in areas with human-scale streets and block sizes under 1.2 hectares (a traditional size in most walkable districts). In fact, the city is dominated by good urban form: 60 percent of the population live in walkable areas with appropriate densities, and only 18 percent live in what generally is considered sprawl—lower-density areas without a walkable street grid.[36]

The scenario process generated three future-growth scenarios for Mexico City for the year 2050, based on four variables: the percent of housing infill, the location of new job centers, the level of transit service, and urban form. In the three scenarios—labeled Trend, Moderate, and Vision—2.7 million new housing units of varying types are accommodated, but some of the scenarios use higher-density types. In all cases, there are 2.9 million new jobs, but the location, regional proximity to housing, access to transit, and location within mixed-use areas of those jobs varies. Transportation varies largely by the quantity of new structured mass transit constructed, but all scenarios use studied and proposed alignments. (See Figure 7–4.)[37]

The Trend scenario explores extending the current direction of urban development, characterized by the inefficient land-use policies that encourage urban sprawl. Employment accessibility still shows a disconnection between the areas that concentrate housing and those that concentrate employment. Regarding access to transport, a moderate expansion of transit is assumed.

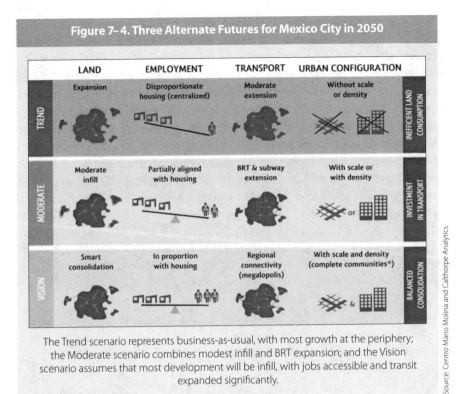

Figure 7–4. Three Alternate Futures for Mexico City in 2050

	LAND	EMPLOYMENT	TRANSPORT	URBAN CONFIGURATION	
TREND	Expansion	Disproportionate housing (centralized)	Moderate extension	Without scale or density	INEFFICIENT LAND CONSUMPTION
MODERATE	Moderate infill	Partially aligned with housing	BRT & subway extension	With scale or with density	INVESTMENT IN TRANSPORT
VISION	Smart consolidation	In proportion with housing	Regional connectivity (megalopolis)	With scale and density (complete communities*)	BALANCED CONSOLIDATION

The Trend scenario represents business-as-usual, with most growth at the periphery; the Moderate scenario combines modest infill and BRT expansion; and the Vision scenario assumes that most development will be infill, with jobs accessible and transit expanded significantly.

Source: Centro Mario Molina and Calthorpe Analytics

Finally, for urban configuration, as housing continues to spread at the edge, a significant percentage of new communities will lack the desired scale, density, and walkability of typical infill projects.[38]

The Moderate scenario posits modest infill within existing urban areas, although sprawl to the periphery does keep increasing. The scenario envisions a large investment in structured public transport that increases its coverage by 50 percent via new BRT lines. It includes a better balance between areas with walkable urban forms and areas lacking human scale or urban density.[39]

The Vision scenario focuses on infill development by strengthening and decentralizing centers of employment. It seeks to establish a balance between the number of jobs and housing units on a sub-regional basis, effectively creating a polycentric metropolis linked by high-quality transit. Public transport service capacity increases, focused mainly on promoting regional connectivity. This scenario assumes that most of the new development will be mixed-use, human-scaled, and walkable—a city of transit-oriented developments.[40]

The study demonstrates that accommodating a population increase of 8 million through 2050 with the Vision scenario will have a dramatic effect on the overall environmental, economic, and social performance of the city when compared to a "Trend" future. (See Figure 7–5.) Land consumption is cut by 78 percent with many tertiary implications, including reducing infrastructure costs cumulatively by a similar amount, even while producing 40 kilometers of new BRT line per year. Water consumption is down 13 percent overall, and household costs and carbon emissions are both down 9 percent. One key metric, average travel time per day, which drives congestion and air quality, is down 23 percent for all auto and transit riders. This, along with a reduction in auto kilometers traveled of about 13 percent, takes stress out of the circulation system and provides better mobility for the poor.[41]

Figure 7–5. Urban Footprint Analysis for Mexico City Showing Positive Impacts for the "Vision Scenario" Across a Range of Economic, Social, and Environmental Measures

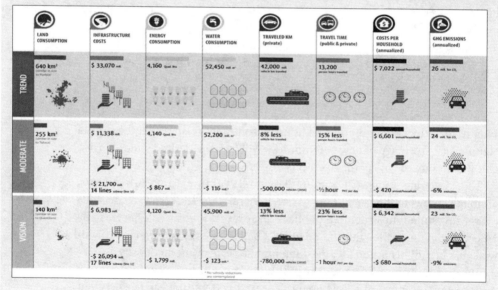

These numbers reflect an important dimension of urbanism as a solution to climate change—the so-called cobenefits. Solving multiple challenges simultaneously with better urban forms is the most cost-effective strategy for carbon reduction, as it results in savings in many other key areas: land, water, infrastructure, travel time for the poor, etc. Carbon reductions, thought of in this context, are more than free. Moreover, such urbanism creates coalitions of currently isolated advocacy groups that can, over the long term, change public policy—a key to implementation.

Conclusion

The Mexico City study highlights the significant differences between most developing economies and the two other forms of global sprawl: high-density and high-income. It is interesting to note that much of Mexico City has walkable urban configurations: human-scaled blocks, pedestrian-friendly streets, and accessible local destinations. This walkable urban form no doubt results from low auto ownership rates and, therefore, the intrinsic need for communities to accommodate local foot trips. But in the developed economies of the world, the need for walkable urbanism has been displaced by low-density sprawl that accommodates cars—generating the classic suburban sprawl typical not only in North America and Australia, but increasingly in Europe. China combines the density of urban places without the walkability, and therefore shares the need for new urban forms with the rest of the developed world.[42]

Cities need transit, but in very different forms. The developed world needs land uses and transit features that are good enough to move people who are rich enough to have a choice out of their cars. China has robust transit systems in its major cities, but in secondary cities it needs affordable transit and more-walkable neighborhoods. The developing world needs massive quantities of affordable high-capacity transit—most likely BRT on auto-free streets.[43]

All of the city types share in the cobenefits of urbanism. As urban form and regional structure improve, all the average per capita metrics studied typically get better: air quality, kilometers driven, fiscal impacts, household costs, infrastructure costs, land consumption, carbon emissions, water consumption, and health costs. All of these metrics are important to each form of global sprawl. However, each city and place will have additional unique metrics that matter: in Mexico, transit travel times for low-income workers jump out; in big Chinese cities, it is smog and gridlock; in many Western cities, it is affordable housing and vehicle miles traveled.[44]

Although varying greatly in degree, the sprawl challenge in the developed world and in China is concerned largely with the quality of life, economics, and environmental impacts of the new and old middle class. Certainly, there are issues in each concerning the needs of the poor, but these pale in comparison to poverty in cities infected with low-income sprawl. Moreover, many low-income cities lack the central control or investment capacity to direct the form, location, or even infrastructure of new growth. So, for China and the developed world, shifting metropolitan forms toward better outcomes is an issue of political will, whereas, for developing economies, it is an issue of capacity. The two developing-world sprawl types described here therefore need differing implementation strategies to deal with different challenges. The means may differ, but the goals converge. Better transit, walkable neighborhoods, higher densities, balanced jobs/housing districts, and more infill provide a good outcome for all futures.[45]

CITY VIEW

Shanghai, China

Haibing Ma

Shanghai Basics

City population: 24.2 million
City area: 6,341 square kilometers
Population density: 3,809 inhabitants per square kilometer
Source: See endnote 1.

Shanghai freeway network at night.

whiz-ka

Sustainability Successes and Challenges in a Chinese Megacity

Over the last three decades, rapid economic growth has transformed China into the world's second largest economy, making it the biggest energy consumer and emitter of greenhouse gases. The central government has made energy conservation and emissions reduction top priorities, a mandate that also has been delegated to the provincial and local levels. In Shanghai, one of the largest cities in China, the movement toward sustainability has been driven mostly by pressure and guidance from the central government, but also by the demands of the 2010 Shanghai World Expo, with its focus on sustainable urban living.

China's Influential Five-Year Plans

The central government's Five-Year Plans lay out the overarching direction for China's economic and social development. The 11th Five-Year Plan (2006–10) was the first to establish mandatory targets for energy conservation and emissions reduction, directing China to cut its energy intensity (energy consumption per unit of gross domestic product, or GDP) by 20 percent and to reduce top pollutants, such as sulfur dioxide (SO_2) and those reflected by chemical oxygen demand (COD), by 10 percent. The 12th Five-Year Plan (2011–15) mandated additional cuts in these pollutants of 16 percent and 8–10 percent, respectively. The Plan also established China's first mandatory target for greenhouse gas reductions, with a goal of reducing the country's carbon intensity (carbon dioxide (CO_2) emissions per unit of GDP) by 17 percent over five years.[2]

The Five-Year Plans also set goals in other areas that are key to sustainability, such as industrial and agricultural water use-efficiency, arable land and forest coverage, and forest reserves. The lower levels of government then release their own plans. In its 11th Five-Year Plan, the Shanghai municipal government stated an overall guiding principle of "fully pursuing a sustainable path." In its 12th Five-Year Plan, the city emphasized resource conservation and environmental protection. Specific policy initiatives include:[3]

- The 11th Five-Year Plan on Energy Development set a wide range of targets for energy security, energy mix, limits on total energy consumption, rate of energy efficiency improvement, and newly installed renewable energy generating capacities.[4]
- The 11th Five-Year Plan on Energy Conservation targeted a wide range of sectors, including industry, buildings, and transport. It mandated that all new residential and public buildings abide by the National Energy Conservation Standard and achieve at least a 50 percent reduction in energy consumption compared to the base house model. The Plan also aimed to raise the share of public transport in the modal mix from 24 percent to 30 percent and to raise the share of public commuting by rail from 14 percent to 40 percent. It set a goal of up to 750 hybrid public buses, 4,000 taxis, and 10,000 other vehicles. The 12th Five-Year

Plan on Industrial Energy Conservation and Comprehensive Utilization set targets across all industrial sectors.[5]

- The 11th Five-Year Plan on Environmental Protection and Ecological Construction set limits on major pollutants such as SO_2 and those reflected in COD. It suggested that environmental investment should account for at least 3 percent of the municipality's GDP. The Plan also divided the municipal land area into different "ecological functions," such as green land, forest, and wetland, and suggested appropriate rules for management and operation.[6]
- Shanghai is one of seven designated pilot cities and provinces for China's CO_2 emissions trading schemes. In November 2013, the municipal government issued a set of management rules and supporting regulations. Shanghai's Emission Trading Scheme covers 10 sectors, including iron and steel, petrochemicals, and electricity. The roughly 200 participating enterprises account for 57 percent of the city's total CO_2 emissions.[7]
- The New Energy Vehicles Promotion Plan, released in 2014, set a target of having 13,000 "new energy" (hybrid and electric) vehicles on the road by the end of 2015. It also requires that at least 30 percent of newly added public service vehicles be new energy vehicles.[8]

Key Achievements to Date

Shanghai's energy consumption and greenhouse gas emissions have yet to peak, but the municipal government has sought to achieve a relative decoupling of economic growth from energy use and environmental impact. From 2006 to 2010, Shanghai recorded average annual GDP growth of 11.2 percent, but the city's energy intensity fell 20 percent, avoiding some 28 million tons of coal equivalent (tce) in energy use. By the end of 2013, Shanghai achieved another 22 percent reduction in energy intensity, surpassing the 18 percent target set in the 12th Five-Year Plan.[9]

The city's energy mix is changing as well. The share of coal dropped from 53 percent in 2005 to 49 percent in 2010 and was expected to reach 40 percent by 2015. Non-fossil fuel sources accounted for 6 percent of primary energy consumption in 2010 and were expected to reach 12 percent by the end of 2015. In 2010, China's first offshore wind farm became operational in Shanghai's Donghai Bridge area. The city's total installed wind power capacity is projected to exceed 200 megawatts in 2016.[10]

Much of the reduced energy demand comes from phasing out or restructuring energy- and emission-intensive sectors. During the 11th Five-Year period, some 2,873 restructuring projects in Shanghai resulted in savings of 4.8 million tce. Some sectors, such as iron alloy processing and plate glass manufacturing, disappeared entirely from the municipality. A total of 149 dangerous chemical factories have either been shut down or relocated. By the end of 2013, the service sector accounted for 63 percent of Shanghai's GDP, close to the goal set in the 12th Five-Year Plan.[11]

Between 2006 and 2010, Shanghai's industrial enterprises with annual revenue larger than 20 million yuan ($3 million) achieved a 28 percent reduction in energy consumption per 10,000 yuan of turnover. Meanwhile, a total of 804 industrial energy conservation projects has reduced energy use by 2.5 million tce. More than 800 energy management contracts contributed to additional savings of 400,000 tce.[12]

During the 11th Five-Year period, Shanghai retrofitted close to 30 million square meters of building space, installing 1.64 million energy-conserving air conditioners and more than 24 million efficient light bulbs in residential buildings. Energy audits of more than 700 energy-intensive industrial facilities and 286 public buildings saved up to 3 million tce.[13]

Shanghai's emission trading exchange was officially launched in November 2013. By September 2015, more than 4 million tons of CO_2 had been traded during the two initial trading periods. Shanghai achieved a more than 8 percent reduction in carbon intensity in 2014, surpassing its annual average goal. With another 2 percent reduction in 2015, the city would meet the emission intensity target set in its 12th Five-Year Plan.[14]

The New Energy Vehicles Promotion Plan projected that 400 new energy vehicles would be put on Shanghai's roads in 2013, another 3,600 in 2014, and 9,000 in 2015. Actual additions ran far lower, at just above 2,000 in 2013–14. To promote greater uptake, Shanghai's government will exempt up to 20,000 hybrid and electric cars from the city's quota on new vehicle licenses (imposed in 1994 to slow the growth in motor vehicle volume). The government also provides financial incentives for buyers of new energy vehicles.[15]

Shanghai's first metro line became operational in 1993, and, by the end of 2014, the network had expanded to 14 lines, with an operational length of 548 kilometers. Within three decades, Shanghai succeeded in building the world's largest rapid transit system by route length and the second largest (after Beijing) by number of stations and annual ridership, with 2.5 billion rides delivered in 2013. Four new metro lines and extensions to five other lines are expected to be completed by 2020.[16]

In another notable achievement, between 2006 and 2010, Shanghai added 6,600 hectares of newly vegetated area, raising the city's green coverage from 37 percent to 38.2 percent. Some 12,000 hectares of forest were added, increasing the city's forest coverage from 11 percent to 12.6 percent. The goal was to raise green coverage to 38.5 percent and forest coverage to 15 percent by the end of 2015.[17]

Problems Encountered

Shanghai has experienced some notable failures in its transition to a more sustainable city, among them the once-acclaimed Dongtan Eco-city project on Chongming Island. In 2005 and 2008, the leaders of China and the United Kingdom signed memoranda aimed at turning the Dongtan area into the world's first "eco-city." According to the plan, Dongtan Eco-city would use 60 percent less energy and 88 percent less water than "regular" cities of its size. It would generate 83 percent less waste, and its citizens would have a 60 percent smaller "ecological footprint" than inhabitants of the Shanghai municipal region. The entire eco-city would be powered by renewable energy sources, achieving zero carbon emissions.

The first phase was to be completed by 2010—in time for the opening of the Shanghai World Expo. However, the planning and construction work was put on hold multiple times. In 2006, dozens of Shanghai-based officials—including the then-mayor of Shanghai as well as the general manager of Shanghai Industrial Investment Co., Ltd., the company that owned the development rights to the Dongtan area—were arrested for economic crimes. Although official reports indicate that the eco-city project is still moving forward, it is unknown when it will be completed and to what extent its original goals will be met.[18]

Another problem relates to energy efficiency in buildings. In November 2010, a high-rise apartment building in Jingan district caught fire, killing 58 residents and injuring dozens more. Although the direct cause of the fire was misconduct by non-certified electric welders, flammable products stockpiled on-site made the fire uncontrollable. Ironically, these products were heat insulators, as the building was undergoing comprehensive renovation for energy efficiency. The incident points to the need for high-quality implementation of desired efficiency goals.[19]

Structural Change and Administrative Powers

Shanghai's transition to a service-oriented city is largely the result of national strategic decisions. As early as the late 1970s, when China began its economic reforms, the central government wished to revamp Shanghai's economy by moving heavy industries out of the municipal area. The industrial sector's share of the city's GDP declined from 77 percent in 1978 to just over 36 percent in 2013, while the service sector's share grew from 19 percent to 63 percent. Given that the service sector generally requires fewer natural resources than other economic activities, this has helped to foster a more sustainable economy in the city.[20]

Dedicated and effective administrative power contributed to Shanghai's quick economic turnaround and to its sustainability achievements. Once policies are decided, China's top-down political system allows for relatively easy and rapid implementation. In Shanghai, many domestic and international experts were invited or contracted to conduct relevant studies, helping to improve policy design and make implementation more efficient.

Both China's central government and the Shanghai municipal government viewed the 2010 World Expo as a window to showcase China's "soft power," or economic and cultural influence. China also wants to demonstrate that it is fully embracing the "advanced" concept of urban sustainability. Many new technologies and practices introduced at the event have been incorporated, or at least considered, in Shanghai's development. The municipal government views the vision that it presented at the Expo as a commitment to the world and feels obligated to retain that spirit.

Citizens Struggling to Make Themselves Heard

Because China remains a largely top-down society, citizens rarely are included in policy making. Although selected citizens participate in proposing and discussing new policies through the municipal-level People's Congresses and People's Political Consultative Committees, this input appears to have made little difference. Many Chinese citizens therefore have grasped social media as a tool to express their collective will. Images of local air pollution readings are readily "retweeted" and create heated discussion. Although relevant government entities, such as the local environmental protection agencies, usually respond verbally in a timely manner, no significant improvement has been achieved.

The Chinese government is more concerned about physical gatherings of citizens. Due in part to the lack of standard feedback channels, citizens rely increasingly on assemblies and demonstrations to influence policy agendas, especially in the case of environmental issues. In June 2015, more than 10,000 residents of Shanghai's Jinshan district came together to protest plans for a new paraxylene production facility that raised pollution and health concerns. Similar gatherings have occurred in several cities across China in recent years. Neither the central nor the local governments have yet found appropriate policy mechanisms to address citizens' environmental concerns expressed through these means.[21]

Lessons Learned

Given the way that China's policy system works—with central government targets subsequently translated into regional and local policies—implementation is, in some ways, easy. Due to the lack of bottom-up approaches, however, very few innovative and unique policies emerge from local needs and are tailored to the local context. A megacity such as Shanghai needs systematic guiding policies that are designed specifically for it, in order to make the city's transition to sustainability more effective.

Although Shanghai's sustainability policies may seem comprehensive, most of them lack real synergy. Implementing measures carried out by different agencies may, in some cases, be repetitive, resulting in a waste of resources. A specially designed, sustainability-themed working group that includes representatives from all relevant agencies would greatly improve policy coordination and efficiency.

Last, but not least, mistakes made as a result of quick implementation of policies in Shanghai reveal an ongoing lack of understanding of sustainability. Both the Dongtan Eco-city and building fire examples demonstrate that sustainable development in Shanghai is often not viewed as a systematic endeavor, but rather as a series of numeric targets with no concrete meaning.

Haibing Ma is China Program Manager at the Worldwatch Institute.

CHAPTER 8

Reducing the Environmental Footprint of Buildings

Michael Renner

In June 2015, the International Renewable Energy Agency (IRENA) moved into its brand new headquarters in Masdar City, Abu Dhabi. The 32,000-square-meter complex, consisting of three interconnected structures, is among the most sustainable buildings in the United Arab Emirates and uses about one-half the water of similar office spaces. Solar thermal units meet three-quarters of hot water demand, rooftop solar panels generate a portion of the electricity, and passive design and smart energy management systems make the structure highly energy-efficient. IRENA's new headquarters is one of the latest high-profile efforts to make the footprint of buildings—offices, commercial structures, residences, and other types—more in line with what is needed for environmental sustainability.[1]

Buildings are among the biggest users of energy, water, and materials, and they contribute substantially to greenhouse gas emissions. Aligning building construction and management with sustainability goals requires a wide range of policies, particularly since the global building stock is growing rapidly. Navigant Research forecasts that the world's building floor area will grow by 24 percent between 2013 and 2023, from 138 billion square meters to 171 billion square meters.[2]

Whereas cities in emerging and developing economies are expanding their building stock rapidly, cities in the older industrialized countries, especially in Europe, are more concerned about improving the performance of existing buildings. Thirty-five percent of buildings in the European Union (EU) are more than 50 years old. And even though 2 million new homes are built in

Michael Renner is a senior researcher at the Worldwatch Institute and Codirector of the *State of the World* 2016 report.

the EU every year, the current housing stock still will account for nearly 70 percent of all buildings in 2050. In the United States, 40 percent of owner-occupied homes in existence in 2013 were built before 1970, and two-thirds were built before 1980. Less than 15 percent of all buildings in New York City are expected to be replaced by 2030.[3]

New buildings such as IRENA's sleek and efficient headquarters offer, in principle, a wide latitude of choices for design, systems, and materials. Cities in emerging and developing economies, in particular, can avoid being locked into high resource consumption by choosing efficient new buildings. Even cities with a mature building stock can reduce their energy and environmental footprint, with the help of retrofits that improve insulation and air tightness or replace inefficient lighting and equipment. Berlin, Germany, has made major strides by nearly halving the energy use of 273,000 apartments in the eastern part of the city.[4]

IRENA

IRENA's new headquarters in Masdar City, Abu Dhabi, United Arab Emirates.

As Mark Roseland, director of the Centre for Sustainable Community Development at Simon Fraser University in Vancouver, Canada, points out, "much of green building is simply good time-tested design practices," such as "building orientation and design, maximizing natural light and ventilation, improved insulation, and sourcing of recycled and sustainable construction materials." But continuous technology development also offers new tools such as sensors that switch lights on or off depending on people's actual use of a given space, and energy dashboards that track energy use and can pinpoint opportunities for savings.[5]

Apartments, single-family homes, and other residential buildings account for three-quarters of global energy use by buildings. But the energy demand of non-residential structures, such as office and commercial buildings, is growing more rapidly: by 22 percent, up from 12 percent in the decade to 2012. During that period, the total floor space of non-residential structures increased by 34 percent worldwide, ahead of population growth at 13 percent and far outpacing improvements in the energy efficiency of buildings.[6]

In industrialized countries, urban buildings typically use less energy per capita than suburban or rural structures, for at least two reasons. They often are attached to surrounding buildings, which reduces heating or cooling requirements, and the floor area per person is smaller. In developing countries, however, the reverse is true: per capita energy use is higher because urban populations tend to have higher incomes, higher comfort expectations, and better access to energy than rural populations.[7]

Municipalities can steer the builders, owners, and users of old and new buildings in greener directions through the use of building codes and permits, zoning regulations, building performance ordinances, and other mandates and regulations. Taxes and other financial policies can provide additional incentives. Requiring building owners or residents to report data on energy and water use can help to establish base-year benchmarks, set goals, and evaluate performance. Retrofits typically involve considerable upfront investment costs, and subsidies likely will be needed to ensure that lower-income residents are not left behind. A variety of social housing and public works programs have embraced the urban sustainability agenda (for example, by mandating the integration of renewable energy solutions such as solar water heaters).

Although cities are pursuing myriad policies, the ambition and comprehensiveness of these measures varies. It is not just city administrations that are acting, or that need to act. National governments and state or provincial authorities need to cooperate with cities and provide support, whether in the form of supplemental funding or by passing the kind of legislation that is outside the authority of mayors and city councils.

Follow the LEED: Green Building Standards

The environmental impacts of building construction, use, renovation, and demolition have prompted the creation of a steadily growing number of green building norms, standards, rating systems, and certifications, developed either by government agencies, industry, or various initiatives and partnerships. By one count, there are nearly 600 relevant green product certifications worldwide. These systems assist in the difficult task of determining how green a given building is.[8]

Private certification approaches frequently rely on design ratings to assess green buildings, whereas government agencies typically focus more on performance ratings. These address either the building as a "package" or specific elements such as the building envelope, lighting, heating, air conditioning, and water use. In the United States, standards for equipment and appliances include the Energy Star and WaterSense labels—one covering appliance energy consumption and the other shower heads, faucets, toilets, and other water-consuming equipment. Another example is Green Seal, which covers a wide range of items including paints, adhesives, and windows.[9]

The earliest green building rating and certification system, the Building Research Establishment's Environmental Assessment Method (BREEAM), was developed in the United Kingdom in 1990. It is still used in Europe, the Persian Gulf region, and other countries and has influenced all subsequent initiatives. Some 450,000 buildings have been "BREEAM-certified" to date, and 2 million more are registered for assessment.[10]

But it is LEED (Leadership in Energy and Environmental Design), developed by the U.S. Green Building Council (USGBC) in 2000, that is more widely known and that has come to be used in most countries worldwide. Other than the United States, the largest areas of LEED-certified building space are found in Canada, China, India, Brazil, and the United Arab Emirates. New York City's Empire State Building and Rio de Janeiro's Maracanã Stadium are among the world's LEED-certified spaces.[11]

Additional rating systems have been tailored to national or regional needs. (See Table 8–1.) LEED and BREEAM consider energy use to be the most important factor, giving it a 33 percent weighting. The United Arab Emirates' "Estidama" system rates both energy use and water use at 25 percent—a critical concern in the water-scarce Middle East—whereas LEED and BREEAM feature water use far less prominently, at 5.5 and 2.5 percent, respectively.[12]

Compared with other rating systems, the Living Building Challenge being used in the United States and Canada is a more ambitious approach that comprises several performance categories, or "petals" (site, water, energy, health and happiness, materials, equity, beauty, and process). A building must, for example, generate all of its own electricity; use only water that falls on the site; incorporate sustainably sourced materials; avoid toxic materials such as asbestos, mercury, and PVC (polyvinyl chloride); and meet livability and social equity criteria. The Bullitt Foundation's headquarters in Seattle, Washington, is among the buildings that have won certification. Another U.S. initiative, the Sustainable SITES Initiative, factors in ecosystem services in addition to the building itself. It was developed in recognition of the limitations of LEED but is able to be used in conjunction with it.[13]

Table 8–1. Selected Green Building Rating and Certification Systems Worldwide

Name	Year Developed or Adopted	Countries Using It	Developing/Administering Organization
BREEAM (Building Research Establishment Environmental Assessment Methodology)	1990	United Kingdom, Europe, Persian Gulf countries, others	Building Research Establishment Global
LEED (Leadership in Energy and Environmental Design)	1998	United States; more than 150 other countries	U.S. Green Building Council
Green Globes (Renamed BOMA BESt [Building Environmental Standards] in Canada)	2000	Canada; adapted for the United States by the Green Building Initiative in 2004	ECD Energy and Environment Canada (administered by BOMA – Building Owners & Management Association)
CASBEE (Comprehensive Assessment System for Building Environmental Efficiency)	2001	Japan	Japan Sustainable Building Consortium
Green Star	2003	Australia, New Zealand	Green Building Council of Australia
Green Mark	2005	Singapore	Building and Construction Authority
Sustainable SITES	2006	United States	Green Business Certification, Inc.
Living Building Challenge	2006	Cascadia (United States and Canada)	International Living Future Institute
GBEL (Green Building Evaluation Label)	2006	China	Developed by the China Building Science Research Institute; administered by the Ministry of Housing and Urban-Rural Development
GRIHA (Green Rating for Integrated Habitat Assessment)	2007	India	The Energy and Resources Institute (TERI)
AQUA (Alta Qualidade Ambiental)	2008	Brazil	Fundação Vanzolini adapted the French HQE (Haute Qualité Environnementale) standard to Brazilian conditions
Pearl Rating System for Estidama	2008	United Arab Emirates	Abu Dhabi Urban Planning Council

continued on next page

	Table 8–1. continued		
Name	**Year Developed or Adopted**	**Countries Using It**	**Developing/Administering Organization**
Green Star South Africa	2008	South Africa	Green Building Council of South Africa (adapted from the Australian system)
LOTUS	2008	Vietnam	Vietnam Green Building Council
Building Environmental Assessment Method (BEAM)	2009	Hong Kong	Hong Kong Green Building Council

Source: See endnote 12.

These standards and certification systems are voluntary in nature. However, cities may require adherence to the approach that best accords with their particular needs and circumstances. Portland, Oregon, and Vancouver, Canada, for example, have mandated that all new public buildings meet the LEED Gold standard, whereas Seattle settled for the LEED Silver standard.[14]

Ambitious standards are critical for achieving urgent sustainability goals. Yet measures to green buildings also must be practical and affordable enough to be implementable. The real-world impact of a handful of standout performers that meet the highest criteria is far more limited than efforts to ensure that the vast majority of the world's dwellings, offices, stores, and factories reduce their environmental footprint substantially. Finding the right balance is often a local task, accounting for circumstances such as the number of heating or cooling degree days, the type and mix of old and new buildings, as well as social and economic conditions.

Formal certification can be a drawn-out, costly process, which may limit the number of buildings recognized under any particular green building standard. LEED and other rating systems, along with changed national and municipal policies and growing environmental awareness, have helped to increase the worldwide share of more-efficient and greener buildings. (See Box 8–1.)[15]

Building Codes

Building codes can be a major driver of more-sustainable practices, particularly for reducing energy consumption. Part of the challenge is not just to

Box 8–1. Green Building Markets and Energy Efficiency Investments

A 2013 estimate put global green building spending at 38 percent of the construction market. However, definitions of what is included in "green" construction vary around the world, making it difficult to come up with meaningful global estimates. In the Asia-Pacific region, green building and construction was estimated to approach $670 billion in 2015, or 40 percent of the region's total construction spending of $1.7 trillion. About a third of the green spending was for retrofits of existing buildings.

According to the U.S. Green Building Council, green construction spending in the United States jumped from $10 billion in 2005 to $129 billion in 2014, or about 13 percent of the $962 billion in construction spending nationwide that year. (LEED-certified construction is a sub-category, with expenditures of some $50 billion in 2014.) The USGBC estimates that green spending will rise to $224 billion by 2018. During the 2015–18 period, green construction in the United States is projected to save $2.4 billion in energy expenditures, save $1.9 billion in materials and other areas, and avoid greenhouse gas releases equivalent to the emissions of 3.4 million passenger cars.

A recent report from the International Energy Agency zeroes in on energy efficiency investments in buildings (excluding for appliances) and estimates the global total to be some $93 billion in 2014. Yet out of a worldwide construction market of roughly $4–5 trillion annually, this would represent only a minor 2 percent. Three countries—China, the United States, and Germany—account for nearly two-thirds of global efficiency spending. Efficiency spending ran to about 5 percent of the total construction market in Germany, 2.4 percent in the United States, and 1.6 percent in China. (See Table 8–2.)

Table 8–2. Building Construction and Energy Efficiency Investments in the United States, China, and Germany, 2014

Country	Total Building Construction	Building Energy Efficiency	Residential Construction	Non-Residential Construction
	billion U.S. dollars			
United States	962	23	10	13
China	1,120	18	11	7
Germany	320	17	13	4

Note: Figures are approximate.
Source: See endnote 15.

The Bullitt Center in Seattle is considered one of the greenest commercial buildings in the world.

© Nic Lehoux for the Bullitt Center

adopt green codes, but to deal with older codes that may inhibit the greening of buildings. In the United States, the Environmental Protection Agency created the *Sustainable Design and Green Building Toolkit for Local Governments* in 2013 to help local governments identify and remove barriers to green building in their permitting processes.[16]

Studies indicate that green building standards can pay for themselves fairly quickly and then generate savings for the rest of a building's life span. Whether adopted by cities, state authorities, or national governments, codes can either be mandatory or voluntary in nature, and their degree of stringency varies considerably. Codes may apply just to municipal structures or also to private (residential, commercial) buildings. Some codes apply only to new buildings, whereas others apply also to old structures. Many cities around the world are establishing codes that exceed the national standards of the countries where they are located.[17]

Some cities have adopted ambitious policies to reduce the greenhouse gas footprint of buildings. In Copenhagen, Denmark, new buildings are required to be constructed according to the country's Low Energy Class ratings; by 2020 the requirement will be for near net-zero energy buildings. This is part of the city's 2012 plan to become the world's first carbon-neutral city by 2025, even as the population grows by a projected 100,000 people. (Copenhagen also relies on moving its energy supply toward greater use of renewable energy and combined heat and power, or CHP.)[18]

As early as 1992, the city of Freiburg in southwestern Germany adopted a Low-Energy Housing Construction standard for all contracts in which the city sold land (the standard was later introduced at the federal level in 2001). Freiburg's Vauban district—a former French military barracks that was turned into residential housing for about 5,000 people—has been a model of green development, and its development plan specified a standard for heating energy consumption of 65 kilowatt-hours per square meter (kWh/m^2) per year. (Some developers in the area decided to exceed this standard, building "passive houses" that lower energy consumption to 15 kWh/m^2 or less.) Equity considerations were part of the planning, with the aim of making one-quarter of the Vauban housing units affordable for lower-income residents; however, because of cutbacks in the state social housing program, funds were insufficient to reach this goal. (See also City View: Freiburg, page 135.)[19]

In the U.S. state of California, San Francisco's Green Building Code sets requirements that apply to all new construction and to certain major alterations and first-time tenant improvements. The code adopts the mandatory measures in California's green building code (CALGreen) and requires compliance with either LEED or GreenPoint Rated standards. In addition, the city's Environment Code requires that all new construction and major alterations of 5,000 square feet (465 square meters) or more in city-owned facilities and leaseholds obtain LEED Gold certification. San Francisco also has an Existing Commercial Buildings Energy Performance Ordinance as well as a Residential Energy Conservation Ordinance, which requires that properties for sale obtain an inspection and install basic energy and water conservation devices or materials.[20]

Seoul, South Korea, launched a Building Retrofit Project in 2008 that initially targeted only public buildings but was expanded to include residences as well as private universities and hospitals. By 2013, some 14,000 buildings were participating. The metropolitan government provides low-interest loans to help finance efficiency installation costs. The project is designed to help Seoul reach its goal of reducing greenhouse gas emissions 40 percent from 1990 levels by 2030. Making a crucial link between environmental and social objectives, the money saved through retrofits is being reinvested in citizen welfare programs.[21]

In Singapore, the Building Control Act (enacted in 2008 and amended in 2012) mandates minimum environmental sustainability standards for all new and existing buildings, with an initial focus on commercial structures. The aim is to green at least 80 percent of the city's buildings by 2030. The Act requires meeting Green Mark-certified standards, conducting energy audits of cooling systems (for buildings with a gross floor area of at least 15,000 square meters) every three years, and annual submission of energy consumption data.[22]

Heating with the Sun

Municipalities in China, Spain, and Brazil have played leading roles in efforts to integrate solar thermal technologies into buildings as a means of heating water for household and industrial use. (By contrast, the promotion of other

CSIRO ScienceImage

Solar hot water units mounted on the roof of an apartment building in Rizhao, China.

forms of renewable energy, such as solar power generation, often requires action by national or state governments.) By 2014, 10 cities in China's Shandong province had adopted mandates for the use of solar water heaters in residential structures. One of them, Rizhao (with a population of 2.8 million), has been promoting solar hot water use for the past two decades. In the city center, 99 percent of households use solar hot water, compared with just over 30 percent in the surrounding suburbs. Support for solar research and development from the Shandong provincial government helped make the solar heaters cost-competitive with conventional electric heaters. All new buildings in Rizhao are required to include solar hot water, and educational campaigns encourage residents to install solar in their homes.[23]

In Spanish cities, water heating accounts for 27 percent of a typical household's energy use. Barcelona was the first European city to implement a Solar Thermal Ordinance, in 2000. Part of the city's wide-ranging policies for climate change mitigation, the ordinance requires that solar energy provide 60 percent of running hot water needs in all new or renovated buildings and in buildings whose primary purpose has been altered. In 2006, Barcelona amended both the ordinance and the scope of buildings to which it applies, leading the area of installed solar thermal panels in the city to increase from just 1,650 square meters in 2000 to 87,600 square meters in 2010.[24]

Still, Barcelona has experienced problems with public acceptance of solar thermal technology as well as with the performance of installed systems. By 2010, when only 46 percent of the total area approved for solar thermal infrastructure was in use, the city redoubled its information campaigns and stakeholder engagement efforts. Architects, engineers, residential building administrators, a consumer association, solar industry representatives, and others came together in Taula Solar, a stakeholder forum, to discuss the objectives and scope of the ordinance. In 2013, an initiative called "Solar Reflection Days" showcased state-of-the-art solar thermal systems. Since Barcelona's pioneering moves, more than 70 other Spanish cities have replicated the ordinance, and, in 2006, a requirement to install solar thermal systems became part of Spain's national Technical Building Code.[25]

Barcelona's experience has influenced decision makers in Brazil as well. Pointing to the Spanish city's positive experience, the Brazilian solar thermal industry association, DASOL-ABRAVA, teamed up with Vitae Civilis, a prominent nongovernmental organization, to promote solar technology. The city of São Paulo's Solar Ordinance of July 2007 mandates that solar technology cover at least 40 percent of the energy used for water heating in new residential, commercial, and industrial buildings (the target may be made more stringent in the future, since solar heaters are believed to be capable of meeting up to 70 percent of energy use). DASOL-ABRAVA projected that if the city installed 580,000 square meters of solar water collectors by 2015, it could reduce its greenhouse gas emissions by 35,000 tons of carbon dioxide equivalent.[26]

São Paulo's ordinance undoubtedly stimulated the market, and the measure is being replicated in cities across Brazil (as of 2014, the country had installed a cumulative 11.2 million square meters of solar collectors, mostly in residential buildings and social housing). Public consultations were a key element in drafting the ordinance. Another important aspect in securing public acceptance of solar water heaters has been product certification to avoid the prevalence of low-quality equipment, via the nationwide labeling program Programa Brasileiro de Etiquetagem.[27]

Social Housing as a Driver

Mark Roseland, with Vancouver's Centre for Sustainable Community Development, comments that, "[a]lthough certified green buildings initially took the form of upscale, architecturally distinct status symbols, the next wave recognized that green building and affordable housing are a natural fit." Roseland points to affordable green housing initiatives in the U.S. cities of Austin, New Orleans, and San Francisco, as well as to government efforts to promote sustainable practices in the context of social housing programs in Brazil, Chile, China, Mexico, South Africa, and the EU. In China, for example, the Ministry of Finance and the Housing Department announced that, from 2014 on, all newly constructed government low-income housing must be Chinese Green Building Label-certified.[28]

Brazil offers a prominent example through its Minha Casa, Minha Vida ("My House, My Life") program, which seeks to reduce the country's massive housing deficit for low-income families. Since the program's launch in March 2009, it has led to the construction of nearly 4 million housing units for low-income families, with the aim of building another 3 million by 2018. The housing units must meet specific environmental requirements, including using rainwater collection systems and certified timber. Solar water heaters were made compulsory for houses in the southern half of Brazil, and close to 900,000 residents of Minha Casa, Minha Vida housing now have them. Since June 2015, the government-owned Caixa bank has offered preferential loan terms for energy-efficient housing built under the program. Developers need to be certified either under LEED, BREEAM, or one of three Brazilian standards (Alta Qualidade Ambiental, or AQUA, based on the French Haute Qualité Environnementale standard; Procel Edifica; or Selo Casa Azul da CAIXA).[29]

The Sustainable Housing Program under Mexico City's Climate Action Plan promotes the inclusion of green building features such as solar photovoltaic (PV) panels, energy efficiency, water efficiency, and wastewater treatment facilities in new and existing multi-family buildings. When Chile's Ministry of Housing and Urban Development introduced an evaluation system to determine the energy efficiency of residential units in 2013, it chose to focus on social housing. As of early 2015, more than 4,000 private and social houses had been certified, and the voluntary program eventually may become mandatory. Meanwhile, the Light Up Good Energy project, established by Chile's Programa País de Eficiencia Energética in 2008, has distributed close to 3 million compact fluorescent light bulbs (CFLs) to low-income residents, enabling them to reduce their electricity use by up to 25 percent.[30]

South Africa has made similar forays. In Johannesburg, buildings in the old city center, such as former hotels and offices, were refurbished for social housing purposes (creating 2,700 homes), with strong community involvement. The project installed solar energy systems, energy-efficient light bulbs, and better-insulated boilers and water tanks, and it introduced energy management systems to avoid use at peak-priced times. Since June 2010, the city also has been implementing a Climate Proofing of Urban Communities Project in 700 low-income households, which involves installing solar water heaters, insulating ceilings, and distributing CFLs. In Cape Town, the Kuyasa pilot project has worked to feature these elements in about 2,300 low-income homes, starting in 2008. Benefits include not only energy savings for heating and cooling as well as carbon emission cuts, but also healthier dwellings and the creation of (temporary) local jobs. Funding limitations have precluded scaling up the initiative, however.[31]

In Europe, between 50 million and 125 million people are estimated to be "fuel poor," meaning that they need to spend more than 10 percent of their household income on heating fuel in order to achieve an adequate standard of warmth. In 2012, members of the European Federation of Public, Cooperative and Social Housing—a network of national and regional federations that together manage more than 26 million homes, or about 11 percent of existing dwellings in the EU—built more than a quarter million new dwellings and refurbished another 155,000. Among the Federation's projects to reduce energy use in its buildings is the "POWER HOUSE Nearly Zero Energy Challenge," which facilitates exchanges and mutual learning about the energy performance of buildings among social housing practitioners.[32]

The EU's Energy Efficiency in European Social Housing initiative (E3SoHo) runs three pilot projects in Italy, Poland, and Spain that advise social housing tenants on how to realize significant reductions in energy use. Another awareness-building pilot project, Saving Energy in Social Housing (eSESH), ran from 2010 to 2013 at 10 sites in Austria, Belgium, France, Germany, Italy, and Spain, involving more than 5,000 tenants.[33]

The Spread of Green Roofs

Roofs play a crucial role in the energy performance of buildings. Conventional roofs that have dark-colored tiles or other materials typically absorb more solar heat, heating the building and requiring greater use of energy for cooling. So-called cool roofs with white or other light colors reflect more sunlight than dark roofs, thereby decreasing the need for air conditioning and lowering energy consumption and emissions of greenhouse gases. In the United States, the Department of Energy indicates that cool roofs can reduce annual air conditioning energy use for a single-story building by up to 15 percent.[34]

Aside from changing the color of roofing materials, other options exist for making roofs more environmentally friendly. The rapidly falling cost of solar PV panels has helped commercial and residential building owners put more panels on roofs. Although the prevalence of rooftop PV in the United States is still relatively limited, the number of homes with solar panels grew more than 10-fold between 2006 and 2013, from about 30,000 to 400,000. The U.S. Department of Energy conservatively projects some 900,000 homes with rooftop PV panels by 2020, although this could jump to as many as 3.8 million if solar costs continue to drop significantly. In addition to PV panels, solar thermal panels (for generating hot water) also can be mounted on roofs.[35]

Another option is green roofs, which are becoming a requirement in a growing number of cities. A green roof is partially or completely covered with vegetation and requires a waterproofing membrane, and perhaps a root barrier, as well as drainage and irrigation systems. The International Green Roof Association distinguishes among three different types of green roofs, depending on the intended use. (See Table 8–3.)[36]

Table 8–3. Typology of Green Roofs

	Type		
	Extensive	Semi-Intensive	Intensive
Use	Ecological protection layer	Designed green roof	Park-like garden
Maintenance	Low	Periodic	High
Irrigation	No	Periodic	Regular
Types of Plants	Moss, sedum, herbs, and grasses	Grass, herbs, and shrubs	Lawn or perennials, shrubs, and trees
Weight	60–150 kg/m²	120–200 kg/m²	180–500 kg/m²
Cost	Low	Middle	High

Source: See endnote 36.

The multiple benefits of green roofs include improved air quality, increased biodiversity, stormwater management, increased longevity of the building's waterproof membrane, assistance with urban food production, and contribution to a more livable city. Green roofs typically also result in lower building energy needs, and thus lesser climate impacts, and can help to mitigate the urban heat-island effect, whereby a city or metropolitan area is significantly warmer than its surrounding rural areas due to human infrastructure and activities. However, green roofs can cost twice as much as conventional roofs or more.[37]

Policies to promote green roofs include stipulations in land-use plans and building codes; green roof statutes; subsidies and other financial incentives (such as reductions in stormwater fees); demonstration projects; and information and awareness campaigns. Cities also can act as role models by greening the roofs of municipal buildings.[38]

By 2012, one-third of all German cities—leaders in this field—had adopted green roof-related regulations. That year, 14 percent of Germany's total roof area consisted of green roofs, an area totaling more than 86 million square meters. As early as 1996, more than 80 cities in Germany were known to offer incentives to building owners to promote green roofs. Stuttgart, with more than 2 million square meters of green roofs, has been the green roof pioneer both in Germany and internationally. The city's first such regulations came into force in 1986, supported by a financial incentive program. Today, all new roofs with a slope of less than 12 degrees must be green.[39]

France has embraced green roofs at a slower pace, but, in March 2015, a law was approved that mandates that all new commercial buildings must be partially covered either in plants or solar panels. In Paris, mayor Anne Hidalgo has called for 1 million square meters of green roofs and walls by 2020, one-third of which are to be dedicated to urban agriculture.[40]

Tokyo, Japan, adopted a Nature Conservation Ordinance in 2001 to require the greening of building roofs and walls in addition to ground-level green-ings for all new construction as well as for buildings that undergo renovation. The initial ordinance mandated 20 percent green coverage for buildings with a gross floor area of more than 1,000 square meters. A 2009 revision raised the requirement to 25 percent for buildings with more than 5,000 square meters of floor area. Altogether, more than 5,700 new or existing buildings in Tokyo have added about 1.8 million square meters of greened surfaces. The city's goal is 10 million square meters by 2016.[41]

Toronto, Canada, was the first North American city to require green roofs on new developments, in May 2009. The Green Roof Bylaw applies to new commercial and institutional, as well as many residential, development appli-cations. Today, at least 444 green roofs exist in Toronto. North American cities, as a whole, installed an estimated 597,000 square meters of green roofs in 2013 and 511,000 square meters in 2014.[42]

Other cities remain in the relatively early phases of green roof development. Mexico City has some 22,000 square meters installed on public buildings such as hospitals, schools, and municipal offices, and the authorities are planning for more. New and existing residential buildings in the city are eligible for a 10 percent reduction in property taxes for installing an approved green roof.[43]

Pushing the Envelope

The big question is whether the collective efforts of cities to minimize the environmental impacts of buildings are making a sufficient contribution to reducing carbon and other pollutants. Cities may need to encourage, or even make mandatory, far-more-efficient building designs. Among these are so-called passive houses, where the inside temperature can be maintained without additional heating or cooling systems. According to the Passive House Institute, the structures' total energy use for all domestic applications (heating, hot water, and electricity) is not to exceed 120 kWh per square meter of floor area per year. The maximum energy use for space heating (or for cooling) is 15 kWh per square meter per year. Passive houses also must meet standards for air tightness and thermal comfort; however, there is no third-party certification for such buildings.[44]

Many European cities are integrating passive-house rules into their regulations, governing either municipal structures or all new buildings. In Germany, major cities such as Bremen, Cologne, Frankfurt, Leipzig, Leverkusen, and Nuremberg, as well as smaller cities like Aschaffenburg, Darmstadt, Heidelberg, Münster, and Ulm, all have passed passive-house legislation in the last decade. Freiburg made the standard mandatory in 2011 for all new residential buildings, and Hamburg decided that from 2012 onward, municipal subsidies for new housing projects will be granted exclusively to passive houses. Hannover is building some 300 residential passive houses, and Munich aims to reduce its heat demand 80 percent by 2058 (relative to 2009) with the help of passive solar design. Belgium's two largest cities, Brussels and Antwerp, have adopted passive-house regulations, making this standard mandatory for all new buildings and retrofits. Oslo, Norway, has required since 2014 that all new public buildings meet the passive-house standard. In the United States, San Francisco and New York City are examining the passive-house standard more closely.[45]

Enter National Governments

National-level policies can support municipal-level efforts or even be the main driver in reducing the environmental footprint of buildings, although the challenges of coordination and varying ambition may exist. Ideally, top-down and bottom-up policies should be combined, as authorities in France have attempted to do. (See Box 8–2.)[46]

In 2012, Chile's Ministries for Public Works, Housing and Urban Development, and Energy and Environment jointly formulated a National Strategy for Sustainable Construction that seeks to link to, and coordinate, the energy and environmental plans of local authorities. South Africa's Department of Energy has created a Municipal Energy Efficiency Demand Side Management program.

> ## Box 8–2. Combining Top-down and Bottom-up Policies for Greening Buildings in France
>
> France's *Grenelle Environment* policy, published in August 2009, established a series of national targets for reducing energy consumption and emissions in different parts of the economy, including cuts in the energy use of new and existing buildings of at least 38 percent by 2020. The Plan Bâtiment Grenelle was drawn up to implement the government's program to improve the energy performance of buildings. However, there also was a recognition that the top-down Grenelle approach needed to be complemented by a bottom-up approach. The French Environment and Energy Management Agency (ADEME) and Alliance Villes Emploi, a network of local authorities, jointly initiated the Employment Centers and Sustainable Development project. Following initial pilot efforts in the cities of Bayonne, Nancy, and Lille to test innovative ways for mobilizing stakeholders around the Grenelle issues, the project was extended to 30 additional locations. Through stakeholder engagement, more than 30 local joint action plans to green the building sector were drawn up and implemented
>
> *Source: See endnote 46.*

And South Korea has developed an eco-friendly building certification program for new buildings known as Green Standard for Energy and Environmental Design, or G-SEED. But whereas the national government has set a target for all new multi-family housing in South Korea to achieve "zero net energy" by 2025, Seoul's Metropolitan Government aims to meet this goal two years earlier.[47]

Among a set of 15 major countries examined by the American Council for an Energy-Efficient Economy (ACEEE), 7 (Australia, France, Germany, Russia, South Korea, Spain, and the United Kingdom) have mandatory national building codes for both residential and commercial structures. Others have mandatory codes in one category but voluntary codes in the other.[48]

The Building Codes Assistance Project finds that, in addition to those countries identified by ACEEE, most countries in Europe, as well as Australia, China, Japan, Jordan, Kazakhstan, Mexico, New Zealand, Tunisia, and Vietnam, all have mandatory national building standards. Standards have been proposed in Algeria, Brazil, Colombia, Iran, Tunisia, and Ukraine, among others, whereas Egypt, India, Indonesia, Pakistan, and South Africa are among those using a voluntary or mixed approach.[49]

In Europe, EU Directives have been an important driver of building-efficiency improvements. The *Energy Performance of Buildings Directive* (first published in 2002 and rendered more stringent in 2010) requires all

EU member countries to introduce national laws to enhance their building regulations. Specifically, countries have to set minimum energy performance requirements for all new buildings and for major renovations of existing buildings. They also have to introduce energy certification schemes for buildings and to conduct inspections of heaters and air conditioners. By December 2020, all new buildings must be "nearly zero energy."[50]

Additionally, the 2012 Energy Efficiency Directive (EED) requires that 3 percent of the floor area of central government buildings be renovated each year, or that alternative measures with at least the same energy savings be put in place. A major objective is to ensure that public buildings showcase the opportunities inherent in building renovations and to pave the way for large-scale renovation of the EU's entire building stock. However, the initial proposal to subject all publicly owned buildings to the EED was pared down to cover only buildings owned and occupied by central governments. Furthermore, as a report by the Coalition for Energy Savings reveals, most EU governments so far have failed to generate adequate inventories of the energy performance of their building stock, presenting the danger of a major opportunity being missed.[51]

How Much Is Enough?

Around the world, cities and national governments are drawing from a broad range of policy options to reduce the energy footprints of buildings. Not only have green standards and rating systems proliferated, but so have the number of city policies that encourage or mandate more-sustainable practices. For both new and existing buildings, these can guide and drive the transition toward greater efficiency in the use of energy, water, and materials.

Green building markets are undoubtedly expanding. But a fundamental question remains: Are current trends and policies enough? Are they getting the cities of the world onto a collective trajectory that will prove adequate in the face of the immense climatic challenge? The likely answer is that more needs to happen—far more. Standout performers such as the IRENA headquarters in Abu Dhabi or the Bullitt Center in Seattle are celebrated, and humanity should embrace these symbols of success during this difficult struggle. But ultimately, what matters most is the performance of the bulk of the world's building stock, which needs to be improved dramatically. Voluntary measures may need to yield increasingly to mandated performance requirements.

Although green buildings, in many cases, will pay for themselves over time, upfront costs are a critical obstacle, and better financing models are needed. Furthermore, the landlord-tenant dilemma needs to be addressed in innovative ways: landlords have little incentive to undertake efficiency and other green investments when the benefit—lower energy bills to run an apartment or office—accrues principally to tenants. Tenants, in turn, are reluctant to share retrofitting costs when it is not clear whether they will remain in a building long enough to reap appropriate cost savings. So-called green leases can help surmount this problem by specifying shared responsibilities and benefits.

The green roof of Chicago City Hall.

TonyTheTiger

Cities will need to step up their efforts to reduce the environmental footprints of buildings. This is particularly critical for new buildings because implementing proper design *before* a structure is built can secure important savings in emissions and other impacts, whereas poor choices upfront lock in decades of unnecessarily high use of energy, water, and materials.

Although much attention is being directed toward the specifics of green design, the rapid growth in the world's overall building floor space seems to be attracting comparatively little interest. If floor space continues to outpace population, as it has in the last decade or so, the task of making buildings more sustainable will be that much harder. Although better design and layout can help make the best use of a given amount of space, building owners ultimately need to ask themselves if they really need as much space as they want.

CITY VIEW

Freiburg, Germany

Simone Ariane Pflaum

Freiburg Basics

City population: 222,203
City area: 153 square kilometers
Population density: 4,800 inhabitants per square kilometer of settled area
Source: See endnote 1.

The city in winter, with the Freiburg cathedral at its center.

A Pioneer in Inclusive, Sustainable Urban Development

Sustainable development has a long tradition in Freiburg, a university city in southwest Germany's Black Forest region. In the 1970s, a strong anti-nuclear movement emerged to protest a planned nuclear power plant in the nearby municipality of Wyhl. The following decade, the Freiburg city council made a landmark decision to divert car traffic from the city center and instead offer public transportation alternatives—a move that ran counter to the mainstream. Today, Freiburg has an extensive transit network, with a tram or bus stop no more than about 300 meters from any residence. The introduction of an environment-oriented public-transit ticket (*Umweltkarte*) in 1982 was the first step toward the development of a regional intermodal ticket (*Regiokarte*).[2]

Freiburg is a signatory to the 1994 Aalborg Charter, an urban environment sustainability initiative inspired by the Rio Earth Summit's *Local Agenda 21* plan and now supported by more than 3,000 local governments. When Freiburg signed the Aalborg Commitments in 2006, it reiterated its political commitment to sustainable development based on 12 overarching policy areas, each with 5 strategic objectives. These were developed in a participatory process by the Freiburg Sustainability Council and adopted by the city council in 2009 as the basis for all political action. In recognition of this commitment, Freiburg became the first major city to be awarded the German Sustainability Award in 2012.[3]

At the invitation of the German government's Council for Sustainable Development, Freiburg Lord Mayor Dieter Salomon serves as one of 20 founding members of the Mayors' Sustainable City Dialogue. Their work has included statements on "strategic principles for sustainable development of municipalities" as well as a brochure detailing the role of cities in ensuring the success of Germany's *Energiewende* ("Energy Transition") policy.[4]

Leadership on Sustainability

To strengthen Freiburg's sustainability profile, the Office of Sustainability Management was set up in early 2011 as a coordination and guidance office reporting directly to the Lord Mayor. Among its tasks are to anchor sustainability as a cross-cutting issue in communal policies, to create an integrated approach to sustainability management, and to coordinate the activities of various urban actors across different issues and organizational boundaries. A sustainability working group was created within the city administration to help set the political course, serve as a multiplier vis-à-vis the various city departments and offices, and share relevant information across administrative structures.[5]

Sustainability cannot be implemented and developed without engaged citizens. Freiburg's Sustainability Council—a 40-member panel of experts drawn from the realms of politics, science, economy, and civil society—seeks to account for the diversity of interests.

Chaired by the Lord Mayor, it advises the city council and recommends ways to implement Freiburg's 60 sustainability goals.

In Freiburg, sustainability is understood as extending far beyond environmental and climate concerns to include matters such as social affairs, education, culture, and generationally equitable fiscal policy. Sustainability is seen as a cross-cutting task, requiring integrated political and administrative action to meet complex and diverse challenges and to enable cooperation among various urban actors. In day-to-day governance, the sustainability mandate demands that social, ecological, and economic change be recognized at an early stage and be taken into account within the framework of an integrated approach.

Freiburg modified the original 10 policy areas of the Aalborg Commitments into 12 overarching policies, each with 5 goals of equal importance, resulting in a set of 60 sustainability targets. In 2012, the city council agreed to prioritize five policy areas: education, social justice, climate and energy, mobility and transport, and urban planning/development. The council's action plan envisions participatory implementation of the sustainability goals and an impact-oriented, step-by-step development of an integrative indicator-based report system to measure local sustainable development.

In December 2014, Freiburg decided to introduce combined financial and sustainability reporting, effective for its 2015/16 budget. Unique in all of Germany, this allows the city to measure how well it is achieving its sustainability goals through a series of monetary and qualitative indicators. Decision makers can direct available resources in support of generational justice and long-term goals, but they also are aware of the limits of municipal sustainability action. Improved monitoring forms the basis for results-oriented policies in all city departments.[6]

Cornerstones of Sustainable Development in Freiburg

Freiburg faces a number of challenges. Among them is the question of how to secure municipal services in the face of rising costs. Other important issues are shifting demographics, climate change mitigation and adaptation, the implications of technological development for municipal infrastructure, tackling a growing city and its social challenges (including providing decent housing), and the integration of an ever-increasing number of refugees. The city is finding ways to deal with these challenges, focused on the following five sustainable development priorities:

Education for Sustainable Development and Lifelong Learning

Equitable access to education and lifelong learning is key to enable children and adults to participate in sustainability action. Under the Learning Life in Freiburg (Leben Erlernen in Freiburg, or LEIF) initiative, the city government set up an educational fund to strengthen sustainability education.[7]

Freiburg's Sustainability Education Cloverleaf program seeks to impart foresight and interdisciplinary thought and action as the basis for sustainable lifestyles for all citizens. The program illustrates key issues, such as energy, food, transport, and consumption, from four dimensions of sustainability: environmental protection, sustainable economic development, social justice, and culture. Bringing together diverse stakeholders sheds light on key problems from different perspectives. In 2011, the German Commission for UNESCO recognized Freiburg as the "City of the UN World Decade of Education for Sustainable Development" for its approach.[8]

Social Justice and Affordable Housing

Freiburg holds considerable appeal for many people. With an average resident age of 40.4 years, it has one of the youngest urban populations in Germany and is one of the country's fastest growing cities. However, housing is scarce, rents are rising, and demand for affordable housing is increasing. Lower-income residents are in danger of displacement, which leads to a further rise in prices, social segregation, and out-migration of families. To address this challenge, the city council adopted a program in 2013 that aims to build 1,000 new apartments per year. The challenge for the city administration is to identify areas for construction in the face of limited available space and the large influx of refugees.[9]

A sustainable society is based on social inclusion, which relates to issues such as education, labor, communications, self-determination, health and rehabilitation, mobility, urban development and housing, recreation, and culture. The *Action Plan for an Inclusive Freiburg* fulfills two functions: to ensure that existing initiatives can continue to unfold in the long term, and to enable prioritization to address new challenges and to reshape existing infrastructure accordingly.[10]

Climate and Energy: A Climate-Neutral Municipality

Freiburg is holding fast to its climate targets, despite the continued rise in greenhouse gas emissions worldwide. The city council voted unanimously in 2014 to halve the city's emissions by 2030 (relative to 1992) and to achieve carbon neutrality by 2050.[11]

Given that about one-third of German energy demand is in the building sector, increasing the efficiency of buildings is an important step for reducing emissions. To reduce building energy consumption sharply and in a sustained manner, Freiburg relies on two different strategies: promoting energy retrofits in existing buildings and requiring that new buildings meet high energy standards.

Freiburg's efficiency standard for new residential dwellings—Effizenzhaus-Standard 55—is considerably more ambitious than the German national standard. It sets the maximum allowable primary energy demand at 55 percent of the standard values established by the federal efficiency decree and sets transmission heat loss at 70 percent of the national value. Commercial buildings used primarily as offices are subject to Freiburg's Effizenzhaus-Standard 70, which exceeds the federal standard by 30 percent.

Long-term lifecycle analysis for residential buildings requires that socially equitable housing objectives be combined with an ecological perspective. In 2011, Freiburger Stadtbau, the municipal housing company, completed the world's first energy retrofit of a residential high-rise from the 1960s, meeting a demanding passive-house standard. The Buggi 50 project, located in an ethnically mixed neighborhood in western Freiburg, gained national accolades because of its positive ecological and economic impact as well as its inclusive, participatory approach. The underlying social concept of *Wohnverwandtschaften* ("living acquaintances") was developed jointly with prospective residents, who were encouraged to become acquainted and to get involved in planning prior to moving in. As a result, they agreed on smaller apartments, communal facilities (such as a senior outdoor sport garden), a concierge to act as a contact person, and activities on each floor to facilitate communal living.[12]

Another smart solution in Freiburg is the development of the city's Green Industry Park. The area known as Industriegebiet Nord (IG Nord) is home to 300 industrial, commercial, and service companies, which employ some 15,000 people and have been drivers of innovation for the "Green City Freiburg" concept. The area also houses science and research institutes, such as the Faculty of Engineering of the University of Freiburg, the Fraunhofer Institutes, the Biotech Park, and the Max Planck Institute of Immunobiology. In a joint initiative of the city, private businesses, and local research organizations, IG Nord is being transformed into the resource-efficient Green Industry Park, funded in part by the National Climate Protection Initiative.[13]

Mobility: City of Short Distances

Early on, Freiburg moved away from a car-oriented transport policy. As early as 1979, the city prioritized environmentally friendly modes of transport in its General Urban Transport Policy. The city's overall traffic concept of 1989 adopted as one of its main goals "traffic avoidance," particularly of motorized individual transport. It pledged to do so through a coordinated urban development and transport policy based on the notion of a "city of short distances." These goals were reaffirmed with the adoption of the VEP 2020 Transport Development Plan in 2008. The focus is on the expansion of tram and bicycle path networks as well as on issues of accessibility and intermodal integration of transport modes.[14]

Attractive Urban Neighborhoods with High Quality of Life: The Example of Vauban

After the end of the Cold War, when French military forces vacated their local barracks in Freiburg in 1989, the city decided to convert the area to new uses. Vauban, a 41-hectare area near the city center, was repurposed into an attractive, family-friendly neighborhood for 5,500 people, marked by active citizen engagement, community building, and environmentally conscious living.

Low-energy building construction is obligatory in Vauban. For most homes, passive house construction and the use of solar technology are standard, and some houses are even net energy producers. Green roofs help with rainwater collection, and green spaces between row houses provide a good micro-climate and serve as playgrounds for children. A social infrastructure has emerged in Vauban that consists of a school, kindergarten, and youth facility, as well as a civic meeting place, marketplace, and leisure and play areas.[15]

Like many parts of the city, Vauban utilizes "traffic calming" (overall, 90 percent of Freiburg's residents live on roads that have a speed limit of 30 kilometers per hour or lower). A large proportion of households are car-free, and residents who do have a private vehicle park them in one of two community garages. Since 2006, Vauban has been accessible by the city's light rail system. Many residents have decided to forego car ownership entirely, using public transit and bicycles instead.[16]

Varied façades and gardening in Vauban.

Outlook

In 2014, Freiburg started developing a new urban development planning instrument, known as Perspektivplan Freiburg. Extensive dialogues were held both within the municipal administration and with citizens to discuss the city's urban development identity, the challenges it faces, and how a city can grow while still maintaining its attractiveness. Through the creation of an informal urban master plan, the outcomes of the dialogues are intended to guide Freiburg's urban development over the next 15 years. The aim is to meet the rising demand for housing while maintaining and enhancing the quality of Freiburg's open spaces. The city is committed to an approach that links these two objectives rather than one that regards them as incompatible. In addition to issues of land use, questions relating to energy, mobility, biodiversity, and social inclusion will play an important role. The planning initiative will be subjected to a comprehensive sustainability rating effort.[17]

Simone Ariane Pflaum is Director of Sustainability Management for the City of Freiburg.

CHAPTER 9

Energy Efficiency in Buildings: A Crisis of Opportunity

Gregory H. Kats

Energy efficiency—using less energy to provide the same or enhanced services—has, for four decades, delivered incremental improvements in energy use around the world. Efforts to increase efficiency in the built environment—including in office buildings, schools, single-family homes, and other structures—have been driven largely by policy mandates, building codes and standards, and a growing private-sector energy efficiency industry. Today, however, more-ambitious policies, enabled by a new set of energy efficiency technologies and new funding tools, are demanding more of energy efficiency and the industry.

Existing buildings represent about one-third of global energy consumption and one-fifth of energy-related carbon dioxide (CO_2) emissions. Achieving deep reductions in CO_2 emissions therefore requires that existing buildings achieve deep energy efficiency savings. The global scientific consensus about the reality and severity of human-caused climate change recently received enormous moral reinforcement from the Pope's clarion call for climate change mitigation and his (and other religious leaders') insistence that addressing climate change is a central moral and religious imperative. Together, these have sharply raised expectations for energy efficiency at the national and local levels, and a doubling of expected energy efficiency improvements is now a globally common target. Meanwhile, the worldwide growth of green buildings relies increasingly on inexpensive energy efficiency gains both to enable cost-effective green retrofits and to deliver a range of related benefits, including pollution reduction and improved occupant comfort.[1]

Gregory H. Kats is President of Capital E and Managing Director of ARENA Investments LLC, a clean-energy investment firm. He is Co-chair of the 2016 ACEEE Energy Efficiency Financing Conference and the first recipient of the U.S. Green Building Council's Lifetime Achievement Award.

At the same time, the cost of renewable energy has dropped, and renewables such as wind and solar have become the largest source of new power generation capacity in both Europe and the United States. Although this has huge health, employment, and security benefits, it also poses increasing challenges to electric grids due to the intermittency of electricity generated from the sun and the wind. Energy efficiency is the lowest-cost way to reshape the electricity load to enable continued rapid expansion of renewable energy. Energy efficiency also is the most cost-effective means to reduce emissions and to provide the "reserve generation" margins that electricity grids require to ensure reliability (particularly as the dirtiest baseload fossil fuel power plants are being shuttered). The world's largest competitive wholesale electricity market, the PJM Interconnection, covers a dozen U.S. states and relies on distributed energy efficiency—rather than on expensive, rarely operated power plants—to provide more than half of its standby capacity.[2]

All of these changes are forcing something of a crisis of opportunity on the energy efficiency industry: it is poised to deliver a far-larger range of benefits and to anchor a global transition to a low-carbon economy. Nevertheless, it continues to be treated as a second-class energy option. For example, emissions trading programs still fail to award the value of CO_2 reductions that result from energy efficiency investments to the firms and cities that make these investments. This failure in market design greatly reduces the financial incentives for energy efficiency and handicaps the efficiency market. (Efforts to address the problem include CO2toEE, an emerging market-based mechanism that would help buildings and companies investing in energy efficiency claim and leverage the value of their associated CO_2 reductions).[3]

Energy efficiency faces enormous obstacles in financing because consumers generally require very high rates of return from efficiency investments. A payback requirement of two years (a 50 percent annual rate of return) is average for investors in the energy efficiency of buildings. (Return requirements are a bit longer for residential buildings and a bit shorter for commercial buildings.) This is a far-higher barrier than the financial return requirements for investments in power generation, such as fossil fuels or renewable energy. Typically, investments in large power plants or on-site generation equipment pay for themselves in five to eight years (a 15–20 percent minimum rate of return).

These challenges and opportunities mean that how we finance and motivate energy efficiency must change rapidly for energy efficiency to deliver on its promise of anchoring the global transition to a low-carbon economy. Energy efficiency is the most cost-effective and largest path to CO_2 reduction in existing buildings and can enable the shifting of the electricity load to accommodate intermittent renewable energy, potentially "anchoring" the world's clean energy options.

The financing requirements for efficiency are substantial. In an August 2015 report, Citigroup used best-estimate energy costs to model a pathway to achieve the deep carbon reductions that are required to limit the worst costs and risks of climate change. To meet the targets for slowed warming that Citi and other financial institutions endorse as prudent and rational, a projected $13.5 trillion in global energy efficiency investment is needed during the period 2014–35. This represents about 26 percent of projected energy investment for the period and is larger than projected investments in oil, natural gas, or renewables. (See Figure 9–1.)[4]

In Citi's analysis, the required transition to a lower-carbon global economy is powered primarily by energy efficiency, which anchors investments in renewable energy and natural gas (oil is used mainly for transportation). Because energy efficiency is very cost-effective, and because renewable energy is increasingly cost-effective relative to fossil fuels, the cost of this transition (according to Citi's analysis) is close to zero—and delivers enormous savings in avoided economic, social, health, and security costs and risks.

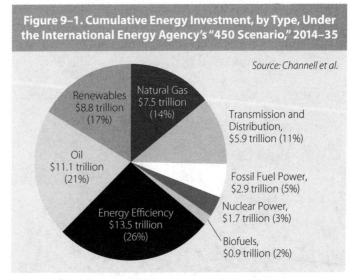

Figure 9–1. Cumulative Energy Investment, by Type, Under the International Energy Agency's "450 Scenario," 2014–35

Source: Channell et al.

Renewables $8.8 trillion (17%)

Natural Gas $7.5 trillion (14%)

Transmission and Distribution, $5.9 trillion (11%)

Oil $11.1 trillion (21%)

Fossil Fuel Power, $2.9 trillion (5%)

Nuclear Power, $1.7 trillion (3%)

Energy Efficiency $13.5 trillion (26%)

Biofuels, $0.9 trillion (2%)

Similarly, the International Energy Agency (IEA) believes that more than 40 percent of greenhouse gas emissions reductions will have to come from energy efficiency in order to keep the global average temperature from rising more than 2 degrees Celsius this century. A recent European policy review found that buildings account for 41 percent of the global energy savings potential by 2035, compared to only 24 percent for the industrial sector and 21 percent for the transportation sector. IEA Executive Director Fatih Birol has observed that "to transition to the sustainable energy system of the future, we need to decouple economic growth from greenhouse gas emissions," and that "energy efficiency is the most important 'arrow in the quiver' to achieve this."[5]

Energy Efficiency Progress and Savings To Date

The energy demand of buildings worldwide increased by around 1 percent annually from 2005 to 2011, with building electricity use increasing 3 percent annually over the same period. Energy efficiency in the European Union (EU) improved 14 percent overall between 2000 and 2013, representing an annual improvement of 1.2 percent, with a slightly higher improvement of 1.7 percent in residential buildings. Annual efficiency improvements by country ranged from a low of 0.6 percent to a high of 3.3 percent during this period, reflecting major differences in the level of support for energy efficiency across Europe.[6]

Andre Carrotflower

This 1896 building in Buffalo, New York, has been converted to apartments and commercial use while being retrofitted with geothermal heat-pump heating and cooling and energy-efficient double hung windows.

Improvements in the United States have been modest as well. The average U.S. building energy efficiency retrofit achieves only about a 15 percent savings in energy use, even when carried out by large, specialized energy service companies operating under long-term contracts. Many of these efficiency improvements are "shallow," resulting primarily from improvements in the performance of appliances and equipment rather than from more "deep" efficiency retrofit measures such as improved building envelopes, ground-sourced heat pumps, or plug load and lighting controls.

Although energy efficiency has improved only gradually, it has delivered large savings on a cumulative basis. Because of energy efficiency investments to date, IEA member countries (i.e., industrialized Europe, North America, and Asia) saved $80 billion in fossil fuel imports in 2014 alone. Globally, energy efficiency improvements over the last 25 years have saved a cumulative $5.7 trillion.[7]

Even so, the current pace of shallow energy efficiency retrofits and incremental improvements in the efficiency of new buildings must be accelerated greatly if we are to meet the climate change targets that the world's scientists view as necessary. The good news is that this is feasible and would be very cost-effective, because large efficiency gains deliver not only large energy-cost savings, but also large health and environmental benefits, enhanced security, and expanded employment.

Doubling the Rate of Improvements

In the mid-1990s, an initiative sponsored by the U.S. Department of Energy led to the establishment of a single, internationally adopted energy efficiency standard called the International Performance Measurement and Verification Protocol (IPMVP). IPMVP allows for the standardization of efficiency-related upgrades and for the reduced cost of retrofits; it also enables expanded financing, including the bundling of IPMVP-compliant projects into single larger, lower-cost financing packages. But although IPMVP was a large and necessary step, it has not proven sufficient to enable the energy efficiency industry to shift from shallow to deep energy efficiency retrofits.[8]

Countries around the world are now setting more-aggressive energy efficiency objectives through the International Energy Code and through standards developed by the American Society of Heating, Refrigerating, and Air-conditioning Engineers (ASHRAE)—for example, ASHRAE 90.1, which governs energy use in commercial buildings. These standards generally are revised every three years. In the last round of code upgrades, energy efficiency targets have roughly doubled relative to prior code upgrades. In countries such as China and India, and in Latin America, national standards for energy efficiency for existing buildings also have become more aggressive, although compliance has lagged, often badly. When countries without building codes adopt and enforce energy efficiency standards, improvements can be relatively rapid. After the EU Directive on the Energy Performance of Buildings came into force in 2006, countries with few existing regulations saw large improvements in energy efficiency: Portugal, for example, achieved a 50 percent reduction in insulation U-values (a measure of heat transfer) in five years.[9]

Financing remains the greatest challenge to increasing energy efficiency, in large part because of the short payback periods that building owners demand for efficiency investments. Private building owners are concerned that promised public incentives for energy efficiency may not materialize because of budget shortfalls, changes in policy, or other unpredictable events. Therefore, efficiency incentives from municipal, regional, or national governments to building owners over multiple years often have relatively little effect on the investment decisions of building owners. Given large discount rates applied to efficiency, incentives for such retrofits need to occur at the time of investment, or at least within the same year that an investment is made.

Achieving Deep Energy Efficiency: New Construction and Retrofits

The last decade has seen rapid adoption of so-called zero or near-zero net energy standards for residential and commercial buildings, as well as the development of various major projects demonstrating cost-effective investment in energy efficiency. The green standards include LEED (Leadership in Energy & Environmental Design) globally, CASBEE (Comprehensive Assessment System for Built Environment Efficiency) in Japan, BREEAM (Building Research Establishment Environmental Assessment Methodology) in Europe and the Middle East, and Green Star in South Africa. More recently, mainly in Europe and then in the United States, more-aggressive energy efficiency and green design standards have emerged, including Passive House (Passivhaus) and the Living Building Challenge, which seek to achieve zero or near-zero emissions buildings.

The German-developed Passivhaus standard has become popular in North America. A passive house typically cuts energy use by three-quarters compared with similar conventional houses. It also does not need installed heating or cooling systems: insulation and weather sealing are so effective that the occupants and appliances alone typically provide sufficient heat, while strategic shading, nighttime purging of heat, and cross-ventilation provide adequate cooling. Passivhaus dwellings are found in every European country, and more than 30,000 passive houses now exist around the world. In 2015, the U.S. Passive House Institute developed a U.S.-specific standard—PHIUS+2015: Passive Building Standard North America—to allow cost-optimized design for a broad range of climate conditions. This new standard, which has served as the basis for passive-house qualification in the United States since spring 2015, is viewed as an important pathway to achieve net-zero and net-positive-energy buildings (the latter of which generate more energy than they consume).[10]

Ideally, energy efficiency and greening efforts should build on existing standards rather than reinventing them. For example, the Energy Star rating system, developed by the U.S. Environmental Protection Agency (EPA) and the U.S. Department of Energy in 1992 to provide energy performance labeling for buildings and appliances, was adopted in Europe in 2001 to cover office equipment that did not already carry an EU efficiency label.

Over the last five years, the Latvian energy service company RenEsco has performed deep renovations of 15 typical Soviet-era apartment buildings in Latvia. These efficiency investments—in building envelopes, heat distribution pipes, heat control, and energy management—have achieved energy savings of 45–65 percent. Financing is provided by local banks (60 percent) in combination with a third-party-guaranteed loan (40 percent) based on project cash flow (and no other collateral). Apartment owners' monthly costs post-retrofit are at the same level as before renovation. All investments are covered by future energy savings over 20 years and by support from a national renovation program. The simple payback period is 9–10 years, and the upgrades are projected to extend the lifetime of the building by up to 30 years. According to energy expert and consultant Steven Fawkes, if the Latvian work were scaled up to retrofit other similar-style housing, it could reduce Latvia's national gas imports from Russia by as much as 40 percent, greatly improving Latvia's security and economic competitiveness.[11]

The U.S. General Services Administration (GSA), which is responsible for most non-military federal buildings in the United States, is several years into a program to deliver deep energy efficiency upgrades of federal buildings. So far, it has achieved average improvement of 38 percent, in a cost-effective manner. In 2014, the U.S. Federal Green Building Advisory Committee developed policy guidance, now being adopted by the GSA, that federal buildings quantify and include the social cost of carbon (SC-CO$_2$) in building and energy designs. Developed by a dozen federal agencies, including the U.S. Treasury and the EPA, the SC-CO$_2$ is an attempt to price some of the damages inflicted by climate change on human health, agricultural productivity, property at risk from increased flooding, and others. The SC-CO$_2$ today is set at about $40 per ton, with annual increases that reflect a consensus, likely conservative (i.e., low), price on the actual cost of carbon emissions. Adopting this type of accounting for carbon helps internalize carbon costs and allows more rational and cost-effective building design decisions. Cities, companies, and institutions such as universities also can incorporate the SC-CO$_2$ metric in their cost-benefit analysis to allow better investment decisions that more fully reflect the cost of climate warming.[12]

Energy efficiency retrofit packages were developed recently that were nearly cost-neutral and that achieved energy savings of 29–48 percent across eight U.S. locations for federal buildings in different climatic zones. These strategies are in line with the U.S. Department of Energy Building Technologies Office's long-term goal of achieving deep retrofits to reduce energy consumption by 50 percent on a whole-house basis, compared to 2010 levels. Conceiving the design and operation of buildings as a single integrated system enables deep, cost-effective reductions in greenhouse gas emissions. The Research Support Facilities building at the National Renewable Energy Laboratory in Colorado achieved a 67 percent reduction in energy use compared to conventional design through an integrated design process, reportedly at no extra cost.[13]

Energy Efficiency Across Multiple Buildings/Campuses

Energy efficiency retrofits of buildings generally have high fixed costs because of the high transaction costs for obtaining an energy efficiency contract, including marketing and sales, contracting, term negotiations, arranging site visits, etc. As a result, it can cost nearly as much to secure a contract for a shallow retrofit as for a much deeper one. This can make it cost-effective to extend the scope of a retrofit to a deeper one based on the marginal cost of equipment and installation. Similarly, because of the cost of scheduling, permitting, etc., it can be more cost-effective to achieve deep improvements in energy and CO_2 reduction on a multi-building, campus, or even town-wide basis. Physical co-location also can be beneficial, simply because neighboring facilities often can be served by a single ground-source heat pump installation. Stockholm's Sjostad development of 10,000 homes achieves deep improvements through an integrated design approach built on integrated planning.[14]

In the United States, the Property-Assessed Clean Energy (PACE) program, an explosively growing residential energy efficiency and renewable energy financing initiative, is using a multi-building, low-cost client acquisition strategy as well. The privately held specialty finance firms Renovate America and PaceFunding are utilizing this standardized-process, multi-building approach with their PACE residential energy efficiency financing program in California and other U.S. states. PACE allows building owners to finance energy efficiency retrofits by adding a line-item surcharge to their property tax bill. Because the energy savings exceed the ongoing financing costs, there is an immediate payback and no out-of-pocket expense.[15]

The U.S. Army and Navy both have net-zero programs aimed at reducing energy use across their military bases, with the Navy targeting 50 percent of its bases to have net-zero energy consumption by 2020. The Army has identified six net-zero pilot installations in each of the categories of energy, water, and waste, with two integrated Army installations striving toward net-zero on all three fronts by 2020.[16]

Implementing New/Expanded Deep Energy Efficiency Incentives, Programs, and Mandates

Because all future population growth is likely to be in cities, city leadership on energy efficiency is essential if global warming is to be limited. Existing buildings account for up to 80 percent of the CO_2 emissions of most cities, so achieving deep emissions reductions requires that existing buildings achieve deep energy efficiency savings. City adoption of innovative and aggressive energy efficiency policies and programs can have enormous social benefits, including a reduced urban heat-island effect, lower health costs, greater livability, and cleaner air and water.[17]

Energy efficiency helps address growing inequity, as well, because the accrued benefits tend to be largest among low-income populations. A 2015 analysis found that applications such as cool roofs (roofs that reflect more sunlight and absorb less heat), green roofs (roofs that are partially or completely covered with vegetation), rooftop solar photovoltaic panels, and solar hot water can be very cost-effective retrofit options for low-income, multi-family properties. These technologies can bring substantial benefits both to tenants and to the broader community and city, including large health benefits. Over a 20-year period, the benefits of green buildings, such as schools and offices, exceed the cost of making these buildings green and energy-efficient by a ratio of about 10 to 1. (See Figure 9–2.)[18]

Given the cost-efficiencies as well as the societal benefits, cities and other governments should be aggressive in adopting city-wide green building and energy efficiency standards. Additional viable city strategies to drive deep energy efficiency include mandating minimum energy and green building standards, providing tax rebates, providing accelerated planning and zoning or flexibility on density, and leveraging public leases.

Mandating Minimum Energy and Green Building Standards

Experience suggests that energy efficiency improvements average around 35 percent in new LEED buildings and 30 percent in retrofitted LEED

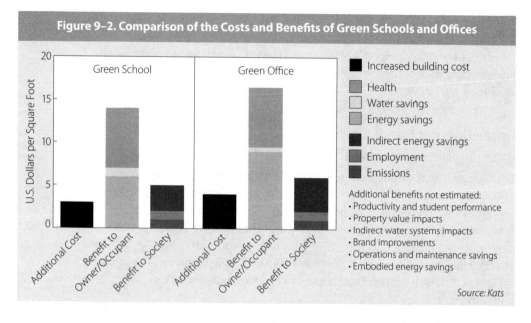

Figure 9–2. Comparison of the Costs and Benefits of Green Schools and Offices

Legend:
- Increased building cost
- Health
- Water savings
- Energy savings
- Indirect energy savings
- Employment
- Emissions

Additional benefits not estimated:
- Productivity and student performance
- Property value impacts
- Indirect water systems impacts
- Brand improvements
- Operations and maintenance savings
- Embodied energy savings

Source: Kats

buildings. To increase efficiency, cities should combine energy performance requirements with green building requirements—for example, by combining LEED Gold with Energy Star 92 by 2018, and LEED Platinum with Energy Star 95 by 2020. LEED and other green building standards, such as BREAAM and CASBEE, are being applied in more than 100 countries worldwide, currently through the national chapters of the World Green Building Council. These groups can serve as ideal partners for city, state, or national governments in implementing combined green and energy efficiency performance standards.[19]

The targeted minimums that cities set for green performance should be higher than the lowest certification levels. Cities often do not keep up with advances made in the building industry. For example, the minimum green performance requirement today in Washington, D.C. is still at the "LEED-certified" level (the basic, lowest level), even though, by 2014, virtually all new construction in the city achieved higher LEED ratings, with LEED Gold being the most common.

Energy Star (or equivalent) performance reporting should be used both to establish minimum energy performance for all buildings and to create an incentive to improve average performance across the entire building stock. Many existing buildings are extremely wasteful, and reducing their energy consumption to meet average performance standards is more cost-effective—and has a larger impact—than improving the performance of buildings that already are efficient. Cities, companies, and other governments should establish an absolute minimum (for example, an Energy Star score of 50) for all buildings as a requirement for sale or acquisition, or to be achieved over a fixed period (such as six or eight years).

Achieving city-set performance objectives for energy, carbon dioxide, and sustainability (for example, through LEED) across the entire building stock for private building owners would provide greater flexibility and would be more cost-effective than doing so on a building-by-building basis. All cities should adopt a minimum energy performance standard (such as Energy Star 85) for new commercial buildings and a higher level for public buildings.

Providing Tax Rebates

Tax rebates should be offered to building owners in exchange for meeting aggressive energy efficiency targets. Financial incentives should be provided within the same year that the upgrade occurs, to allow for faster payback. For example, a city's tax rebate schedule might offer $5 per square foot the year that a building achieves 40 percent energy savings, $10 per square foot for 55 percent savings, and $15 per square foot for 70 percent savings. This would be expensive to a city in terms of forgone revenue that year from complying buildings, but it would yield large savings over many years from reduced air pollution, improved health, increased employment, and other benefits.

Providing Accelerated Planning and Zoning, or Flexibility on Density

In exchange for meeting aggressive energy efficiency targets, cities should provide building developers with access to accelerated planning and zoning processes and/or allow for flexibility in the density of construction (for example, with regard to a building's "floor area ratio," or FAR, the ratio of the total floor area to the area of the lot). In cities like Washington, D.C., where buildings are height-restricted, FAR flexibility for developers is limited. One promising option is to provide greater flexibility on penthouse construction, so that the mechanical penthouse that normally is located on the top of a building can instead locate mechanical equipment on a lower floor, freeing up the area on top for use as leasable space (e.g., office space or a penthouse apartment). This type of FAR and zoning flexibility would create considerable value at no cost to cities. According to D.C. Building Industry Association President Sean Cahill, this may be sufficient to motivate developers to achieve LEED Platinum or similarly aggressive energy efficiency/green levels of design.[20]

Leveraging Public Leases

Another pathway to drive deep energy efficiency is to require coordinated aggressive efficiency standards for space leased by public agencies. In the case of Washington, D.C., the federal and city governments, combined, lease approximately 2.8 million square meters of space that should be used to create a large program of deep energy efficiency performance on a forward schedule. Because of the importance of long-term federal or city leasing contracts to anchor a development, such contracts are enormously valuable to developers and can motivate large changes in design—including commitments to deep energy efficiency or CO_2 reduction for the entire development.

Ultimately, achieving very deep energy efficiency in new and existing buildings depends largely on whether these investments are recognized as valuable for the building owner/occupant and for the city, corporation, or other government jurisdiction. As noted earlier, the benefit-cost ratio for green buildings is about 10 to 1 over a 20-year period. Many of these benefits accrue to building owners and occupants, but many also accrue more broadly, including hard-to-capture benefits such as improved air and water quality, reduced illness, lower summer peak temperatures, and slowed global warming. Like green building designs, deep energy efficiency also delivers critical benefits beyond energy cost savings (see Box 9–1), but these often are ignored or misunderstood.[21]

Conclusion

The energy efficiency industry faces a crisis of opportunity. Energy efficiency is increasingly recognized as indispensable if we hope to avoid the most-severe climate change costs. But energy efficiency is underfunded and is treated as a second-class energy choice. To meet emissions reduction targets, new buildings must become far more energy-efficient, with a rapidly growing portion of new buildings achieving zero or near-zero net emissions—primarily through energy efficiency. Even more importantly, retrofits need to go from being relatively shallow today to being deep—achieving 40-plus percent reductions in building energy use rather than the 10–20 percent reductions that are the norm today.

How we finance and motivate energy efficiency must change rapidly for

Box 9–1. Supplemental Benefits of Energy Efficiency

Enabling renewable energy. Multiple studies have examined the impact on grid operations of generating higher shares of electricity from intermittent renewable energy sources, such as the sun and wind, as well as the benefits of using energy storage technologies to enable greater consistency of power output. Reflecting a broad consensus, these studies recommend expanding storage solutions but focus fairly exclusively on "hard storage," such as batteries, compressed air, flywheels, capacitors, and other expensive hardware. Likewise, regulators, utility executives, and venture capital firms are investing primarily in these hard power-storage technologies.

But such technologies are far more expensive than "virtual storage": distributed, intelligent energy efficiency and the use of building thermal mass and time-of-energy-use measures to reshape power consumption to enable far greater use of intermittent renewable energy. Virtual storage using energy efficiency includes smart meters, low-cost monitoring and sensor equipment, inexpensive distributed intelligence, and real-time energy management services. It offers a much lower-cost load-shaping ability to accommodate large additions of renewable energy capacity and to enhance grid reliability.

Increasing economic competitiveness. By cutting long-term energy costs, energy efficiency increases competitiveness by allowing more-productive investments. A 2012 U.S. National Academy of Sciences report highlighted the importance of energy efficiency to enhancing U.S. global competitiveness. In Germany, energy efficiency improvements in the national economy boosted the country's trade surplus 12 percent by reducing energy use and thus avoiding $30 billion in energy imports in 2014. For companies, cities, and countries, energy efficiency increases competitiveness.

Enhancing security. The U.S. Department of Defense, the largest energy consumer in the world, understands that investments in energy efficiency strengthen security. Annual energy costs of $21.3 billion and the fragility of the electrical grid leave the U.S. military "vulnerable to service disruptions" and put the continuity of critical missions "at serious and growing risk." The Defense Department therefore has set ambitious targets to reduce energy use and to develop renewable energy sources.

In its *Vision for Net Zero*, the U.S. Army states: "Addressing energy security and sustainability is operationally necessary, financially prudent, and essential to mission accomplishment." Energy is, in the words of former Chairman of the Joint Chiefs of Staff Admiral Mike Mullen, about "not just defense but security, not just survival but prosperity. . . . Saving energy saves lives." The Department of Defense realizes the value and practicality of energy efficiency, officially codifying it as "a force multiplier" in the 2010 *Quadrennial Defense Review*.

Source: See endnote 21.

energy efficiency to deliver on its promise of anchoring the global transition to a low-carbon economy. This is eminently doable, as there is a wide range of identified pathways for scaling energy efficiency financing. Making these policy choices also will provide very large financial savings and other benefits. The policy options and rationale to achieve deep energy efficiency improvements are available. All that is lacking is initiative.[22]

CITY VIEW

Melbourne, Australia

Robert Doyle

Melbourne Basics

City population: 122,207
City area: 37.7 square kilometres
Population density: 3,242 inhabitants per square kilometre
Source: See endnote 1.

Rowing on the Yarra River, with the city skyline in the background.

David Hannah

Tackling Climate Impacts, Shoring Up Livability

Focused around a central business district, metropolitan Melbourne extends over 8,806 square kilometres of suburbs; however, the City of Melbourne comprises the 40 square kilometres that make up the city centre. The City's residential population, estimated at 122,000 in 2014, is expected to nearly double to around 230,000 by 2035. The current daytime population (including visitors and workers) approaches one million people. The City's economy accounts for 27 percent of Victoria's gross state product and 6 percent of Australia's gross domestic product.[2]

Like most Australian cities, Melbourne faces diverse challenges related to a changing climate, urban sprawl, water and energy security, and population growth. In recent decades, a host of extreme weather events such as drought, flooding, and storm surge has affected the health and quality of life of the community. Heat now causes more deaths annually in the city than road accidents. While 173 people died during the Black Saturday Bushfires of 2009, the five-day heat wave preceding that catastrophic event resulted in 374 deaths in metropolitan Melbourne due to extreme heat.[3]

Scientists predict that by 2030, Melbourne's climate will become warmer and the city will face increased impacts from heat waves, lower rainfall, intense storm events, and flash flooding. By 2070, the city is projected to experience more than twice as many heat waves, an 11 percent decrease in rainfall, and a significant increase in storm events. The City of Melbourne is looking at new ways to adapt to climate change while maintaining and enhancing its position as one of the world's most sustainable and livable urban areas.[4]

Open Space and Urban Forest Strategy

The City of Melbourne recognises that urban open spaces and a healthy urban forest will play a critical role in maintaining the health and livability of Melbourne. The City directly manages 480 hectares of parkland and, through its Open Space Strategy, aims to increase this area by 7.6 percent, providing 20 square metres of open space per person. In recent decades, 46 hectares of asphalt from central city streets and parking lots has been converted to parkland and pedestrian paths, and another 6.5 hectares is earmarked for conversion in the coming years.[5]

Some 27 percent of Melbourne's tree population (70,000 council-owned trees) is under threat in the next decade, and 44 percent in the next 20 years. The City of Melbourne has worked closely with the community to develop an Urban Forest Strategy, which aims to double the canopy cover to 40 percent by 2040 to help cool the city by four degrees Celsius. To boost the urban forest's resilience to climate change, the strategy aims to increase tree diversity, with no more than 5 percent of any one species, 10 percent of any one genus, and 20 percent of any one family present in the city.[6]

Healthy Catchment, Healthy City

In the coming decades, reduced rainfall and more-frequent and severe droughts are expected to strain Melbourne's water supply. The City of Melbourne's vision of a "healthy city in a healthy catchment" aims to involve the entire community—residents, workers, and businesses—in thinking about water and its role in the city's future. As a drainage authority and as one of the biggest water users in the municipality, the City has a leadership role to play in implementing integrated water cycle management and has developed a plan called *Total Watermark – City as a Catchment.*[7]

The City of Melbourne has invested AUD$20 million (US$14 million), and the state and federal governments AUD$5 million (US$3.5 million), in a network for harvesting stormwater, which delivers 25 percent of the City's annual landscape water requirements and is reducing reliance on potable water. In the coming years, the City aims to source half of its water requirements from rainwater tanks. In 2016, a 2 million litre underground water tank will be installed to capture and treat stormwater for use in irrigation. The AUD$4 million (US$2.8 million) project also will reduce downstream flows, minimising the chances of flooding and increasing the capacity of the drainage network without having to replace the current pipes.[8]

Carbon-Neutral by 2020

In 2002, the City of Melbourne adopted an ambitious target of zero net emissions by 2020, both for the Council's operations and for the municipality as a whole. At the current trajectory, annual greenhouse gas emissions are projected to total 7.7 million tonnes by 2020, a 60 percent increase from 2010 levels. Even if Melbourne were to implement all currently viable emissions reduction opportunities by 2020, municipal emissions would still exceed those of 2010.[9]

Achieving zero net emissions requires substantial structural, economic and policy changes to increase energy efficiency, decrease the use of carbon-intensive fuels, and offset any remaining emissions. The Council is providing city residents, business owners, building owners, workers, and visitors with information on reducing their emissions. Through its Zero Net Emissions strategy, the City of Melbourne is focusing on six areas where it can achieve the most effective and viable emissions cuts: council operations, commercial buildings and industry, residential buildings, non-transport electricity and gas use, transport and freight, and waste management.[10]

Council Operations

The Council must lead by example. The City of Melbourne became a certified carbon-neutral organization in 2011/12 and has since maintained that status. However, the City's operations contribute less than 1 percent of Melbourne's total municipal emissions. Engaging and activating the community is paramount to achieving carbon neutrality for the city.[11]

Commercial Buildings and Industry

Commercial buildings and industry have the biggest impacts on the city's emissions, with electricity and gas use in these sectors alone contributing more than 70 percent of municipal emissions. The City of Melbourne, through its 1200 Buildings Program, is encouraging the retrofit of 1,200 commercial buildings, two-thirds of the building stock that contains office space. Tools include a Website with advice sheets and case studies, seminars to improve knowledge around retrofitting, a panel of lighting providers to facilitate access to discounted upgrades, and links to state and federal grants. In 2014, the program was recognised with the inaugural C40 and Siemens City Climate Leadership Award for its development of "environmental upgrade finance," which allows owners to borrow funds for retrofits that are then paid back via a Council-levied charge on the property.[12]

Melbourne also participates in the CitySwitch Green Office program, a nationwide initiative that supports office-based businesses in measuring, managing, and reducing their energy use and improving their sustainability. CitySwitch provides free advisory services along with Web resources, discount offers, industry case studies, research and thought leadership, and annual recognition awards. The program uses the industry-accepted standard for operational energy efficiency, NABERS, to measure and verify emissions reductions. Over the last 10 years, CitySwitch has become a well-respected program securing partnerships with state and federal governments, industry media, and NABERS, and has reduced emissions in the sector by more than 400,000 tonnes of carbon dioxide-equivalent.[13]

The City of Melbourne is setting an example through its own green building landscape. In 2014, the City inaugurated the Library at the Dock, Australia's first public building to receive a "Six Star" design rating from the Green Building Council of Australia. The City also is home to 138 "Six Star" new office buildings, the largest number in Australia, and has the largest urban renewal project in the country, Docklands.[14]

Residential Buildings

Melbourne is experiencing rapid population growth, leading to a transformation of its residential sector. During the last two decades, the population of the municipality has almost tripled, to more than 100,000 in 2011, and, by 2031, an estimated 190,000 residents will be living in more than 115,000 homes.[15]

Approximately 80 percent of City of Melbourne's residents live in apartments, which accounted for 93 percent of new homes built between 2006 and 2012. Studies have found that apartments in mid- and high-rise buildings consume 25 percent more energy than detached dwellings due largely to the energy use of shared services in common areas such as foyers, corridors, pools, gyms, and parking lots.[16]

The City's Smart Blocks program is designed to help apartment owners and managers improve energy outcomes in common areas and reduce energy costs. The Smart Blocks

Website offers a toolkit of smart energy projects to guide buildings through the process, as well as case studies and measurement and evaluation capability. The program also offers free solar assessment and rebates, reimbursing half of the cost, up to AUD$3,000 (US$2,100), for solar panels and lighting upgrades that reduce energy use in common areas. Smart Blocks has won national awards and was nominated for an international C40 Cities Award.[17]

The High Rise Recycling Program, one of 10 initiatives within the City of Melbourne's *Waste and Resource Recovery Plan 2015–18*, works with the managers and committees of 101 apartment buildings to improve the waste and recycling systems available to residents. The program has increased recycling by some 35 kilograms per apartment, and the collection of unwanted items for charity has been implemented in 76 buildings. As a result, some 180 tonnes per year of unwanted household goods was diverted from landfills.[18]

Non-transport Electricity and Gas Use

Melbourne's grid power supply, provided by the electricity grid of Victoria and sourced primarily from brown coal, is the most emissions-intensive in Australia. A major focus is to move the city to a renewable energy supply, sourcing 25 percent of the municipality's electricity from renewables by 2018. Meeting this target requires an innovative, scalable approach that appeals to all levels of business and to the community.[19]

In an Australian first, the City of Melbourne has teamed up with a group of local governments, cultural and educational institutions, and private companies to explore group purchasing of renewable energy. By enabling large energy users to sign contracts that directly link their electricity consumption with new and identifiable renewable power stations, this approach can help guarantee the financing needed for these projects. The City will go to market with the initiative in 2016.[20]

Transport and Freight

Melbourne is one of Australia's fastest growing municipalities. The City of Melbourne's *Transport Strategy 2012* establishes the key directions, goals, and actions to ensure that the city is prepared to meet this anticipated growth. Walking, cycling, and public transport are prioritised as the dominant modes of transport in inner Melbourne, although the car will continue to play a role given the city's sprawl and the lifestyles and habits of citizens.[21]

The new Melbourne Metro Rail Project, when finished, will alleviate pressure on the existing public transport network by allowing an additional 20,000 people to access inner Melbourne at peak times. The project involves building two nine-kilometre rail tunnels beneath the city and creating five new metro stations.[22]

Some 146,000 trips are taken by bike on a weekday within the municipality, and the aim is to increase this to 200,000, or 7 percent of total trips, by 2020. To meet the high demand for bicycle parking in popular destinations, the City of Melbourne is installing 200 on-street bicycle hoops each year as well as bike hubs with parking, showers, and changing facilities. The City's Good Wheel Project, in collaboration with partners, works with unemployed residents to refurbish unwanted bikes as a means to help people from culturally diverse backgrounds access cycling.[23]

Waste Management

If current trends continue, the municipality of Melbourne is expected to send 208,000 tonnes of waste to landfill by 2020—84,000 tonnes more than the city produces today. The City's *Waste and Resource Recovery Plan 2015–18* aims to increase resource recovery, reduce waste to landfill, and improve local amenity. The plan's 10 initiatives focus on both residential and commercial waste management.[24]

The Degraves Street Recycling Facility uses traditional recycling and high-tech machinery such as a food waste dehydrator and a cardboard baler to reduce landfill waste and to turn organic food waste into a soil conditioner for use on city parks and gardens. The GreenMoney recycling project issues rewards to participating households based on the quantity of materials recovered. Other projects include a trial of on-site technology to process food waste within residential buildings, working with food rescue organizations to capture unwanted food from the commercial sector, and educational programs to reduce food waste going to landfills.[25]

Looking Forward

The City of Melbourne's adaptation journey began in 2008 with the publication of *Future Melbourne – City of Melbourne*. This long-term plan for the city, developed through collaborative public engagement, recognises the strategic importance of tackling climate change. The City is now sponsoring the development of the *Future Melbourne 2026* plan to engage diverse stakeholders in creating a community strategy for shaping Melbourne's next decade.[26]

In 2013, the City of Melbourne was among the first 32 cities invited to participate in the 100 Resilient Cities Challenge, a Rockefeller Foundation program dedicated to helping cities around the world become more resilient to the physical, social, and economic challenges of the twenty-first century. In November 2014, Australia's first Chief Resilience Officer was appointed to lead the development of a resilience strategy on behalf of the 31 local government areas that comprise metropolitan Melbourne. The objective is to strengthen the city's ability to identify and manage shocks, including natural and human-caused disasters as well as social and economic stresses.[27]

Robert Doyle is Lord Mayor of Melbourne. At the author's request, Australian spellings were maintained throughout this City View.

CHAPTER 10

Is 100 Percent Renewable Energy in Cities Possible?

Betsy Agar and Michael Renner

In March 2015, the City of Vancouver, Canada, tabled a commitment to source the city's energy from 100 percent renewable sources by 2050. The plan passed unanimously in November, and Vancouver has been generating international buzz ever since. (See also City View: Vancouver, page 171.) With this commitment, Vancouver joins the ranks of other forward-thinking cities such as Copenhagen, Denmark, and Oslo, Norway, in transitioning to 100 percent renewable energy community-wide. Toward this end, collaborative efforts among like-minded cities—such as the 100% RES Communities and RES Champions League in Europe, and the Global 100% RE initiative worldwide—also have sprung to life.[1]

Committing to 100 percent renewable energy means significantly more than flipping a few switches. It requires making strong commitments to energy efficiency as well as to renewables in the three major urban energy-use sectors: electricity, heating and cooling, and transportation. In addition to the energy sources that it includes and the sectors to which it applies, committing to 100 percent renewable energy drives social change, animates a diversity of actors, demands innovative policies, transforms economies, and develops knowledge and skill capacities.

Vancouver's energy plan is gaining international recognition for being comprehensive and ambitious, as well as achievable. Around the world, many cities are taking steps to put their energy supplies on a more sustainable footing, primarily in the electricity sector but also in many other areas. This is driven

Betsy Agar is Research Manager at Renewable Cities, Simon Fraser University Centre for Dialogue, Vancouver, Canada. **Michael Renner** is a senior researcher at the Worldwatch Institute and Codirector of the *State of the World 2016* report.

not only by resource and environmental concerns, but also by the realization that the cost and reliability of conventional energy supplies—which often are imported—may fluctuate and even undergo severe swings, compromising local economies.

According to the Renewable Energy Policy Network for the 21st Century (REN21), "cities and municipalities are on the leading edge of integrating renewable energy into power infrastructure, buildings, and transportation systems." They are using their purchasing power, regulatory authority (through target setting), and various support mechanisms to encourage renewable energy deployment. Some cities have put in place feed-in tariffs for renewable power that are comparable to the national-level policies that have been adopted in nearly 100 countries worldwide. Action by city administrations is important not only in setting deployment targets, providing support measures, and removing policy barriers, but also in serving as a positive model for private residents and businesses.[2]

Goal Setting for Renewables

In broad strokes, "renewable energy" describes hydro, wind, solar, biomass, geothermal, and tidal, wave, and ocean energy, but "renewable" does not necessarily mean zero-impact. Although hydropower fits within the definition of renewable energy, most large hydropower projects permanently alter local ecosystems. Similarly, wood can be burned for energy as a biofuel, but there is growing evidence that it is more effective at reducing carbon emissions when used as a structural material in place of concrete and steel. Nuclear power, in contrast, does not meet the definition of renewable energy because its fuel source is non-renewable, yet it is considered "carbon-free"—as emphasized by the American Nuclear Energy Institute—a tension that divides environmentalists.[3]

Support for nuclear energy dipped in 2011 after a tsunami severely damaged the Fukushima Daiichi nuclear plant in Japan. Described as the greatest nuclear power plant disaster since the Chernobyl accident in 1988, the effects of the "Great East Japan Earthquake" rippled worldwide. Three months later, the German government pledged to retire all of the country's nuclear facilities within a decade. This, in turn, precipitated a shift in Germany's planning framework, with the country transitioning from being simply "low-carbon" to using 80 percent renewable energy by 2050, a strategy known as the *Energiewende*. Most countries have yet to show such a high level of commitment to renewables at the national level.[4]

In contrast, many cities are taking steps to increase the use of renewable energy sources, with many of them committing to high shares of renewables in the energy supply. Dozens of cities worldwide, including 74 in Germany alone, already have reached a goal of 100 percent renewable electricity. Other local governments have committed to reaching that goal in future years, including 140 additional cities in Germany, 50 cities in the United Kingdom, the cities of Aspen and San Francisco in the United States, and the city of Malmö in Sweden. (See Table 10–1.)[5]

Other cities have adopted less-ambitious targets for renewable electricity but probably could do far more. The Institute for Local Self-Reliance estimates

Table 10–1. Community-wide Renewable Energy Targets Set by Selected Local Governments					
City	Population	Target Date	Electricity Target	Heating/Cooling Target	Transportation Target
Vancouver, Canada	603,500	2050	100%	100%	100%
Aspen, Colorado, USA	6,600	2015	100%	–	–
San Francisco, California, USA	805,235	2020	100%	–	–
Malmö, Sweden	318,107	2030	100%	–	10%
San Jose, California, USA	960,000	2022	100%	–	–
Munich, Germany	1,388,000	2025	100%	–	–
Ulm, Germany	120,714	2020	100%	100% (by 2030)	100% (by 2030)
Wellington, New Zealand	204,000	2025	78–90%	–	–
Austin, Texas, USA	912,791	2025	55%	–	–
Amsterdam, the Netherlands	779,808	2040	50%	–	60–90%
Tokyo, Japan	35,682,460	2024	20%	–	–
Cape Town, South Africa	3,400,000	2020	10%	–	–

Note: Targets and achievements are continually evolving. Please refer to updates from local governments for accurate details and progress updates.
Source: See endnote 5.

that installing rooftop solar panels on just the municipally owned buildings of the roughly 200 U.S. cities that have 100,000 people or more could generate more than 5 gigawatts of electricity; 1 gigawatt can power 750,000 homes, the equivalent of about two coal-fired power plants. New York City alone could support more than 400 megawatts (MW) of solar capacity on its public buildings, a figure that exceeds the city's current 10-year overall goal of 350 MW.[6]

Many cities are promoting not just renewable electricity, but also renewable heating and cooling technologies—an area where national governments have lagged. China remains the world leader in the use of solar water heaters, although the technology is also popular in countries like Brazil and Spain. (See Chapter 8.) Israel has the oldest mandatory solar water heating ordinance, with a law from 1980 requiring that the heaters be installed in virtually all new homes. Today, nearly 90 percent of Israeli households use solar thermal energy to heat their water. The Indian city of Chandigarh made solar water heating mandatory as of 2013 in industries, hotels, hospitals, prisons, canteens, housing complexes, and government and residential buildings. Even in Austria, a country with far less sunshine, the capital city of Vienna has established an ambitious goal to cover half of its heat energy demand with solar thermal energy by 2050.[7]

The use of renewables in district energy systems is a promising solution in many cities. In such systems, steam, hot water, or chilled water is produced at a centralized location and then sent through a network of pipes to interconnected buildings to provide space and water heating and/or cooling. District energy networks are more efficient than having to install isolated equipment in each building. By developing the world's largest district heating network, Copenhagen was able to cut its carbon emissions 21 percent between 2005 and 2011. The city now is exploring ways to transition from fossil fuel use altogether by 2050. Dubai, in the United Arab Emirates, has developed the world's largest district cooling network, as an efficient, low-carbon alternative to conventional air conditioning. Paris is home to Europe's first district cooling network, and other European cities such as Helsinki, Finland, and Vilnius, Lithuania, source nearly all of their heating and cooling from district energy networks. In many European cities, particularly in Austria, Denmark, Germany, and Sweden, these networks are fed by large solar thermal heating plants.[8]

Some local governments are seeking to develop local renewable energy manufacturing industries as a way to support their broader renewables goals. Since the late 1990s, Dezhou, China, a city of some 5.8 million inhabitants in northwestern Shandong province, has worked to create a local solar industry cluster. The 2005 Dezhou Solar City Plan provided incentives such as tax waivers and reductions, rebates, preferential land use policies, and low-interest loans. Dezhou succeeded in building a strong local economic base, with more than 120 solar energy enterprises and some 30,000 jobs. The city's Million Roof Project, launched in 2008, required that all new residential buildings be equipped with solar water heating. Today, solar thermal or solar photovoltaic (PV) technology is integrated in 95 percent of new buildings in Dezhou.[9]

Creative Local Solutions: Procurement, Benchmarking, and Zoning

Germany's *Energiewende* epitomizes the strength of political and public cooperation on energy issues. Although it is a national strategy, it is centered on building city and regional capacity to take action on energy production, procurement, and demand at the local level. That said, such strong national-level support is not a necessary condition for local governments worldwide to transition away from fossil fuels.

In May 2015, Vancouver-based Renewable Cities (see Box 10–1) held a Global Learning Forum to discuss how the world's cities can lead the way toward implementing renewable energy and energy efficiency measures. Through this dialogue, participants learned that local governments in countries that are unsupportive of (or even opposed to) non-fossil energy strategies are finding creative ways to leverage the operational and development tools that do lie within their control. For example, although the City of Melbourne,

---

Box 10–1. About Renewable Cities

Renewable Cities is a new global program of the Simon Fraser University Centre for Dialogue, based in Vancouver, Canada, that aims to accelerate the adoption of 100 percent renewable energy by cities globally. The five-year program has been developed through dialogue with leaders in local government, the private sector, key innovators and thought leaders, and utilities. Renewable Cities grew out of the Carbon Talks program, which has worked with municipalities and utilities since 2010 on transitioning to low-carbon policies.

Renewable Cities launched with a Global Learning Forum in May 2015, which included among its participants more than 300 city staff, elected officials, members of the private sector, individuals from civil society, and researchers. Renewable Cities continues to leverage its expertise as a research-based dialogue convenor in support of cities through their transition to 100 percent renewable energy and energy efficiency.

Australia, is situated in a brown coal-exporting country and does not control its energy utility, it does control its energy procurement program. Recognizing this leverage point, Melbourne is piloting a procurement model that aggregates its energy needs with those of local businesses so as to achieve enough scale to warrant construction of a new renewable energy power plant, which is now out for tender. (See also City View: Melbourne, page 155.)[10]

In India, energy policies are overseen mostly by state and national governments, and city governments have limited decision-making authority in this area. However, the Solar Cities Program of the country's Ministry of New and Renewable Energy (MNRE) is helping local authorities achieve a minimum 10 percent reduction in conventional energy use over five years through a combination of renewable energy and energy efficiency measures. MNRE has proposed supporting 60 cities with financial, technical, and planning assistance, and, as of August 2015, it had approved the master plans of 27 cities.[11]

In places where local utilities are in private hands, city governments may face a powerful obstacle to shaping their own energy destiny. Private owners may not share a city's desire to embrace renewable energy and energy efficiency. One solution is to put the utility into the public realm—potentially against the express wishes of the private investors. In recent years, cities have engaged in a growing number of "remunicipalization" efforts, including of utilities that once had been publicly owned but subsequently were privatized. (See Chapter 16.)

According to REN21, efforts to take over control or ownership of local utilities generally have had "a positive impact on the deployment of renewable electricity at the local level." In Europe, the Community Power (CO-POWER) project was launched to support the creation of community power systems and to enable greater uptake of renewables. In the United States, by early 2015, more than 2,000 communities had created community power systems to enable the uptake of renewable energy. At least 800 electricity cooperatives also exist in the country.[12]

Cities can influence the market-based decisions of individuals and businesses as well. One approach is to "benchmark" the energy used by buildings, for example by asking the owners or operators of residential or commercial structures to voluntarily measure and report the energy performance of these buildings. Unfortunately, local governments have found that voluntary energy benchmarking programs tend to plateau: the initiatives appear to attract building owners who already are inclined to carry out energy retrofits, but fail to inspire less-progressively minded building owners.

The City of Seattle is responding to this pattern by taking its long-established benchmarking program to the next level: mandatory public disclosure. At present, owners are obliged to report to the city upon request, but publishing the energy performance of buildings empowers potential buyers with data on the actual energy use and associated costs, enabling them to factor this information into real estate purchases. Seattle anticipates that building owners will reap the rewards of investing in energy retrofits, which, in turn, will spur market drivers that encourage deeper efficiency measures. (See Chapter 9.)[13]

Not all local governments have the authority to mandate energy performance monitoring and disclosure; some opt for creative zoning that prioritizes district energy (or similar systems such as district heating, combined heat and power, etc.). Existing buildings that reach the end of their service lives can present unique opportunities for energy-efficient retrofits. During the scoping phase of a retrofit, it is not unusual to find that the existing heating systems are oversized for the average needs of the occupants—as was the case for the University of British Columbia. This leads to inefficient low firing temperatures and to cycling losses.

By zoning for district energy, cities enable building owners to take advantage of the existing mechanical systems and ultimately optimize operational capacity. Often thought feasible only for new builds, existing steam or hot water pipes in buildings can be adapted to accommodate district energy systems. Växjö, Sweden, has been a pioneer in using biomass and cogeneration for district heating purposes, even in old buildings, which has been rare in other cities due to the high costs of installing new piping. Coupled with waste heat recovery from other urban systems—such as the London "Tube" or Vancouver's sewer system—district energy has also been successful in recovering energy that otherwise would be lost to the atmosphere.[14]

Addressing Grid Defection and Social Equity Concerns

Scaling down to the household level, local governments are developing innovative financial incentives, policy levers, and behavior modification strategies to spark energy efficiency and/or distributed (on-site) energy production by private homeowners. To explore the challenge that cities face in changing homeowner behavior, Renewable Cities invited representatives from the municipally owned electrical utilities of Durban, South Africa, and Austin, Texas, to discuss the impacts of distributed solar power on centralized utilities, regardless of the local context.[15]

Both Durban's and Austin's municipally owned utilities have found that as the cost for solar PV declines, the number of people installing solar has increased significantly, in turn reducing electricity sales from the utility. In Durban, some of those customers are defecting from the grid altogether, which is especially detrimental in a city like Durban where electric rate structures are set up to provide cross-subsidies (charges are set in proportion with the household income). Cross-subsidies are needed to provide residences in poorer, informal settlements with basic services such as flush toilets and reliable electricity. Unfortunately, it is the city's wealthiest households that can afford solar, and their defection disproportionately affects the utility's ability to provide affordable service to other customers. (See City View: Durban, page 337.)

Halfway around the world, Austin's utility is managing similar energy inequities. Under net metering, most of the fixed costs of running Austin Energy and operating the grid were recovered in the upper tier of energy rates. This meant that as solar customers reduced their grid consumption, non-solar customers were paying greater shares of those fixed costs. To combat this and other inequities, Austin Energy implemented a Value of Solar (VoS) rate structure: instead of billing customers for the net amount of energy passing to or from the grid, households are charged for the total electricity they consume (at standard energy rates), but solar-producing customers are credited for the amount of electricity they produce on-site (at the current VoS rate). By teasing apart energy consumed from energy produced, VoS compensates homeowners for solar electricity that they send back to the grid, but it also motivates them to reduce their net energy use and ensures that the cost of maintaining the grid is distributed equitably.[16]

Another way to address potential energy inequity challenges is to ensure that renewable energy solutions are available and accessible to lower-income communities. In the United States, where tax credits for residential solar installations tend to benefit wealthier homeowners, one alternative is a so-called solar power purchase agreement (SPPA). The developer owns, operates, and maintains the PV system, and the homeowner agrees to site the system on the home's roof or elsewhere on the property. The homeowner purchases the electricity produced, rather than the PV system itself. This arrangement enables the homeowner to avoid barriers such as high upfront capital costs, complex design and permitting processes, and other risks.[17]

In addition, solar purchasing cooperatives can bring together several households to negotiate affordable prices with solar installation firms. In the United States, solar worker cooperatives such as Evergreen Energy Solutions in Ohio,

Namasté Solar in Colorado, and PV Squared in Massachusetts have a strong focus on the local communities in which they operate.[18]

Community solar gardens allow access to solar even without home ownership. For people who have few assets or savings to invest in energy alternatives, so-called on-bill financing relies on projected future electricity bill savings as a revenue stream to fund investments in renewables or energy efficiency. In New York State, this option is available for energy efficiency investments by low-income households. Another option is to provide subsidies, as California's SASH and MASH programs (Single-/Multiple-family Affordable Solar Homes, respectively) are doing.[19]

These arrangements underline that cities need to go beyond general target setting and similar policies and to incorporate policies that allow people of various income groups to benefit from renewable energy development.

Cleaner Energy for Transportation

A third energy-use sector that is key to cities transitioning to renewable energy is transportation. Like buildings and utilities, elements of urban transportation systems are both private and public. Under its Carbon Neutral planning framework, the City of Copenhagen is investing in infrastructure that favors active multi-modal transportation: walking, cycling, and public transit. Commuting by public transit is considered active transportation because it involves walking or cycling to and from the service. Studies show that active transportation has manifold health benefits, including lower rates of diabetes, heart disease, and mental health issues associated with the stress of driving in rush-hour traffic. In economic terms, fewer personal vehicles on the road means that infrastructure has greater capacity to move goods efficiently, thereby reducing the times that transport trucks idle in stop-and-go traffic, for example.[20]

Transportation alternatives, such as hydrogen-fueled vehicles, are gaining traction, but electric vehicles are expected to play a key role in curbing our addiction to fossil fueled-mobility. Cooperation with local authorities is essential for ensuring that charging infrastructure is well-distributed and serves electric vehicle owners sufficiently. One of the most interesting arguments in favor of electric vehicles is that they can be designed to charge when electricity production is high in the middle of the day—helping to drive prices down—and to send electricity back to the grid when demand is high. This has the effect of flattening the belly of the so-called duck curve (a supply-demand plot of the net hourly electricity load that projects the risks of over and under-generation throughout the day and that happens to take the shape of a duck when renewables contribute to the energy mix). Using electric vehicles to flatten this curve helps to lessen the impacts of variable loading associated with renewable sources and stabilizes electricity prices.[21]

Conclusion

With a little creativity, cities are finding countless ways to overcome the many obstacles they may face in integrating renewable energy and energy efficiency into their systems and operations. If local governments are still looking for reasons to commit to 100 percent renewable energy, they can look to municipalities like Greensburg, Kansas, which was flattened by a tornado in 2007. In the heart of America's coal country—where the politics also happen to swing right—the entire town was rebuilt using renewable energy for the benefits of resilience and energy self-sufficiency. Cities like Greensburg succeed when the community has a sense of ownership over decisions about their energy systems and residents benefit directly from the revenues and jobs generated locally. This shows that commitments to 100 percent renewable energy are both politically neutral and locally beneficial.[22]

Ty Nigh

Built as part of Greensburg's reconstruction, the 5.4.7. Arts Center is LEED Platinum. It takes its name from the date the tornado struck.

The world is teetering on the cusp of a global energy shift. Technologically, solutions such as district energy systems—and even innovations like the "Powerwall," Tesla's solar-based home energy storage unit—swing 100 percent renewable energy from the realm of the possible to the realm of the preferable. Now, more than ever, cities have the planning tools, financial incentives, technical know-how, and public support to transition to 100 percent renewable energy. All that this movement needs is leadership from cities like Greensburg and Vancouver to lend their political, legislative, and financial weight. The world is ready.

CITY VIEW

Vancouver, Canada

Gregor Robertson

Vancouver Basics

City population: 603,502

City area: 115 square kilometers

Population density: 5,248 inhabitants per square kilometer

Source: See endnote 1.

View of Vancouver Harbour and Convention Center from the seawall promenade in Harbour Green Park.

Guilhem Vellut

A City with a Bright Green Future

In 2009, Vancouver's Greenest City Action Team was created. This group of local experts researched best practices from leading green cities around the world to make recommendations and establish goals and targets that would not only keep Vancouver green, but also make it the world's greenest city. Their work—*Vancouver 2020: A Bright Green Future*—outlined 75 quick-start actions and tasked City staff with developing a robust implementation strategy.

The result—the Greenest City 2020 Action Plan (GCAP)—was approved by the city council in 2011. Vancouver's GCAP is a vision and strategy to create opportunities today while building a strong local economy, vibrant and inclusive neighborhoods, and an internationally recognized city that meets the needs of generations to come. The plan outlines 10 goal areas and 15 measurable targets, including doubling the number of green jobs from 2010 levels, requiring all buildings constructed from 2020 onward to be carbon-neutral in operations, and having 51 percent of trips taken by bike, walking, or transit.

The goals, targets, and more than 150 actions all work together to form one integrated plan. Actions with cobenefits are given priority. For example, increasing composting and gardening helps to achieve the Green Economy, Zero Waste, and Local Food targets. Improving transit service supports Climate and Renewables, Green Transportation, and Clean Air targets.

Now halfway through the implementation of the GCAP, the City is looking beyond 2020 and has committed to transforming Vancouver into a city powered completely by renewable energy before 2050. Vancouver has joined 16 other cities in the Carbon Neutral Cities Alliance, committing to these aggressive long-term carbon-reduction goals. Vancouver also helped found the ICLEI–Local Governments for Sustainability 100% Renewable Cities Network.

To achieve the vision of a city where all power for buildings, heat, and transportation is provided by renewable sources, the City of Vancouver has developed the *Greenest City: A Renewable Future* strategy, a guide for long-term actions necessary to take advantage of renewable energy opportunities in Vancouver's building, transportation, and waste systems.

Key Policies

More than 80 percent of the highest priority actions listed in the initial GCAP are now complete, and the next set of priority actions and strategies has been identified for 2015–20. Vancouver also has begun to look beyond 2020, with plans to green transportation through 2040 and to power the city completely with renewable energy before 2050.

In 2014, the City updated the Vancouver Building By-Law with additional requirements and revisions, including requiring new buildings on rezoned land to use 22 percent less energy than specified in North American standards (ASHRAE 90.1 2010), while also requiring buildings undergoing renovations to incorporate a range of energy-saving retrofits. The City also has developed a retrofit strategy to reduce greenhouse gas emissions from energy use in existing buildings.

The most significant change to energy use in buildings and greenhouse gas emissions has been the City's world-leading focus on establishing and expanding low-carbon neighborhood energy systems. The flagship Neighbourhood Energy Utility in South East False Creek, built to heat the athletes' quarters during the 2010 Winter Olympics, uses heat from sewer waste to heat homes, reducing the amount of heating-related greenhouse gases by 70 percent.

Building on this success, in 2012 the City adopted a Neighbourhood Energy Strategy, a roadmap for building other low-carbon district energy systems. In district energy zones, City policy has shifted the design of new condominium and apartment towers away from natural gas and electric baseboard heating to hot water-based heating systems that enable buildings to connect to, and benefit from, neighborhood systems.

Vancouver continues to be a leader in North America for sustainable transportation. As of spring 2015, 50 percent of all trips originating in the city were by foot, bike, and/or transit, up from 40 percent in 2008. During the same period, the number of daily bike trips doubled from 50,000 to 100,000.

Land use and urban design play an important part in changing transportation mode shares. Vancouver builds mixed-use, walkable communities that are well-served by transit. The City also has a city-wide *Transportation 2040* plan that builds on the approved high-level direction and detailed ideas generated through the Greenest City planning process. This plan reaffirms the GCAP mode share and distance-driven targets (see below) and outlines a 2040 target for at least two-thirds of all trips originating in the city to be made by foot, bike, and/or transit. *Transportation 2040* also includes a target for zero traffic fatalities.

Building on the original Greenest City targets for quantity of green space, the Vancouver Board of Parks and Recreation has made plans to further enhance Vancouver's natural spaces, including *Rewilding Vancouver, an Environmental Education and Stewardship Action Plan* (2014), the *Bird Strategy* (2015), and a *Tree Protection Bylaw* (2014). The City has partnered with Metro Vancouver, the regional governance body, in pushing toward the goal of zero waste. Policies include banning organics from landfills and introducing several new programs to collect and divert compostable food scraps. On January 1, 2015, a Metro Vancouver ban on the disposal of organic waste with garbage came into effect region-wide.

In 2014, the province launched an extended producer responsibility (EPR) program for packaging and printed paper from residential properties, increasing the types of materials that can be recycled. EPR programs shift the burden of dealing with materials from taxpayers to the producers and users of products.

Key Achievements and Outcomes

Since 2011, a dozen strategies, policies, and plans have been approved that are complementary to the *Greenest City 2020 Action Plan*—from the *Food Strategy* to the *Transportation 2040* plan and the *Urban Forest Strategy Framework*.

In the past 25 years, Vancouver's population has grown by 34 percent, with jobs increasing by 30 percent and energy use increasing by about 15 percent, making Vancouver the fastest growing economy in Canada. Over the same period, Vancouver's carbon emissions have dropped by 7 percent and are expected to keep falling, showing that the city can continue to grow and be economically strong while removing the burden of carbon pollution. Vancouver's green building policies and support of the local food movement and green transportation have contributed to the creation of 3,200 green jobs, up 19 percent since 2010.

Of the high-priority actions identified in the GCAP, 80 percent are now complete. Beaches, shorelines, and water bodies throughout the city have been cleaned up, and wildlife such as salmon, beavers, and even whales have been returning to Vancouver's waters. A program to install electric vehicle charging stations throughout the city has supported growth in electric vehicle ownership and use, resulting in cleaner air quality. The City has increased the number of farmers markets and community gardens.

Other key GCAP achievements include:

• Reduced waste going to the landfill by 18 percent, from 480,000 tons in 2008 to approximately 395,000 tons in 2013.

• Passed one of the greenest building codes in North America; new homes built in Vancouver will now use 50 percent less energy than those built elsewhere in British Columbia.

• Increased the proportion of trips made by sustainable transportation within the city to 50 percent of all trips.

• Reached the target of reducing the average distance driven per resident by 20 percent from 2007 levels.

• Expanded the walking and cycling network, creating a network of protected cycling infrastructure that allows people of all ages and abilities to enjoy cycling for both recreation and transportation.

• Opposed industries that will damage Vancouver's health, ecology, and reputation, such as the creation of a new coal export terminal on the Fraser River and the Kinder Morgan oil pipeline, which would see a massive sevenfold increase in oil tanker traffic in Vancouver's harbor, putting the city's shoreline and the planet's climate at risk.

• Approved a ban on coal shipments through Vancouver to protect residents from toxic dust.

What Made These Policies Possible?

Community and stakeholder engagement is key to the success of Vancouver's Greenest City Action Plan. More than 35,000 people from around the world participated in the development of the GCAP through social media, online, and in face-to-face workshops or events. More than 60 City staff, 120 organizations, and thousands of individuals contributed to the creation of the GCAP. The City then formed External Advisory Committees for each goal area, with representatives from key partner organizations such as business and industry associations, other levels of government, nongovernmental organizations, and academia. Staff consulted with these groups and included community input from the engagement process as they developed their implementation plans.

In many cases, the City faces limited jurisdictional control. The City's success, therefore, relies in part on action taken by other levels of government, residents, businesses, and community partners. For example, the City has developed collaborative relationships with local utilities, building organizations, and many others to help owners of existing buildings and homes reduce both energy costs and greenhouse gas emissions. The City also supports residents and businesses to help Vancouver reach its Greenest City vision through the CAD$2 million (US$1.4 million) Greenest City Fund. The fund has supported projects ranging from small neighborhood gardens to large community-wide education programs.

A community garden thrives near downtown construction.

The City also launched a Green and Digital Demonstration Program to allow businesses the opportunity to accelerate the pace of innovation, commercialization, and job growth in Vancouver's clean technology and digital sector.

Social Dimensions

Vancouver's vision for sustainability comes in three parts: Greenest City, Healthy City, and the Economic Action Strategy. All of these complement global sustainable development goals.

Vancouver's Healthy City Strategy is a long-term, integrated plan for healthier people, healthier places, and a healthier planet. The Healthy City Strategy also provides a framework for a healthy environment as a right. The Strategy further ensures that the City is working to create an environment in which workers are paid well and housing is fairly priced. This includes a Living Wage Policy, a social procurement framework that includes social enterprises, Community Benefit Agreements with developers to include low-income residents in building construction, and a Poverty Reduction and Advocacy Strategy.

Early in the development of the Greenest City Action Plan, the City sought to prioritize projects that supported both green and social goals at the same time. One example is the grants provided to Save On Foods and Sole Food, two organizations that focus on providing both employment to low-income residents and local, healthy food.

As Vancouver builds a green economy, it also looks at how to promote inclusive economic development. The City specifically includes local food as part of its definition of a green economy, as it fosters a growing local food industry and helps create jobs for people who face barriers to employment. Vancouver is a world leader in social impact businesses, with more than 400 social enterprises.

Scalability, Replicability, and Lessons Learned

Every city is a unique geopolitical ecosystem, so there is no single template to guide a city in becoming green, sustainable, and economically viable. However, the process by which Vancouver created the *Greenest City 2020 Action Plan*, and the associated systems of accountability, can be replicated. Vancouver's programs, policies, and goals are designed specifically to do two things: to inspire others to act, and to be duplicated and scaled.

The inspiration comes from the GCAP's success, and the duplication potential comes from four essential ingredients: leadership, plan, action, and partnerships. It sounds simple, and really, it is, but not without hard work, perseverance, and being able to understand that the targets and sometimes the plan need to be adjusted when the context changes.

In 2015, Vancouver became the lead city for the C40 District Energy Network. Vancouver thus plays a critical role as a thought leader and communicator to help accelerate the uptake of district energy systems in cities around the world. Vancouver's district energy leadership has been recognized globally, and its wastewater recovery system (the Neighbourhood Energy Utility) is the first of its kind in North America. Over the last 10 years, Vancouver has learned many lessons—about developing business models, engaging with businesses and utilities, and understanding the role of technology—that the City is eager to share with others.

Gregor Robertson is Mayor of the City of Vancouver.

CHAPTER 11

Supporting Sustainable Transportation

Michael Renner

Transportation—the movement of people and goods—is the lifeblood of a city. Inadequate transport systems constrain a city's economy and vitality. But making a city too dependent on motorized transport can cause a host of other problems: traffic jams and deadly accidents, debilitating air pollution, and the loss of valuable land to streets, highways, and parking lots. Car- and truck-centered transportation systems run the risk of becoming like clogged arteries: they are bad not only for the vitality and attractiveness of cities, but also for urban residents' health, local environmental quality, and the global climate.

At first glance, transportation policy appears to be principally about the "modal mix"—the types of vehicles being used and the supporting infrastructure that they require. But although vehicle choice (private cars versus trams, subways, or bicycles) is a central aspect, fuel efficiency is another important variable: the less energy a vehicle needs to move a given distance, the fewer air pollutants or greenhouse gases it emits. Hybrid gasoline-electric technologies can make engines more efficient. Pure electric vehicles eliminate the air pollution that results from the use of internal combustion engines, and, if the electricity that they use is generated from wind and solar, they do not contribute to carbon emissions while operating. An analysis by the Union of Concerned Scientists finds that even with the current fuel mix of power plants in the United States, battery-electric passenger vehicles sold today produce less than half the greenhouse gas emissions of comparable gasoline-powered models across the entire life cycle.[1]

As important as these considerations are, transportation choices are influenced and shaped by a broader set of issues. Land-use policies, zoning codes,

Michael Renner is a senior researcher at the Worldwatch Institute and Codirector of the *State of the World* 2016 report.

and the resulting degree of density in a given city determine the type of transportation system that is feasible. In principle, cities can undertake a wide range of measures to make their transportation systems more sustainable. However, urban areas that were designed with the car in mind, or whose structures were changed to accommodate cars, have to overcome a deeply ingrained structural problem. They typically suffer from sprawl that makes automobiles the only practical mode of transportation. (See Chapter 7.)

The cities of Atlanta in the United States and Barcelona, Spain, make for a telling contrast. The two cities have a comparable number of inhabitants, yet Atlanta's built-up area exceeds Barcelona's by about 12-fold. (See Figure 11–1.) The much greater distances in Atlanta sharply limit the practicality of travel modes other than the private automobile. Walking and biking are virtually impossible (and often too dangerous) in many parts of the city, and public-transit systems cannot adequately serve the many far-flung destinations that result from sprawl. Once a city's DNA is synonymous with the car infrastructure—roads, highways, parking lots, and so on—it is extremely difficult to reorient it. This path dependence affects all decisions and likely will take decades to overcome, even with dedicated effort.[2]

Figure 11–1. Population and Urban Area in Atlanta and Barcelona, 2014

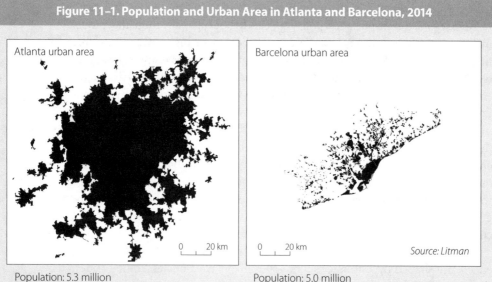

Atlanta urban area

Barcelona urban area

0   20 km   0   20 km   *Source: Litman*

Population: 5.3 million
Urban area: 7,692 km²
Transport carbon emissions per capita: 6.90 tons

Population: 5.0 million
Urban area: 648 km²
Transport carbon emissions per capita: 1.16 tons

Professor Stephen Wheeler at the University of California at Davis studies the patterns of built landscapes in metropolitan regions. Historically, he says, cities were characterized by compact settlement areas (grids or quasi-grids that were home to "mixed use," a combination of residential, commercial, cultural, institutional, or industrial uses). Only after the advent of the automobile did longer city "superblocks" and so-called degenerate grids, where streets do not readily interconnect, become possible. Many cities became more stretched out, less walkable, and more dependent on individual motorized transport. Superblocks and large apartment complexes still allow public transit-systems to work, but they often are sterile, monotonous, and excessively large (as in many Chinese cities; see Chapter 7). Many suburban residential areas follow circuitous patterns that Wheeler calls "loops and lollipops," artificially creating distance.[3]

Mixed use is a rarity in sprawled cities where homes, workplaces, schools, hospitals, and shops are segregated from each other. A feature found predominantly in the United States are the shopping malls and "big box" stores rising far from where people live, girded by huge parking lots. A consequence of various forms of sprawl is that traffic is funneled onto a limited number of collector and arterial streets that are easily subject to congestion. A city like Atlanta is an extreme exponent of this unsustainable pattern. Sprawling forms of land use account for 82 percent of the city's metropolitan area (with "loops and lollipops" alone accounting for 55 percent), whereas compact settlements represent only 1 percent.[4]

The website associated with Wheeler's work observes that "built landscapes often correlate with livability and sustainability variables (walkability, motor vehicle use, greenhouse gas emissions, demographic diversity, urban heat island effects, etc.)." From the perspective of sustainability, livability, and equity—and the type of decision making needed to advance these goals—the development of cities as described above has potentially fatal implications. Wheeler writes: "The privatized character of many suburban neighborhoods also discourages the sort of mixed-use neighborhood centers and public spaces that have traditionally served as locations for community gatherings and political protests, and thus may work against social dimensions of sustainability."[5]

Density, Equity, and Transportation

Compared with North America and Australia's sprawling metropolises, European cities generally are far denser, although not as dense as Asian cities. Atlanta's density is only 636 persons per square kilometer; other large North

American and Australian cities score somewhat better but fall in a range far below that of many European cities, such as London (5,907). Cities in the Middle East and Africa are even denser. In Asia, the density of major cities ranges from 8,000–9,000 persons per square kilometer in Tokyo and Singapore to about 20,000 in Shanghai and Seoul and more than 30,000 in Hong Kong, Mumbai, and Ho Chi Minh City. Figure 11–2 shows that the less dense a city is, the higher its car reliance and transportation energy use. With it come more-extensive road networks and low shares of public transit, walking, and cycling. But the figure also shows that medium density is quite sufficient to attain lower energy use levels.[6]

Sprawl also is becoming more prevalent in some developing countries. In Mexico City, a combination of ill-fated national economic and housing policies led to massive dispersal of residential areas along the periphery. (See Chapter 7.) During 1980–2010, the city's population doubled but its area grew sixfold, inflating its transport needs and greenhouse gas emissions.[7]

A low-density city may look deceptively green—with parks and landscaped

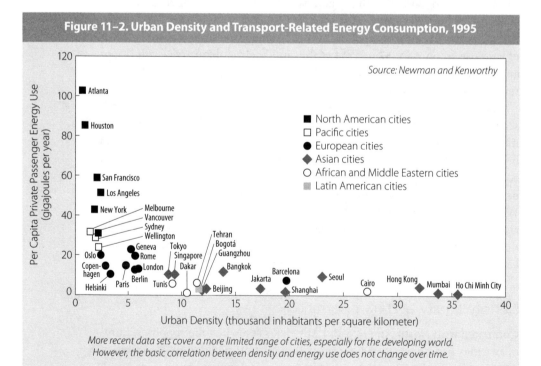

Figure 11–2. Urban Density and Transport-Related Energy Consumption, 1995

Source: Newman and Kenworthy

More recent data sets cover a more limited range of cities, especially for the developing world. However, the basic correlation between density and energy use does not change over time.

spaces—but the underlying sprawl is anything but green. Sustainable cities need to be far denser than many of today's urban agglomerations. They can take steps to reduce distance by encouraging mixed-use, transit-oriented development—where each street features a mix of homes, shops, restaurants, offices, and other places of urban activity and where each borough or similar unit offers adequate housing, jobs, schools, hospitals, and amenities—so that few people have a need to traverse vast expanses of urban spaces on an everyday basis. Unlike urban areas that are functionally segregated, such a mix provides for a vibrant kind of density that allows for well-functioning public transportation systems.

Sustainable cities need to integrate a social equity perspective into their policy making. Joan Clos, executive director of UN-Habitat, has argued that a car-centered, sprawling city "creates a heavy demand on workforce mobility, with large numbers of people needing to travel great distances each day." In cities of the developing world, car-centered transportation is not affordable to poorer people, and public transit is often inadequate or even non-existent, so they are stuck with informal vans, jeepneys, matatus, and similar unregulated vehicles that abide by no set schedule and are typically old, unsafe, and highly polluting. The poorest people end up walking along the margins of dangerous, congested roads. Affordable public transport systems are needed to provide better access to jobs and livelihoods for the majority of urban residents.[8]

Metropolitan areas in richer countries are not immune to inequities in transportation and housing. In New York City, members of low-income households tend to face longer commuting times than wealthier residents. A 2010 study by the Applied Research Center found that "black New Yorkers have the longest commute times of all, 25 percent longer than the average commute time for whites." And the gap is widening. The escalating cost of housing in areas near transit stations is pushing lower-income people into more-remote areas of New York that are not well-served by public transit. In the 25 largest metropolitan areas of the United States, combined transport and housing costs rose faster than incomes in the 2000s. In rich and poor cities alike, sustainable housing and sustainable transportation policies must go hand in hand.[9]

From Congestion Pricing to Car-Free Days and Car Sharing

Many cities have come to understand the serious problems of pollution, congestion, noise, health impacts, and, more recently, carbon emissions that come with heavy reliance on motor vehicles.

Various measures seek to dissuade drivers from entering certain parts of the city (such as downtown areas) or to otherwise reduce the number of cars on the roads. They include: congestion pricing, implemented in cities like London, Milan (see Box 11–1), Singapore, Stockholm, Tehran, and Washington, D.C.; vehicle quotas through auctions or lottery systems (in Chinese cities such as Beijing); license plate restrictions, such as Mexico City's *Hoy No Circula* ("Today Don't Drive") program and initiatives in other Latin American and Chinese cities; low-emission zones (adopted in 226 European cities as of 2013); and parking restrictions (in Singapore as well as cities in Europe, Japan, and the United States). As a result of congestion pricing, the number of cars entering central London dropped by one-third after 2002.[10]

Box 11–1. Congestion Pricing in Milan, Italy

With one of the highest rates of car ownership worldwide, Milan predictably suffers from heavy traffic congestion and dangerous air pollution. Two successive pricing schemes—"Ecopass" (introduced in 2008) and "Area C" (launched in 2012)—succeeded in reducing city center traffic, raising public-transit ridership, and improving air quality. Ecopass imposed a charge on the most-polluting vehicles entering the city center and succeeded in persuading motorists to switch to less-polluting cars, but it failed to solve the issue of congestion. Voters endorsed replacing it with "Area C," a congestion charge that applies to motorists entering an eight-square-kilometer Low Emission Zone (but that exempts electric and hybrid cars).

During its first year, the Area C scheme reduced traffic by 30 percent. Milan's public-transit system, Azienda Trasporti Milanesi (ATM), used the funds raised through the congestion fee to finance needed upgrades to the city's subway cars, trams, buses, and signaling system, and to extend "BikeMi," the city's bike-sharing system. Between 2005 and 2013, the share of private motorized transport in Milan decreased from 44 percent to 37 percent.

Source: See endnote 10.

More than 100 big cities, many in Latin America and Europe, now close some roads on weekends. Indian cities have caught on fairly recently to this movement with "Raahgiri Day," a weekly car-free Sunday first tried in 2013. Delhi is among the cities doing this, and so far the approach has influenced the discourse about public spaces and traffic in 30 other Indian cities. In a bold move that would be far more daunting in megacities (cities with 10 million people or more), the city council of Norway's capital, Oslo—where 650,000 residents have about 350,000 cars—announced a plan in late 2015 to ban cars completely from the city center by 2019. To make this possible, the city intends to expand public transport and bicycle lanes substantially.[11]

Other alternatives are emerging as well. Car-sharing ventures offer short-term access to a vehicle rather than requiring ownership. Most of the programs are membership-based and involve networks of stations and vehicles. In principle, car sharing offers environmental and anti-congestion benefits by greatly reducing the need for private car ownership (and thus requiring fewer parking spaces) and reducing vehicle kilometers traveled; however, real-life experiences vary widely. Other benefits include reduced emissions of air pollutants and greenhouse gases, especially if shared vehicles are efficient models, hybrids, or electric cars.[12]

From faint beginnings in the 1980s and early 1990s in a few European cities, car-sharing systems have spread to more than 1,000 cities in over 30 countries. Between 1995 and 2006, the ranks of car sharers increased from some 15,000 members to just under 350,000 members, with close to 11,700 vehicles being shared. By 2014, these figures reached 4.9 million members and 92,200 vehicles. Navigant Research projects that participation in car sharing may surpass 12 million by 2020.[13]

Different sharing models have evolved. A two-way model requires members to return the vehicle to the pick-up location. A more convenient one-way model allows drop-off elsewhere. Peer-to-peer systems enable users to share privately owned vehicles rather than those of a provider like Zipcar (with services in many cities worldwide) or Car2Go (in Europe and the United States). The basic appeal of car sharing is that it can help reduce the number of vehicles in use. In Philadelphia, the nearly 500 vehicles in the PhillyCarShare system serve some 50,000 members, displacing about 20,000 cars, reducing driving by an estimated 50 million miles, and avoiding 46,000 tons of carbon dioxide (CO_2) emissions.[14]

It is critical that all urban residents have access to cleaner mobility options. In the United States, the Shared-Use Mobility Center (SUMC) has helped introduce car sharing services for low-income communities in Chicago and the cities of Albany and Buffalo in New York. In Los Angeles, SUMC is working with the city and the California Air Resources Board on a three-year pilot project to make 100 electric and hybrid vehicles and more than 100 charging stations available to low-income residents. Funded through revenues from the state's cap-and-trade emission trading program, the goal is to help reduce greenhouse gas emissions by replacing some 1,000 private vehicles.[15]

Several car manufacturers and rental companies have jumped in, setting up pilot projects and joint ventures and acquiring leading car-sharing providers. In 2011, for example, PhillyCarShare was acquired by Enterprise, one of the largest U.S. car rental companies. Although this may bring financial muscle and scaled-up opportunity to this budding field, the impact of corporate-driven—as opposed to community-led—car sharing remains to be seen. Profit interests (revenues per vehicle) may result in car sharing becoming a supplemental mode of transportation layered on top of the existing system, rather than a means to greatly reduce the number of vehicles on urban streets.[16]

Despite the promise of car sharing, it represents only a very small share of the world's car fleet. Services remain centered largely in North America, Europe, Japan, and Australia, although car-sharing programs are making small inroads in cities elsewhere, such as Bangalore, Beijing, Mexico City, and São Paulo.[17]

Recent years also have seen the rise of companies like Uber, Lyft, and Sidecar, whose ride-hailing smartphone applications make it possible, in principle, for any driver to offer services. Such initiatives have gained supporters among people who are discontented with traditional taxi services. They also could help address the so-called last-mile problem, bridging the distance between homes and remote transit stops in suburban areas.

Uber has attracted growing criticism for its business practices, however, including violations of local laws and regulations (such as licensing and insurance requirements) and labor practices that some observers regard as exploitative. In Germany, Italy, the Netherlands, and Spain, courts have banned Uber from operating, and the company's practices have prompted protests in London, Madrid, Paris, and elsewhere. The Uber model is driven by profit motives, adding another layer on to existing transport options and thus intensifying car use rather than reducing car dependence. Other models, such as BlaBlaCar—which operates in a dozen European countries, including France, Germany, and Russia—focus on creating cost-sharing opportunities for people traveling to the same destination, rather than on a lucrative investor proposition.[18]

Although measures such as car sharing and congestion pricing can help to reduce the number of cars on the road, they cannot, on their own, make a city's transportation system sustainable. Car sharing, for example, may simply attract additional drivers hoping to cash in—people who previously had not owned a private car, perhaps because they could not afford one. Car sharing needs to be integrated into a well-planned and reliable multi-modal transport system to ensure that it helps to reduce overall private car use. In Bremen, Germany, the public-transit agency and the private car-sharing service Cambio successfully partnered to introduce an extension to the regular public transit ticket: for an additional €30 ($33) a year, customers gain access to car-sharing vehicles.[19]

Public Transit: Metros and Trams

To become sustainable, cities need to sharply reduce reliance on automobiles and to work to ensure a better mix of well-integrated transportation modes. Expanding and improving public-transit systems is a key ingredient of a more balanced system. The U.K.-based Light Rail Transit Association lists a total of 718 subway, light rail, and tram systems worldwide.[20]

The world's first subway system opened in London in 1863. By October 2015, some 157 metro systems were in operation, up from just 17 in 1950. The number has accelerated sharply since the 1970s. (See Figure 11–3.) Since 2000 alone, 56 new systems have been added. Another 35 are under construction and are projected to open between now and 2020.[21]

Historically, Europe has had the most metro systems (it now has 60), but Asia and the Middle East have since overtaken it, with 63 systems today. The subways in Beijing and Shanghai, built in 1969 and 1993, respectively, are now the longest and most heavily used systems in the world. Cities in the Americas

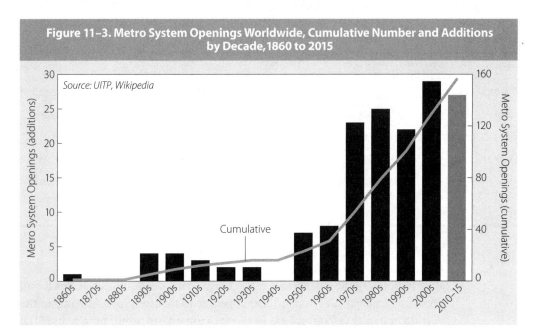

Figure 11–3. Metro System Openings Worldwide, Cumulative Number and Additions by Decade, 1860 to 2015

have 33 systems in operation. Collectively, the world's metro systems carry some 150 million passengers daily, including 71 million passengers in Asia and 31 million in Europe. The nearly 540 individual subway lines in operation worldwide have a combined length of 11,000 kilometers, with 9,000 stations.[22]

One problem inherent in metro systems is that they are expensive and time-intensive to build—something that many cities, especially in the developing world, can ill afford. Light rail or tram systems are far cheaper because they do not require digging tunnels; building underground stations, elevators, and escalators; or installing costly ventilation, lighting, and air conditioning. One study found that the per kilometer cost for selected subway systems in cities around the world ranged from a low of $71 million in Helsinki to a high of $684 million in Singapore, whereas the light rail systems examined ranged from $39 million in Strasbourg (France) to $68 million in Melbourne. The median cost for subways was $288 million, nearly nine times the median cost of $33 million for light rail projects. However, subways achieve higher speeds than light rail and tram systems and are able to carry more passengers. (See Table 11–1).[23]

Table 11–1. Characteristics of Light Rail and Metro Systems			
System	Speed	Peak Capacity	Segregation from Other Traffic
		(Passengers per hour)	
Streetcar or Tram	Low	Low (5,000 or less)	No significant segregation
Light Rail Train	Low to medium	Low to medium (10,000 to 12,000)	Partially segregated
Electric Commuter Train	Very high	Medium (~30,000)	Completely segregated
Light Metro	High	Medium (15,000 to 30,000)	Completely segregated
Heavy Metro	High	High (60,000 or more)	Completely segregated

Source: See endnote 23.

BRT: New Kid on the Block

In addition to light rail, another attractive alternative to subways is Bus Rapid Transit. So-called BRT systems first appeared in the late 1960s, but only in the last decade has the concept gathered real momentum around the world. The unique features of BRT systems—including dedicated right-of-way lanes, bus-only corridors, off-board fare collection, platform-level boarding, and stations that typically are aligned to the center of the road—make service comparable to light rail or metro systems in reliability, convenience, and speed.[24]

BRT also can greatly reduce greenhouse gas and air pollutant emissions. Passengers who switch from single-occupancy cars to high-occupancy BRT buses help reduce overall vehicle-kilometers traveled. Modern BRT bus fleets also are far more fuel-efficient and cleaner than the private automobiles and informal vans that they typically replace.[25]

According to the BRT Data website, just 21 cities had adopted a BRT system prior to 1990. The pace has picked up substantially since then, with 20 additional cities building BRT corridors in the 1990s, 104 cities during 2001–10, and 50 cities since 2011. (See Figure 11–4.) The combined track length rose from only 625 kilometers before 1990 to 5,229 kilometers (across 402 individual BRT corridors) in 2015. In addition to the 195 cities known to be operating a BRT system today, 48 cities (mostly in Latin America) have expansion plans, and 141 more cities either are planning a BRT system or have one under construction.[26]

The picture varies strongly by region. The city of Curitiba, Brazil, popularized BRT systems in the 1970s, and Latin America still has the most BRT

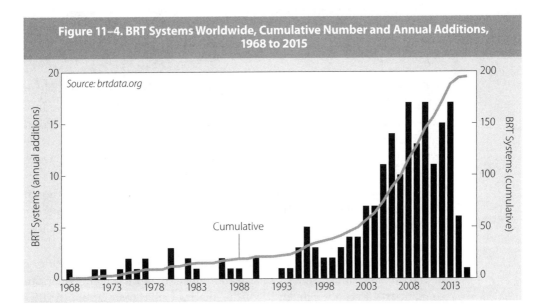

Figure 11–4. BRT Systems Worldwide, Cumulative Number and Annual Additions, 1968 to 2015

Source: brtdata.org

systems, the largest fleets, the greatest length of dedicated BRT lanes, and by far the largest ridership. (See Table 11–2.) Brazil's 33 BRT cities alone have close to 12 million riders daily, half the region's total. A 2011 survey among C40 cities showed that every one of them in South America either already had a BRT system or was planning one.[27]

Table 11–2. Number and Characteristics of BRT Systems, by Region				
Region	Number of Cities	Number of Vehicles	Length	Passengers per Day
			kilometers	millions
Africa	3	807	83	0.262
Asia	40	7,839	1,429	8.735
Europe	56	1,312	935	1.982
Latin America	63	44,283	1,745	20.036
North America	27	1,097	942	1.045
Oceania	6	593	96	0.430
World	**195**	**55,931**	**5,230**	**32.490**

Note: The data for individual systems are drawn from slightly varying years. For 70 out of the 195 cities, no vehicle fleet information is available; thus, the "world" figure understates the size of the global fleet.

Source: See endnote 27.

One of the most successful BRT systems is Bogotá's TransMilenio, which has a daily ridership nearing 1.9 million. TransMilenio is estimated to reduce CO_2 emissions by nearly 1 million tons per year and has led to a 43 percent cut in sulfur dioxide (SO_2) emissions, an 18 percent decline in nitrogen oxide (NO_x) emissions, and a 12 percent reduction in particulate matter. Buenos Aires, Argentina, instituted a major change in 2013 when it transformed several center lanes of its 20-lane Avenida 9 de Julio into a BRT corridor. Mexico City's Metrobus BRT now serves about 800,000 passengers per day; another 1.2 million people use regular buses or minibuses (typically of poor quality), and 4.8 million travel by subway, but 5 million still drive their own cars.[28]

Many cities in Asia and Europe also have BRT systems, but those in Europe in particular are much smaller in size and impact than their Latin American counterparts. Even though fewer North American cities than European cities have such systems, the total track length is about the same in both regions. But North American ridership is small, surpassing only that of cities in Oceania and Africa.[29]

The BRT system in Guangzhou, China, launched in 2004, carries more than 850,000 passengers a day. Guangzhou's experience has inspired similar systems (and the urban renewal that they often allow) elsewhere in China, as well as in Southeast Asia. Ahmedabad, India, introduced its Janmarg ("the people's way") BRT system in 2008, as part of a broader 2006–12 Comprehensive Development Plan. The plan introduced policy changes in favor of dense, mixed-use development, public transit, walking, and cycling, with the aim of increasing the public transport share from 17 percent to 40 percent over 10 years. Janmarg serves low- and higher-income communities equally and has encouraged urban regeneration. (See City View: Ahmedabad and Pune, page 231.)[30]

The Rea Vaya in Johannesburg, South Africa, was Africa's first full BRT system. Before its launch in 2009, more than two-thirds of public transit between Soweto and downtown Johannesburg was carried by crowded, unreliable, poorly maintained (and thus highly polluting) minibus taxis. The BRT not only saves travel time and operating costs, but also increases road safety and reduces CO_2 emissions. Many minibus operators became BRT drivers. A critical element was the inclusion of key community stakeholders in designing the system. Citizen engagement similarly was essential to BRT's success in Lagos, Nigeria, where the system replaced reliance on old and polluting private passenger buses. Its design was influenced by the experience of cities in Brazil, Chile, and Colombia. A primary aim was to meet the mobility needs of the urban poor, reducing their transportation expenditures and travel times.[31]

As BRT systems proliferated, the Institute for Transportation and Development Policy (ITDP) and its partners developed a "BRT Standard" that seeks to establish a common definition and to ensure that BRT systems "more uniformly deliver world-class passenger experiences, significant economic benefits, and positive environmental impacts." (See Box 11–2.)[32]

ITDP finds that many cities "have, on their own initiative, brought about significant long-term shifts away from private car use." Still, in many cities (including in India, Indonesia, and the United States), the expansion of mass transit infrastructure has not kept pace with growing populations and falls short in addressing the climate challenge. A useful yardstick is measuring kilometers of transit tracks per million urban residents—what ITDP calls the RTR (rapid transit-to-resident) ratio. France, for example, has achieved a high RTR ratio by expanding mainly its light rail systems (and subways to a

Box 11–2. The BRT Standard

ITDP has created a scoring system to offer recognition to high-quality BRT systems world-wide. In 2013, 11 corridors (6 of them in Bogotá, Colombia) were certified as satisfying criteria for the highest, or Gold, standard; 27 corridors were accorded Silver status, and 24 corridors received Bronze status. Many of the BRT systems in China and the United States, by contrast, are of relatively low quality. Scoring for the BRT Standard includes points for meeting certified emissions standards. But reducing emissions from all bus services remains a major challenge in cities. Some cities are switching to natural gas-fueled buses, which generate lower emissions of air pollutants. Over the years, the European Union and the United States have tightened bus emissions standards for particulate matter and NO_x, which requires the adoption of low-sulfur diesel fuel and stricter tailpipe controls.

Source: See endnote 32.

lesser degree). But many developing countries cannot afford the high investments required. Since 2000, China has built more than half of all mass transit (measured by length of lines) worldwide, but at a high financial cost due to its emphasis on metro systems. China's RTR ratio remains lower than that of Colombia and Mexico, two countries that have shown how high-quality BRT systems can help cities render their transportation systems more sustainable, in an affordable way.[33]

Walking and Biking to Save the Planet

Another key piece of the sustainable transportation puzzle is providing safe, attractive spaces for bicyclists and pedestrians. This requires a range of measures, including promoting density, deceleration (the slowing down of motorized traffic), pedestrianizing core areas of cities (closing streets for motorized traffic either entirely or on certain days), and building supporting infrastructure. In the United States, Portland, Oregon, has promoted the concept of "20-minute neighborhoods" to enable residents to meet all of their non-work needs by walking or cycling. And numerous urban areas have reduced speed limits to make streets safer for cyclists and pedestrians.[34]

Freiburg, a city in southwestern Germany with about 220,000 inhabitants, has been a leader on many of these fronts since the early 1970s, when it established a pedestrian zone and issued its first bicycle plan. Since then, its network of bike paths has expanded from 30 kilometers to 420 kilometers. Freiburg also was the first German city to introduce an integrated monthly public transport ticket that allowed the use of all trains, trams, and buses in the city and the surrounding region.[35]

Continued efforts to reduce motorized transport and boost alternatives in Freiburg (already, the city has a much lower car density than most similar-sized cities) will play a big part in plans to make the city climate-neutral by 2050. Freiburg has sought to accommodate population growth within city limits by transforming Rieselfeld, a formerly polluted industrial area, and Vauban, a former French military base, into compact and attractive areas that promote car-free living. These examples show that transportation and housing policies should go hand in hand. Short distances and good public-transit systems are key elements of success: because of Freiburg's efforts, the number of motor vehicles per 1,000 residents in Vauban, at 250, is half the German average. (See City View: Freiburg, page 135.)[36]

Cycling offers social, health, and environmental benefits, coming as close to "zero-carbon" as any mode of transportation other than walking. It also improves urban livability and invigorates local business. Biking can flourish with the help of dedicated bike paths and lanes (especially if they form a continuous and coherent network), bike parking, and safety measures such as restricting vehicular access and speeds in parts of the city. Thanks in part to a supportive infrastructure, bicycles have been outselling cars in many European countries.[37]

Copenhagen, Denmark, is famous for its high share of bike use: more than one in three trips is made by bicycle, and the city boasts some 400 kilometers of bike paths. In Münster, Germany, cycling has a comparable modal share thanks to farsighted policies dating back to the 1950s; two-thirds of all trips today are made by bike, on foot, or by public transit. Utrecht, in the Netherlands, is building the world's largest bicycle parking facility, with space for 12,500 bikes. Malmö (Sweden), Sevilla (Spain), Strasbourg (France), Antwerp (Belgium), and Glasgow (Scotland) are among the many cities that have invested heavily in bicycle infrastructure.[38]

The "Copenhagenize Index," a comprehensive ranking of the world's bicycle-friendly cities, is based on 13 criteria that include bicycle culture, facilities, infrastructure, bike-sharing programs, modal share, safety, politics, advocacy and social acceptance, gender split, as well as the broader issues of urban planning and traffic calming. Copenhagen and Amsterdam typically vie for the top spot on the list, and European cities occupy the first dozen ranks. But in 2015, some surprising cities made the Index's "Top 20," including Minneapolis, which rose past Montreal, and Buenos Aires, which ranked as the top non-European city. Biking culture in Brazilian cities is growing: in Rio de Janeiro's *favelas*, or slum areas, cycling has a 57 percent modal share despite the challenging topography of the hillside communities (but with many of the city's wealthier residents eschewing bicycles, Rio dropped out of the Top 20).

Tokyo and Nagoya were on previous editions of the list, but, in 2015, no Asian city made the cut.[39]

Policies to promote cycling are no longer limited to avant garde cities, as even cities with a history of heavy car dependence have joined the fray—often due to visionary leadership in city halls. In Bogotá, the support of successive mayors since the late 1990s has led to more than 350 kilometers of bike paths, helping bicycle use rise from just above zero to about 5 percent of all modes by 2010. In Buenos Aires, an alternative transportation policy begun in 2009 has led to 138 kilometers of protected bike lanes known as *bicisendas*, bike sharing has attracted more than 140,000 registered users, and the city aims to pedestrianize more than 100 blocks of the city center. Since 2010, Mexico City has embraced similar policies, including the pedestrianization of several neighborhoods, the Programa de Corredores de Movilidad No Motorizada ("Non-Motorized Lanes Program"), and the Ecobici bike-share program.[40]

Bike-sharing programs thrive when they go hand in hand with a reliable and safe urban biking infrastructure. As recently as 2000, the world's bike sharing fleet was limited to 4,000 bicycles in six European countries, with Copenhagen alone accounting for half. As of late 2013, 639 cities in 53 countries had a combined fleet of nearly 643,000 bicycles, and, by 2014, the number had risen to 806,000 bicycles in 712 cities. The largest number of shared bicycles is found in the Asia-Pacific region (460,000 in 108 cities in 2013). Europe has by far the largest number of sharing programs (472), even though its overall fleet, at some 147,000 bikes, is smaller than Asia's. Cities in the Western

Terrestrial and floating bike parking in Amsterdam.

Michael Renner

Hemisphere are playing catch up, but New York, Chicago, and others have ambitious expansion plans.[41]

Bike sharing can be scaled to small towns as well as large cities. The number of C40 megacities with bike-sharing programs increased from 6 in 2011 to 36 in 2013. In addition, 80 percent of C40 cities have introduced bike lanes. The specific designs and goals of bike-sharing systems vary tremendously by city, however. Mexico City, Montreal, Barcelona, and Lyon are among the cities with the best bike-sharing performance, measured relative to trips taken per resident population and trips per available bicycle.[42]

During the past decade, bike sharing docking stations have become more sophisticated—with smartphone apps in more than 100 cities now indicating bike availability—and smart-card payment systems have been introduced. Moreover, global positioning system (GPS) technology has led to systems that allow users to park bikes almost anywhere, rather than having to return them to a fixed location. Electric bicycles are included in some systems to enhance the programs' attractiveness: Birmingham, Alabama, is the first city in North America to include bikes with battery-powered pedaling power.[43]

Several cities are emphasizing equity aspects, working to make bike sharing available and affordable for low-income residents. Chicago's "Divvy for Everyone" provides discounted memberships for qualified applicants. And Philadelphia's "Indego" system is putting one-third of its 600 bicycles in low-income neighborhoods.[44]

Moving Beyond the Winners

To make transportation systems more sustainable, cities need to shift their modal mix away from car dependence and toward a much more balanced group of options. Depending on the city's circumstances, alternatives include car sharing, congestion pricing, public transit, and promotion of bicycling and walking. But these efforts can bear fruit only in the broader context of strategies to increase density and limit sprawl.

Since 2005, an alliance of organizations has awarded the annual Sustainable Transport Award to cities that demonstrate leadership and innovation in improving mobility for residents, reducing emissions of greenhouse gases and air pollutants, and improving safety and access for cyclists and pedestrians. The award committee brings together the World Bank, ITDP, ICLEI–Local Governments for Sustainability, the World Resources Institute, Germany's Agency for International Cooperation (GIZ), and other organizations.[45]

As experience worldwide shows, wide-ranging policy options are available to cities wanting to reduce the footprint of their transportation systems. Although car-dependent metropolitan areas face a tremendous challenge, many cities, including some that struggle with too many vehicles on their streets, have been at the forefront of efforts to create more-sustainable transportation policies. It is important to recognize and celebrate these efforts, as the Sustainable Transport Award does. Part of the function of such awards is to encourage other cities to take up the challenge. It is critical that success stories be replicated as broadly and as quickly as possible, and that the lessons learned be shared around the world. The opportunities are matched by the urgency with which cities everywhere need to act.

CHAPTER 12

Urban Transport and Climate Change

Cornie Huizenga, Karl Peet, and Sudhir Gota

Energy use in the global transport sector is poised to double by 2050 despite ongoing improvements in vehicle technology and fuel economy, and urban transport accounts for 40 percent of total transport-related energy consumption. The demand for mobility is growing particularly rapidly in cities in the developing world. Under a business-as-usual scenario, the total number of urban passenger-kilometers traveled could triple in the period 2010–50. Cities must find ways to meet these mobility challenges while also reducing overall greenhouse gas emissions from transport.[1]

A crucial way to do this is to capitalize on existing opportunities in cities. The transport sector represents one-third of the global potential to reduce urban greenhouse gas emissions in the period leading to 2050, so there is a strong need to implement sustainable transport strategies for their emissions-reduction potential. Cities offer an immense opportunity to scale up sustainable, low-carbon transport solutions to contribute to climate change mitigation, to achieve positive health outcomes through non-motorized transport, and to create more-compact developments that increase residents' access in addition to improving mobility. Many cities already have the necessary ingredients to reduce climate change impacts, and the current need is to prioritize these existing factors for success.[2]

These factors can be strengthened by optimizing the mitigation potential of urban transport; enhancing coverage of urban transport in the United Nations climate change process; scaling up and accelerating urban transport measures

Cornie Huizenga is Secretary-General of the Partnership on Sustainable Low Carbon Transport (SLoCaT). **Karl Peet** is SLoCaT's Research Director, and **Sudhir Gota** is SLoCaT's COP-21 Senior Transport GHG Consultant.

proposed in the Intended Nationally Determined Contributions (INDCs) of countries; and expanding sustainable transport commitments under the Lima-Paris Action Agenda (LPAA) and other city initiatives.

Mitigation Potential of Urban Transport

To meet projected mobility demands under a business-as-usual scenario, urban infrastructure (especially roads) would have be expanded by 129 percent in the next 40 years, which would necessitate increasing urban transport investments by a factor of seven. However, if cities improve and invest in public transport and non-motorized travel, a huge sum could be saved. More than $100 trillion in cumulative public and private infrastructure spending and 1,700 million tons (40 percent) of annual carbon dioxide (CO_2) emissions from urban passenger transport could be eliminated by 2050. (See Figure 12–1.)[3]

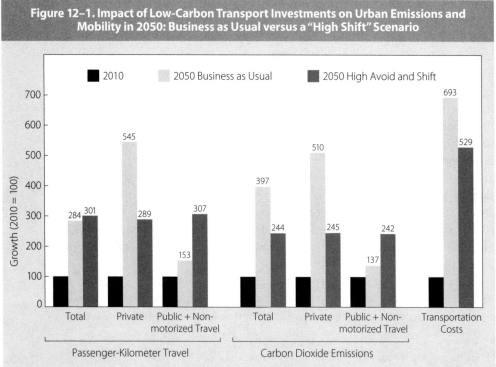

Figure 12–1. Impact of Low-Carbon Transport Investments on Urban Emissions and Mobility in 2050: Business as Usual versus a "High Shift" Scenario

Legend: 2010 | 2050 Business as Usual | 2050 High Avoid and Shift

Y-axis: Growth (2010 = 100)

Passenger-Kilometer Travel:
- Total: 284, 301
- Private: 545, 289
- Public + Non-motorized Travel: 153, 307

Carbon Dioxide Emissions:
- Total: 397, 244
- Private: 510, 245
- Public + Non-motorized Travel: 137, 242

Transportation Costs: 693, 529

Source: Replogle and Fulton

In addition, research from the New Climate Economy project suggests that an incremental investment of $10.6 trillion over the period 2015–50 in public, non-motorized, and low-emission passenger and freight transport could yield an annual abatement of up to 2.8 gigatons of CO_2-equivalent by 2050 relative to business as usual, with an average payback of less than 12 years.[4]

For the developing world, the International Transport Forum (ITF) projects that cities in Latin America, China, and India that have more than 500,000 inhabitants will more than double their share of world passenger transport emissions—from 9 percent in 2010 to 20 percent in 2050—based on current urban transport policies. Under a business-as-usual scenario, 38 percent of the growth in surface transport passenger emissions worldwide to 2050 will come from big cities in these three regions. These projections underscore a critical choice for policy makers: whether to pursue urbanization models that prioritize public versus private transport policies, which will lead to very different mobility futures. (See Figure 12–2.)[5]

In Latin America, private-transport-oriented policies would lead to an 82 percent share for cars, whereas a public-transport-oriented policy scenario—which supports the combination of low sprawl, high public transport expansion, and high fuel prices—would result in a 50 percent share for public transport, compared to only 44 percent for cars and 6 percent for two-wheelers. In China, an urban policy with restrictions on new roads and car ownership would lead to a 55 percent share for cars, 34 percent for public transport, and 10 percent for two-wheelers; in the absence of these measures, cars would account for 78 percent of urban modes, and public transport for only 9 percent. In India, a private-transport-oriented policy would lead to 67 percent of urban mobility being covered by car traffic, whereas, with pro-public-transport policies, the share of buses and other public-transport modes could reach 39 percent (roughly equivalent to the auto mode share). Thus, urban transport policies will be crucial in establishing transport sector-wide low-carbon trajectories in the coming decades.[6]

Coverage of Urban Transport in the UN Climate Change Process

Many cities around the globe are undertaking their own initiatives to mitigate greenhouse gas emissions from urban transport. Internationally, however, the primary arena for discussion and negotiation of such reductions is the United Nations Framework Convention on Climate Change (UNFCCC), an international treaty established in 1992 to limit average global temperature

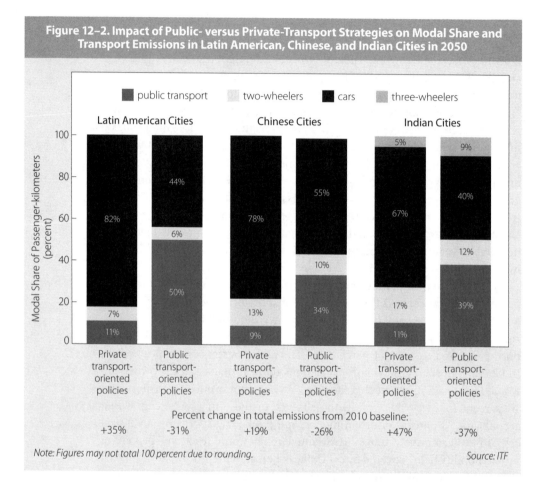

Figure 12–2. Impact of Public- versus Private-Transport Strategies on Modal Share and Transport Emissions in Latin American, Chinese, and Indian Cities in 2050

Note: Figures may not total 100 percent due to rounding.

Source: ITF

increases and the resulting climate change impacts. In 1997, to accelerate efforts to reduce global emissions, countries adopted the Kyoto Protocol, which legally binds developed countries to emission reduction targets. The Protocol's current commitment period began in 2013 and will end in 2020. There are now 195 Parties to the UNFCCC and 192 Parties to the Kyoto Protocol.[7]

The transport sector has traveled a winding road within the UNFCCC process. Transport traditionally has been viewed as a subsector of energy in the UNFCCC framework, which has led to a failure of governments to significantly scale up transport projects to reduce climate change impacts. As a consequence, the UNFCCC's Clean Development Mechanism (CDM)—a mechanism that enables developed countries to lower their own emissions by supporting emission reduction projects in developing countries—has proposed methodologies that do not fully consider the characteristics of transport. As of December 2015, only 33 of 7,685 CDM projects approved were in the transport sector.[8]

More recently, national commitments to greenhouse gas reductions—so-called Intended Nationally Determined Contributions (INDCs)—have played an integral role in international climate discussions, particularly in the December 2015 UN climate talks in Paris. INDCs communicate country targets and strategies to reduce carbon emissions for the post-2020 period, and each country faces a unique set of circumstances influencing reduction strategies, including socioeconomic development patterns, historic emission trajectories, and varying financing requirements.

Because INDCs represent a bottom-up, nationally determined process, they have the potential to drive progress in countries (particularly developing countries) that are shaping emerging climate policies. Starting in 2016, countries will have to operationalize the transport components of their INDCs and thus will need a robust kit of data, tools, and analytical methods to ensure that INDC targets ultimately are realized. Likewise, urban transport investment strategies must be spelled out clearly in the scope of existing and forthcoming INDCs.[9]

Urban Transport in the INDCs

As of November 2015, a total of 133 INDC submissions representing 160 countries had been submitted to the UNFCCC. These countries account for nearly 93 percent of global greenhouse gas emissions and roughly the same share of global transport greenhouse gas emissions. Although nearly all of the INDCs acknowledge the transport sector, only 12 countries specifically translate the 2030 economy-wide target for emission reduction into a transport sector target. Among the INDCs submitted, roughly 35 percent make specific reference to urban transport improvements. Other modes mentioned in the INDCs provide indirect support for urban transport (for example, freight transport, railways, and waterways). (See Figure 12–3.)[10]

Urban transport measures included in the INDCs are allocated among a variety of *direct* measures, which include public transport (mentioned in 24 percent of the INDCs), walking and cycling (8 percent), compact land use (8 percent), and parking (3 percent). (See Figure 12–4 and the examples in

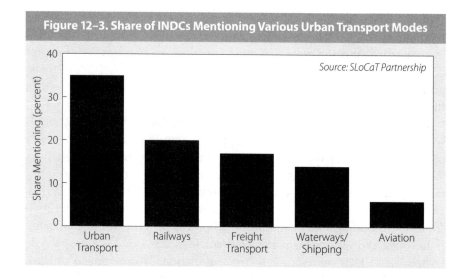

Figure 12–3. Share of INDCs Mentioning Various Urban Transport Modes

Source: SLoCaT Partnership

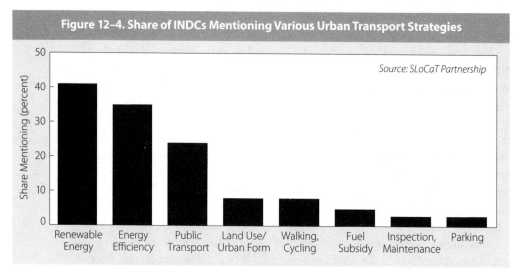

Figure 12–4. Share of INDCs Mentioning Various Urban Transport Strategies

Source: SLoCaT Partnership

Table 12–1.) In additional, several overarching strategies in the INDCs can contribute *indirectly* to urban transport, including renewable energy (41 percent), energy efficiency (35 percent), and vehicle inspection and maintenance (3 percent).[11]

Transport sector-related *targets* (in contrast to *measures*) emphasize a reduction in the magnitude of emissions in comparison with a base year and/

Table 12–1. Transport Sector Measures in Selected INDCs	
China	Increase fuel quality and promote alternative fuels; increase the mode share of public transport in large- and medium-sized cities to 30 percent by 2020; promote dedicated pedestrian and bicycle infrastructure in cities; accelerate development of green freight
Côte d'Ivoire	Integrate climate considerations in territorial planning to limit travel distances; propose efficient policies in urban transport plan development (e.g., Abidjan urban train); accelerate uptake of low-emission vehicles through standards and incentives
Ethiopia	Promote clean rail transport and compact development
Gabon	Increase infrastructure investments and public transport services (e.g., congestion reduction in Libreville); restrict importation of vehicles that are more than three years old
Japan	Promote modal shift to public transport and railways; develop traffic safety facilities and improve traffic flow through Intelligent Transport Systems (ITS); promote driverless cars, eco-driving, and car sharing
Jordan	Increase public transport mode share to 25 percent by 2025; reduce vehicle fuel emissions and vehicle travel, particularly in densely populated areas
Macedonia	Increase electrification of transport, use of railways, and use of bicycles and walking; renew vehicle fleets; introduce a parking policy
South Korea	Continue to expand infrastructure for environment-friendly public transport while introducing low-carbon standards for automobile fuel efficiency and emissions; provide incentives for electric and hybrid-electric vehicles

Source: See endnote 11.

or with a business-as-usual baseline. Emission targets in two INDCs submitted to date highlight urban transport specifically: the Democratic Republic of the Congo aims to reduce CO_2-equivalent emissions by 10 million tons compared to a business-as-usual scenario through urban transport improvements, and Trinidad and Tobago intends to reduce greenhouse gas emissions in the public transport sector by 30 percent by December 31, 2030, compared to a business-as-usual scenario.

On an economy-wide scale, mitigation measures proposed in the INDCs are expected to fall short of meeting the internationally agreed target under the Paris Agreement of keeping global temperature rise below 1.5–2 degrees Celsius, with a more likely scenario in the range of 2.7°C. It also is unlikely that the transport sector (and, in turn, proposed urban transport measures) will attain a 1.5–2°C scenario by 2030 through the proposed targets and measures, based on existing policies and on the levels of ambition expressed in the INDCs. To achieve the deeper emission cuts that are necessary to put the transport sector on track for a 2°C scenario, the level of ambition would need to be intensified, implying the need for transformational rather than merely incremental change.

LPAA Transport Initiatives on Urban Transport

The Lima-Paris Action Agenda (LPAA) is a joint undertaking of the Peruvian and French presidencies of the UNFCCC Conference of the Parties, the Office of the Secretary-General of the United Nations, and the UNFCCC Secretariat. The UN Environment Programme's (UNEP) 2014 *Emissions Gap Report* asserts that business-as-usual emissions in 2020 are projected to exceed emission levels required to achieve a 2°C scenario by roughly 12 gigatons of CO_2-equivalent, and the LPAA intends to contribute to closing this emissions gap by further increasing pre-2020 ambition to support the 2015 Paris Agreement, and to emphasize the need for greater action to strengthen resilience to climate impacts.[12]

Building on the UN Secretary-General's Summit in September 2014, the LPAA is committed to scaling up regional, provincial, and city-level climate initiatives to advance sustainable development. Among these initiatives are numerous urban transport-focused commitments for scaling up sustainable urban mobility, which account for 9 of the 15 transport commitments under the LPAA. These include the following:[13]

MobiliseYourCity, a coalition that helps local governments in developing countries plan sustainable urban mobility, reduce greenhouse gas emissions, and develop more-efficient cities. Adequate transport-related activities at the national and subnational levels could yield a 50 percent reduction in urban emissions by 2050 compared to business as usual. MobiliseYourCity aims to engage 100 cities (including Amman, Casablanca, Ouagadougou, and Tunis) in integrated mobility policies by 2020 and to spur national governments to create comprehensive urban mobility frameworks (including legislative frameworks, funding schemes, and evaluation methodologies).[14]

C40 Clean Bus Declaration of Intent, a declaration announced during the C40 Latin American Mayors Forum in March 2015. The ultimate goal is to incentivize and help manufacturers and other stakeholders, such as multilateral banks, develop strategies to make electric, hydrogen, and hybrid bus technologies more affordable for cities. As of November 2015, 26 cities across Africa, East Asia, Europe, Latin America, and North America had signed on to the declaration, including Addis Ababa, Bogotá, Mexico City, Oslo, and Seoul.[15]

Action Platform on Urban Electric Mobility, an initiative to increase the market share of electric vehicles in cities to at least 30 percent of all new vehicles (including cars and motorized two-or three-wheeled vehicles) sold each year by 2030, while developing the enabling infrastructure for their effective use. The initiative aims to reduce CO_2 emissions in urban areas through increased use of electric mobility for passenger transport (both private and public) as well as freight transport, combined with measures to reduce transport demand and increase the use of public and non-motorized transport.[16]

International Association of Public Transport (UITP) Declaration on Climate Leadership, a declaration that encourages UITP members to make commitments to reduce carbon emissions and strengthen climate resilience within their cities and regions. As of November 2015, UITP had stimulated around 350 commitments and actions from 110 public transport organizations (e.g., bus fleet renewal in Dakar, metro expansion in Moscow, bike sharing in Munich, efficient transport infrastructure lighting in Rio). Actions aimed at giving a greater role to public transport in mobility will help decrease carbon footprints in metropolitan regions. They also will support UITP's goal of doubling the market share of public transport by 2025 (compared to 2005 levels), which would prevent half a billion tons of CO_2-equivalent compared to business-as-usual projections.[17]

A public bike share station in Milan, Italy.

World Cycling Alliance (WCA) and European Cyclists' Federation (ECF) voluntary commitment, a commitment that seeks to boost a modal shift to cycling worldwide and to double the cycling mode share in Europe by 2025 (compared with the current share) within various countries. This will be achieved by advocating for the importance of cycling in achieving the new UN Sustainable Development Goals through collaboration with the UN, the Organisation for Economic Co-operation and Development (OECD), the Partnership on Sustainable Low Carbon Transport (SLoCaT), and the Transport, Health and Environment Pan-European Partnership (THE PEP), and by mobilizing the support of WCA and ECF members to enable local, national, and international governments and institutions to scale up action on cycling. In September 2015, UITP and ECF signed an agreement to support each other's missions to double both cycling and public transport mode share and to establish a stronger lobby position when talking to European institutions and the UN. The agreement marks a more-intensive collaboration process for the development of policy messages on the economic benefits of sustainable mobility, public health, transport policy, and urban mobility data collection.[18]

Vehicle Fuel Economy Energy Efficiency Accelerator, a project led by the FIA (Fédération Internationale de l'Automobile) Foundation that calls for a doubling of the efficiency of all new vehicles by 2030 and a doubling of the efficiency of the entire global vehicle fleet by 2050, relative to a 2005 baseline. These fuel economy numbers would save more than 1 gigaton of CO_2 per year by 2025 and more than 2 gigatons per year by 2050, thus reducing more than $300 billion in annual oil imports in 2025 and more than $600 billion in 2050. The Global Fuel Economy Initiative—a partnership of the International Energy Agency, UNEP, ITF, the International Council on Clean Transportation, the Institute for Transportation Studies at the University of California at Davis, and the FIA Foundation—works to secure improvements in fuel economy, has expanded its network of pilot countries through a range of outreach processes, such as training workshops and meetings, and has achieved global recognition as the leading fuel economy initiative.[19]

Climate and Clean Air Coalition-coordinated Global Green Freight Action Plan, an action plan that brings together more than 20 committed governments and dozens of nongovernmental organizations and companies to expand, harmonize, and scale up freight programs that reduce black carbon, particulate matter, CO_2, and other emissions from global freight transport. These goals are to be accomplished by enhancing existing green freight efforts through peer-to-peer partnerships and government industry exchanges (for example, as modeled on the SmartWay Transport Partnership in the United States and Canada, with more than 3,000 partners, and the Clean Cargo Working Group in the marine sector), and by expanding green freight practices in interested countries to build bridges among policy makers, business leaders, and civil society at the global level (for example, as modeled on the World Bank and Netherlands Government Sustainable Logistics Trust Fund).[20]

International Zero-Emission Vehicle Alliance, a collaboration of national and subnational governments, coordinated by the California Environmental Protection Agency, that aims to accelerate the adoption of zero-emission vehicles (ZEVs), including electric vehicles, plug-in hybrids, and fuel-cell vehicles (which produce very-low to zero tailpipe emissions but do produce indirect emissions from electricity generation and manufacture), and to foster collaboration on policies to advance investment and innovations required to achieve ZEV targets. The alliance was formally launched in August 2015, and 13 North American and European governments announced a target at the Paris climate talks to make all new passenger vehicles in their jurisdictions ZEVs no later than 2050.[21]

Two electric car-share vehicles plug into a charging station in Berlin.

© Tony Webster

Intelligent Transport Systems for the Climate, an emerging initiative from ATEC-ITS France and TOPOS Aquitaine (two French organizations focused on using "smart" technologies to improve transport) and other partner organizations that works to facilitate the deployment and operation of Intelligent Transport System (ITS) services. ITS services use digital technologies to enable users to be better informed and to make safer, more coordinated, and more efficient use of transport networks in order to reduce CO_2 emissions in the transport sector. The aims of the initiative are to facilitate the increased integration of transport modes for people and goods; to promote efficient navigation of vehicles; to encourage local authorities to optimize intermodal investments in infrastructure, vehicles, and training; and to share best practices for deployment of ITS to reduce transport greenhouse gas emissions.[22]

Other City Initiatives and Commitments on Transport

To complement the LPAA-endorsed transport initiatives, city governments are taking steps to expand actions and strengthen partnerships, which—similar to the submission of INDCs—indicate a growing willingness on the part of local authorities to prioritize action on sustainable low-carbon transport. These city actions include:[23]

Civitas, an initiative cofunded by the European Union with the objective of helping cities redefine their transport policies to create cleaner transport systems. So far, Civitas has helped some 60 demonstration cities to implement innovative measures to develop greener transport (e.g., electro-mobility in Stuttgart, hybrid and clean natural gas buses in Ljubljana, improved goods distribution in Kraków) by maintaining networks and working groups on transport topics and compiling best practices for broader dissemination. Civitas also provides funding for the transfer of smart measures from one city to another. At present, the Civitas initiative has a database of more than 700 mobility-related commitments.[24]

EU Covenant of Mayors, a joint initiative developed and administered by five of the largest city networks in Europe. Covenant signatories aim to meet and exceed the EU's objective of a 20 percent reduction in CO_2 by 2020. The transport-related submissions are generally local pledges, which range from improving public transport to increasing accessibility for cyclists. Of the signatories' planned actions toward 2020 that the Joint Research Centre of the European Commission has already assessed and approved, 24 percent relate to sustainable transport, with an estimated total reduction of 117 terawatt-hours per year, equivalent to the total annual energy consumption of the Netherlands.[25]

Sustainable Urban Mobility Campaign, an initiative launched in 2012 to support sustainable urban mobility campaigners in the EU's 28 member states, plus Iceland, Liechtenstein, and Norway. Known less formally as "Do the Right Mix," the campaign advocates the use of different modes of transport to help reduce the cost and impact of each journey. "Do the Right Mix" recently joined forces with the annual **European Mobility Week**, which encourages European cities to promote the use of sustainable transport and invite local residents to try alternative forms of transport. The event is organized each September to promote innovative mobility measures by local authorities, encourage exchanges with citizens on urban mobility themes, and find concrete solutions to related issues (e.g., urban air pollution). In 2015, more than 1,700 cities participated in European Mobility Week.[26]

These examples suggest a continued willingness by the transport sector to engage in voluntary commitments to reduce the impact of sustainable transport infrastructure, services, and policies (for example, through the Secretary-General's Climate Summit Initiatives and other emerging transport commitments), creating, in essence, a set of "supply-side" commitments. At the same time, there is a growing interest within cities and countries to engage in sustainable low-carbon transport initiatives and implementation measures (e.g., city commitments, business-sector commitments, and transport-focused INDC targets and measures), creating a set of complementary "demand-side" commitments. Analysis shows a remarkably good fit between the areas where cities and countries would like to take action on transport for both the sustainable development and climate change-oriented transport commitments. For example, INDCs highlighting urban transport from India, Japan, and Senegal can be matched to the UITP public transport commitment, and INDCs highlighting cycling from Azerbaijan, Cambodia, and Ghana can be matched to the ECF cycling commitment.

Recommendations

Urban transport requires additional attention within the UNFCCC framework, which can be achieved in several ways. First, technology transfer discussions under the UNFCCC offer the potential to scale up urban transport solutions, which should be implemented through balanced "Avoid/Shift/ Improve" approaches. (See Box 12–1.) For example, "Shift" strategies should incorporate non-motorized transport enhancements as well as public transport improvements, and they should be complemented by "Avoid" approaches that encompass both compact development and travel demand management (TDM). "Improve" approaches should be used as a supportive strategy rather than as a primary or sole strategy.

Second, fossil fuel subsidy reform efforts under the UNFCCC should be accompanied by efforts to allocate dedicated funding streams for the implementation of urban transport infrastructure and services and supportive compact development patterns. Third, because transport systems worldwide are vulnerable to the increasing impacts of extreme weather, and because rapid urbanization and motorization increase the potential for catastrophic impacts, the UNFCCC should provide guidance and support for sustainable urban transport systems to adapt to climate change and thus to achieve their full mitigation potential.

A growing number of the INDCs submitted by UNFCCC Parties mention urban transport among planned mitigation measures, and these generally are dominated by public transport. Yet many INDCs that define transport measures make little or no explicit mention of urban transport, focusing solely on sector-wide measures such as fuel decarbonization and energy efficiency.

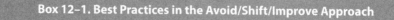

Box 12–1. Best Practices in the Avoid/Shift/Improve Approach

Avoid:
- Vehicle registration quotas allocated through auction (Singapore)
- Congestion charging (Bergen, London, Milan, Oslo, Singapore, Stockholm)
- Emission-based road use charges for heavy goods vehicles (Germany's national road system, London's Low Emission Zone for trucks)
- Mixed-use, public transport-dependent development (Curitiba, Hong Kong, London, Stockholm)

Shift:
- Bus rapid transit (Ahmedabad, Bogotá, Brisbane, Cambridge (U.K.), Capetown, Cleveland, Guangzhou, Johannesburg, Ottawa)
- Public bicycle systems (Barcelona, Brisbane, Hangzhou, Montreal, New York, Paris, Shanghai)
- Rail-based mass transit (Berlin, Hong Kong, London, Melbourne, Montreal, New York, Tokyo, Toronto)
- Pedestrianization, greenways, and cycling networks (Copenhagen, Guangzhou, Sydney, Toronto)
- Parking management and pricing (Paris, San Francisco, Tokyo, Zurich)
- Intermodal freight system management for optimizing rail and water freight (Germany)

Improve:
- Fuel efficiency regulation (California, EU, Japan)
- Electric bikes (20 million-plus produced annually in China)
- High-efficiency cars and trucks: hybrids, neighborhood electric vehicles, biogas buses (Sri Lanka, Stockholm)
- Time-of-day road charges (keep traffic at optimal speeds 85 percent of the time in Singapore)

Although such approaches are valid, they overlook the fact that the transport sector accounts for one-third of the global urban potential for emissions reductions, which can relieve pressure on other sectors to make economy-wide reductions. Urban transport measures in particular should maximize mitigation potential through a balanced set of Avoid/Shift/Improve strategies. Finally, urban transport measures in INDCs should incorporate adaptation strategies to ensure resilience to more-frequent extreme weather events, which will help increase mode share and thus maximize the potential of the transport sector for climate change mitigation.[27]

The UN Sustainable Development Goals (SDG) framework provides ample opportunity to further low-carbon transport on a global scale. SDG 11 is dedicated to urban issues, and associated Target 11.2 is focused exclusively on sustainable transport; thus, these are key avenues for advancing the urban transport agenda. In addition, SDG 13 is focused on reducing climate impacts, with corresponding targets focused on both mitigation and adaptation. Efforts focused on scaling up low-carbon urban transport should be linked closely to the forthcoming implementation of the SDGs in the 2015–30 period.[28]

The SLoCaT Partnership has made a number of recommendations for solidifying the position of urban transport within the proposed SDG indicators, including establishing indicators on urban access (proportion of the population that has convenient access to public transport); road safety (number of road traffic fatal injury deaths per 100,000 people); energy efficiency (rate of improvement in energy intensity measured in terms of primary energy and gross domestic product); and air quality (annual mean levels of fine particulate matter in cities). Although sustainable transport is reasonably well covered in those indicators (with the notable exception of walking and cycling), the cross-cutting nature of transport continues to be underemphasized. Furthermore, there is no clear consensus within the transport community on how best to track urban transport-related targets (for example, some proposed indicators focus on access to transport while others focus on the essential services that can be reached via transport). Thus, building internal consensus is a first step to ensure that transport targets are being tracked with the active support of the sustainable transport community.

A biogas bus on the road in Stockholm.

In October 2016, Habitat III—the UN Conference on Housing and Sustainable Urban Development, to take place in Quito, Ecuador—will set the agenda for urban development over the next decade. Scaling up sustainable transport infrastructure and services within the world's cities will be a critical component in the sustainable urban development process. Thus, it is crucial to link Habitat III (and associated issue papers) more closely to global processes on sustainable development, climate change, and financing for development, specifically as related to the sustainable transport sector.[29]

Transport has been mainstreamed as a cross-cutting sector in the sustainable development process, and the sector needs to be addressed in a similar manner under Habitat III, as it cuts across several relevant conference topics (e.g., Safer Cities, Urban-Rural Linkages, Jobs and Livelihoods).

Conclusion

As global cities continue to grow, and as business-as-usual patterns of transport become increasingly unsustainable, the big question is whether incremental steps toward sustainable urban transport are sufficient, or whether achieving sustainability will require more-disruptive changes. There are rays of hope on the topic of post-2020 climate change mitigation and adaptation ambitions, based on the overall spirit of the Paris Agreement and the overall structure of mechanisms being put in place, to ensure that such ambitions have the potential to be scaled upward as we near a new 2020 starting line.

However, a transformational change in transport is not likely to happen purely on the basis of climate change goals. It is more likely to be driven by sustainable development concerns (for example, as a cobenefit of reducing urban air pollution as a primary policy thrust). For this reason, the transport sector (and urban transport in particular) could benefit from a stronger linkage between the post-2015 development agenda and the emerging climate change agenda to improve the chances of translating mitigation and adaptation ambition into long-term implementation of sustainable low-carbon transport measures.

CITY VIEW

Singapore

Geoffrey Davison and Ang Wei Ping

Singapore Basics

City population: 5.5 million
City area: 719 square kilometers
Population density: 7,967 inhabitants per square kilometer
Source: See endnote 1.

The Singapore Festival of Biodiversity attracted 27,000 members of the public over a single weekend in 2015.

National Parks Board

A Pragmatic Approach to the Environment and Quality of Life

Singapore lies at the southernmost tip of the Malay Peninsula, less than 140 kilometers north of the Equator. The nation has been independent since 1965 and is a parliamentary republic. Its population is moderately dense, at more than 7,900 inhabitants per square kilometer. Because Singapore is a small city-state with effectively no hinterland, there is no distinction between the city proper and outlying regions.

When Singapore became independent in 1965, economic survival was an absolute necessity. Many laws had to be put in place quickly and at short notice, and administrative systems had to be overhauled. Singapore is now a global economic hub, relying on banking and financial services, foreign exchange, refining and trade of petrochemicals, shipping, and aviation. The nominal per capita gross domestic product in 2014 was $56,319.

Without economic survival, Singapore's sustainability in the broader sense would have been moot. Even in 1965, the need to balance different imperatives—from national defense to food security, education, and employment—was fully recognized. There was to be no compromise in housing standards or provision of public transport in order to finance other sectors. It was recognized that continual improvement in every aspect of livability would feed back to support improvement in others. Sustainability of the economy could not be separated from the sustainability of social capital (including health, education, skills, and harmony in a multi-ethnic, multi-religious community) and sustainability of the environment.

Livability for a growing population within strictly limited space has been a constant theme. For more than 50 years, Singapore has pursued the goal of being the cleanest and greenest city in Southeast Asia. As a former official in the Ministry of Environment and Water Resources noted: "Singapore is not a green utopia with zero carbon emissions, large-scale renewable energy sources, or cutting-edge, zero-energy buildings. What it does have is a practical, cost-effective, and efficient approach towards sustaining its environment, which contributes to the high quality of life."

Key Sustainability Policies

Approximately 80 percent of Singapore's population lives in publicly built apartment buildings (although many of the units are now owned by their occupants). The small size of Singapore and the high value of land resulted in a policy to maintain a mix of landed property and apartment dwellings, as well as private and public ownership. This has been tied closely to infrastructure policies. The high share of public housing has allowed an unprecedented integration of services, including public transport by road and rail, as well as natural gas, electricity, and water utilities.

Although Singapore is a city-state, with very few tiers of government, it is not administratively structureless. Residential areas are concentrated as municipalities, each managed by a town council (16 as of late 2015), with land use throughout the nation guided by 10-year Concept Plans and 5-year Master Plans, using a long, forward-looking time horizon.

Small size has perhaps been an advantage in planning and implementing the national water supply. Sustainability in water use depends on four national "taps": rainwater capture, storage, and treatment; desalination, in which Singapore is a world leader; recycling of used water, or "Newater"; and water imports. These water sources rely on the storage capabilities of Singapore's 17 reservoirs. The four inland reservoirs are located in a forested catchment that aids rainfall capture, soil permeability, groundwater recharge, and maintenance of water quality. Good management of human activities throughout Singapore helps to minimize pollution of the 13 coastal reservoirs downstream and to maintain their viability.

Maintenance of the forested catchments contributes to the sustainability of Singapore's rich native plants (2,145 species) and animals (more than 40,000 species). In addition to the remaining natural forest, there is much secondary woodland, 3 million roadside trees, and greenery on vacant land. This greenery provides a matrix for biodiversity, facilitates genetic exchange among populations, and increases the diversity of available habitats.

The mosaic of greenery, guided by the Concept Plans and Master Plans, is a key to livability in Singapore, as it provides exercise and recreational space, offers educational and research opportunities, and contributes to the mitigation of urban heat-island effects. Based on a target in the 2015 *Sustainable Singapore Blueprint*, 90 percent of households should have a park within 400 meters (or an estimated 10 minutes walking time), and this will be achieved by 2030.

In considering the integration of policies for sustainability, it is important to recognize feedback loops. Singapore promotes trade liberalization and has a clear, straightforward tax regime. Attractiveness to foreign businesses—whether in the form of direct investment, company headquarters, or as a global and regional hub, along with supporting banking, insurance, and investment services—helps to maintain economic sustainability, enabling the government to allocate revenue to greenery and the environment.

Although certain spots, such as the Singapore River, have suffered from pollution, overall pollution levels in the nation have never been high. Plenty of greenery, clean air, and safe drinking water—and, consequently, high standards of living, with abundant recreational opportunities in pleasant, comfortable surroundings—encourage inflows of foreign investment.

Success in attracting business has led to population growth and to pressure on land space. Limits on the number of private vehicles, electronic road pricing, and a policy target for 75 percent of all peak-hour travel to make use of public transport by 2030 (the share was

64 percent in 2013) are ways of minimizing pollution and saving energy. Such reliance on public transport is possible only because of the nature of land-use planning and the concentration of public housing. Hence everything comes full circle.

Key Achievements and Outcomes

In 2010–11, Singapore ranked third (out of 139 countries) in the World Economic Forum's Global Competitiveness Index and ranked first in the category of "Efficiency Enhancers," an economic measurement reflecting value-added to raw materials. In 2014, Singapore ranked 9th in the United Nations Development Programme's Human Development Index, up from 25th in 2005. It ranked 25th in the 2014 Mercer Quality of Living survey and consistently has been ahead of Japan's four top cities and Hong Kong. On the Siemens Green City Index 2014, Singapore ranked first out of 22 cities assessed in Asia, scoring above average in all eight sub-categories and well above average in two categories: waste and water.[2]

Singapore is responsible for an estimated 0.14 percent of global greenhouse gas emissions and is a party to the United Nations Framework Convention on Climate Change. The nation maintains a comprehensive online data set of Millennium Development Goal indicators, including carbon dioxide emissions as well as shares of forested land, protected areas, and threatened and endangered species. Singapore maintains a national *Red Data Book* for plants and animals, used as a basis for measuring changes in their status.

What Made These Policies Possible?

Powerful political will, adherence to rule of law, and minimization of corruption have been key tenets of Singapore's philosophy and development. Although some view the city-state's long-term political stability as authoritarian, others see it as the necessary basis for a high and sustained quality of life. Relative prosperity has created a positive feedback loop facilitating environmental care. During 50 years of independence, environmental sustainability has evolved from being the personal vision of a few government leaders to being engrained in all national policies on land, water, and greenery.

Singapore's approach derives from the strong direction provided by Parliament, as well as from the strong public demand for high-quality services. The government structure helps make policy implementation more efficient, with a large number of statutory boards (such as the Public Utilities Board, the Housing Development Board, and the National Parks Board). They operate with highly trained staff, receive adequate funds, are backed by strong business models and national commitment to anti-corruption, and report to equally resourced ministries.

Public feedback is encouraged. During the Concept Planning process, committees and subcommittees are formed that either include nongovernmental organizations and their representatives or are tasked to solicit such views. Although there is no law requiring environmental impact assessment, land allocations cannot be completed without due regard for the EIA process. Full EIAs are made public, and there is an increasing trend to consult civil and environmental organizations prior to development.

Social Dimensions

Social aspects include the modular urban design of Singapore's component townships, each with an administrative town council. The society is multi-cultural and multi-lingual, with four official languages (Malay, English, Mandarin, and Tamil). Social integration exists in housing and in the educational system. As a social leveler, all public parks have free entry, including Singapore Botanic Gardens, which attracts more than 4 million visitors annually; exceptions include specific attractions, such as the national Orchid Garden and the indoor domes and super-trees at Gardens By The Bay.

All government agencies have public feedback channels, and speed and efficiency in handling public feedback are incorporated into standard operating procedures. For many years, government agencies have practiced a "no wrong door" policy in which any civil servant is required to deal with public feedback and is responsible for ensuring that the consultation is passed on correctly and is completed. A Municipal Services Office works with key government agencies to improve feedback management and customer service for six core municipal services.

Singapore strives to innovate wherever possible and leads in the development of skyrise greenery. Since 1992, 300 kilometers of the Park Connector Network in seven loops have been estab-

National Parks Board

Office buildings and apartment blocks are set within a matrix of tree-covered parks that include a network of bicycle and pedestrian trails.

lished nationwide to facilitate the movement of pedestrians and cyclists between parks. The network continues to expand and will be followed by a Round Island Route circling the entire city-state.

In 2005, the National Parks Board initiated "Community in Bloom," a nationwide gardening movement of more than 400 citizen groups that aims to bring residents together. A follow-up "Community in Nature" initiative aims to muster additional support from nature enthusiasts, photographers, and citizen scientists. Singapore Botanic Gardens and the four nature reserves rely heavily on volunteer guides and wardens for their successful management and public outreach. Every child within the Singapore education system can expect to visit a nature reserve, the botanic gardens, or a major public park at least once, and usually several times, in the course of curricular as well as co-curricular activities.

Scalability, Replicability, and Lessons Learned

Singapore is unusual in being a single city-state and is notable for its location close to the Equator. Replicability cannot occur domestically because no other cities exist within its jurisdiction. Scale can be measured either by replication across municipalities or by repeated examples of nationwide implementation sector by sector. The various municipalities strive for common practices, and Singapore's small size eases the nationwide implementation of single systems. Nevertheless, sustainability challenges are likely to increase because of the growing population within a limited land area, the changing labor supply, reliance on imports (particularly food), global economic shifts, natural events such as El Niño years, and continuing climate change. It will not be possible to take sustainability for granted.

Singapore has unique circumstances, including poor access to domestic renewable energy sources such as solar (because of cloud cover), tidal (because of low tidal range), wind (low average wind speeds), and geothermal. The lack of a significant hinterland necessitates food imports as well as economic reliance on trade, services, and international transport links. Singapore is highly dependent on shipping and aviation; together with high educational standards and the use of English, these contribute to powerful international outreach.

Rather than scalability within Singapore, replication of Singapore's ideas occurs overseas. China has been keen to take up lessons from Singapore's successful drive toward modernization. For example, China has adapted Singapore's themes of water and waste management, integrated transport, public housing, and distribution of green recreational spaces in the planning for Tianjin Eco-city, a major urban development outside of Beijing. China and Singapore also are collaborating in the design of Guangzhou Knowledge City, a 6,000 square kilometer sustainable city for knowledge-based industries, which will incorporate green connectors and water bodies as well as integrated residential, business, and recreational areas built around a transport-oriented model.

Geoffrey Davison is Senior Deputy Director of the Terrestrial Branch of the National Biodiversity Centre, and **Ang Wei Ping** is Deputy Director of the Policy & Planning Division—both at the National Parks Board of Singapore.

CHAPTER 13

Source Reduction and Recycling of Waste

Michael Renner

As more people move to urban areas and as consumption levels rise, cities are producing ever-growing volumes of waste. In 2012, worldwide flows of municipal solid waste (MSW, known more commonly as trash or garbage) totaled some 1.3 billion tons, a figure that could rise to 2.2 billion tons per year by 2025. Much of this waste ends up in landfills, which generate serious air and water pollution and contribute to greenhouse gas emissions. MSW is the third largest source of human-caused methane emissions, and the open burning and transport of waste release significant amounts of black carbon particulates and carbon dioxide (CO_2). Leachate from landfills contaminates groundwater and poses a risk of vector-borne disease.[1]

Cities generate large volumes of waste because they are home to high concentrations of people. But many other factors—from lifestyle choices to systems of production—influence how much waste, and what kind of waste, is generated. As Mark Roseland observes in his book *Toward Sustainable Communities: Solutions for Citizens and Their Governments*, "the dilemma for local governments is that the most desirable options in the waste management hierarchy . . . are behavioral choices that are largely outside of [the city's] realm."[2]

Although the consumption choices of city residents are an important factor in waste generation, larger, unsustainable patterns of production and consumption exist across entire economies. Much of the responsibility lies with corporate decision makers, who see opportunities for making a profit by urging people to buy more "stuff" and by manufacturing overly packaged, short-lived

Michael Renner is a senior researcher at the Worldwatch Institute and Codirector of the *State of the World* 2016 report.

products that cannot easily be repaired—all without having to shoulder the financial and other consequences of such strategies. National governments can take steps to curb these practices through laws that minimize unnecessary packaging, eco-taxes that discourage wasteful practices, and mandatory "take-back" regulations that compel manufacturers to re-assume responsibility for products at the end of their useful lives (thus creating an incentive to design products in more-sustainable ways).

Because municipalities generally are in charge of waste collection, it is in their economic and environmental self-interest to take action to reduce the waste streams entering landfills—and thus to limit the share of their own budgets absorbed by waste-management operations. Cities alone may not be able to act on the full range of policies needed, given the roles played by national governments and corporations. But they need to be conscious of the sustainability "hierarchy" of options—ranging from conventional waste collection and disposal, to waste-to-energy plants (which reduce the burden on landfills but generate their own problems, such as emitting dioxin when they burn chlorinated plastic), to "source separation" by individual consumers or at centralized facilities so that recycling and composting, as well as reuse and refurbishing of materials, become viable. (See Chapter 14.)

Most important, however, are efforts to reduce the generation of waste in the first place. This can be achieved by redesigning and rethinking products, by creating more-circular materials flows (such as through a cradle-to-cradle approach), and by moving from the demolition of buildings to deconstruction that allows materials recovery. Policies need to transition from waste management to waste avoidance.

Recycling and waste reduction generate substantial environmental benefits, including reductions in air, water, and land contamination; high energy and water savings relative to virgin production of metals, plastics, and finished products (see Table 13–1); and reduced greenhouse gas emissions. Recycling and refurbishing of products also can generate local jobs, an objective at the heart of many city policies.[3]

Pay As You Throw

Among the policies that municipal governments can pursue are financial incentives such as higher landfill charges (which reflect the costs of disposal more fully and are intended to encourage recycling or composting), tax credits (to encourage businesses to rely more on recycled materials and refurbished

Material	Share of Scrap in Global Supply	Energy Savings Relative to Virgin Production
	percent	
Aluminum	25	95
Copper	>40	85
Plastic	—	80
Steel	44	74
Paper	—	65
Lead	45	65
Zinc	30	60

Table 13–1. Energy Savings from Recycling versus Virgin Materials Production

Source: See endnote 3.

products), and deposit/refund systems (to encourage the recycling of beverage containers and other items). Cities may pass recycling ordinances or adopt a "pay as you throw" system—as pioneered in Zurich, Switzerland—that rewards residents who generate less garbage. Reward systems and other market-based approaches also need to be paired with regulatory policies, such as disposal bans or procurement policies that mandate the purchase of products that contain recycled content.[4]

Recycling rates are influenced heavily by national-level policies and vary widely among countries. In most countries in eastern and southeastern Europe, less than 30 percent of the material in MSW streams is recovered, compared with a recovery rate of more than 50 percent in Austria, Belgium, and Germany. The United States recovers only about one-third of its MSW stream. Many developing and emerging economies continue to lack appropriate laws as well as the institutional and market infrastructure needed to ensure high recycling rates.[5]

Either individually or acting jointly, a number of cities are showing the way forward. Among so-called C40 cities—a group of megacities that are taking action to reduce greenhouse gas emissions—the Sustainable Solid Waste Systems Network aims to strengthen alternatives such as source reduction, improved collection and transportation, and recycling, as well as organics

Leon Brocard

The Spittelau waste incineration plant in Vienna, Austria, was redesigned and given its present colorful, irregular structures by artist Friedensreich Hundertwasser following a fire in 1989.

utilization and landfill diversion (although C40 also promotes waste incineration for "energy recovery," a problematic approach; see below). Network participants have pledged to share their experiences and expertise, and they gather advice from technical experts.[6]

At the United Nations Climate Summit in New York in September 2014, the Climate and Clean Air Coalition, a national government-led initiative, announced its intention to have 50 cities worldwide committing by December 2015 to develop and implement plans of action for the waste sector. Targets include reducing short-lived climate pollutants (black carbon, methane, tropospheric ozone, and hydrofluorocarbons) from the waste sector by 2020, expanding the number of participating cities to 150 by 2020, and disseminating best practices to 1,000 other cities.[7]

The Zero-Waste Challenge

A growing number of cities in North America and Europe are seeking to boost recycling rates and to reduce the amount of waste going to landfills. Portland, Oregon, has adopted a set of procurement policies intended to reduce waste flows, including guidance that instructs the city to ensure, "to the maximum extent economically feasible," the purchase of environmentally preferable products or services that are "durable, recyclable, reusable, readily biodegradable, energy efficient, made from recycled materials, and nontoxic." The city aims to have no more than 10 percent of waste go to landfills and to reduce waste from city operations by 25 percent below the 2009–10 level. Portland prohibits restaurants, grocery stores, and other retail vendors from using polystyrene foam containers. After the city introduced a weekly compost pickup and reduced garbage collection to biweekly in 2011, it saw a 37 percent drop in trash production. (See City View: Portland, page 291.)[8]

San Francisco figures prominently on the U.S. and Canada "Green City Index," having achieved a recycling rate of 77 percent by 2010. A California state law, the Integrated Waste Management Act of 1989, provided the initial impetus for reducing MSW flows to landfills, but the city subsequently passed several waste ordinances of its own. In 2002, San Francisco set the goal of zero waste disposal by 2020, and, in 2009, it made recycling and composting mandatory for all residents and businesses.[9]

New York City, which generates 14 million tons of waste annually, has established a goal of diverting 75 percent of its solid waste from landfills by 2030. Buenos Aires, Argentina, has adopted a more ambitious pace. Through its Solid Urban Waste Reduction Project, the city wants to reduce waste sent to landfills by 83 percent by 2017, with the help of measures such as source separation, waste recovery and recycling, and valorization (the process of converting waste materials into more useful products, including chemicals, materials, and fuels). In Canada, Metro Vancouver (made up of 22 municipalities) adopted a Zero Waste Challenge in 2007 aimed at diverting 80 percent of materials such as food scraps and wood from landfills by 2020. (See City View: Vancouver, page 171.)[10]

Small towns can sometimes be pioneers and laboratories for larger cities. The Italian town of Capannori (46,700 inhabitants) committed itself in 2007 to send zero waste to landfills by 2020. By 2013, it had reduced per capita waste generation 39 percent (compared with 2004 levels), and just 18 percent of waste went to landfills, putting the town on a firm trajectory toward its target. Tax incentives in the town encourage local small businesses to stock food items that customers can refill using their own containers, eliminating the need for throwaway packaging. This "short chain" model of food distribution allows for lower prices and gives farmers a higher return. Capannori's approach has been emulated elsewhere in Europe, as some 100 cities in Spain's Catalan and Basque regions have adopted similar policies.[11]

In the Basque region, municipalities in Gipuzkoa, a province with more than 700,000 inhabitants, have successfully adopted the zero-waste vision. By the end of 2014, 60 of the 88 municipalities had signed on to a zero-waste policy and achieved recycling rates of 70 percent, up from just 5 municipalities in early 2013. In Hernani, a town of 20,000 inhabitants, source separation led to a dramatic reduction in waste disposal, and disposal costs dropped from 74 percent of the municipal budget to just 17 percent. In another town, opposition to a planned incinerator led to a community participatory process to design an alternative plan, and the political party in favor of incineration was voted out of office. Between 2011 and early 2015, total waste production in Gipuzkoa province fell 7 percent, and the overall recycling rate rose from 32 percent to 51 percent, well on the way to the target of 70 percent by 2020. The creation of "Ecocenters" to encourage second-hand sales and the reuse of recovered materials has brought social benefits, including providing jobs for people at risk of social exclusion.[12]

Ljubljana, Slovenia, was the first European capital city to commit to zero waste. The city recycles more than 60 percent of its MSW and aims to reach 78 percent by 2025. Strong citizen opposition led Ljubljana to abandon its planned construction of waste incinerators, and the city's zero-waste goals have largely eliminated the need for the facilities. Between 2004 and 2014, the average annual quantity of recovered material per resident increased from 16 kilograms to 145 kilograms. The city's overall per capita waste generation, at 283 kilograms per year, is well below the European Union average of 481 kilograms.[13]

The Flanders region of Belgium is perhaps the leading example of a broad range of forward-looking anti-waste policies in Europe. The regional authorities are responsible for environmental issues, including legislation and policies concerning waste management. Most of Flanders's 308 municipalities are grouped in 27 associations to provide MSW services. The Flemish government has adopted mandatory source-separated collection, has passed restrictions and taxes to discourage landfilling and incineration, and relies on pay-as-you-throw laws. The region requires municipalities to conduct waste prevention education campaigns and has created tools to promote cleaner production and sustainable product design—with the aim of waste avoidance. It also provides subsidies to second-hand (reuse) shops and has adopted "extended producer responsibility" legislation that requires manufacturers to take back products at the end of their useful lives.

As a result of these initiatives, recycling and composting have increased in Flanders over the past few decades, reducing the waste sent to landfills to marginal amounts, even though the region's incineration capacity has not expanded for 25 years. Overall, the region, home to 6.2 million people, has achieved the highest landfill diversion rate in Europe, with three-quarters of residential waste being reused, recycled, or composted. Waste generation has been decoupled from economic growth. However, as in most of the world's cities and regions, Flanders has not yet succeeded in reducing absolute levels of waste.[14]

Waste-to-Energy: Friend or Foe?

Oslo, Norway, has had mixed success in reducing waste. Between 2006 and 2015, the city increased the share of household waste that is recycled from 27 percent to 37 percent, and it has set a goal of 50 percent by 2018. In 2013, 60 percent of Oslo's recycled waste went to waste incinerators—used to power the city's district heating system—while 6 percent went to landfills and 1 percent was reused. Oslo stopped dumping biodegradable waste in landfills in 2002, seven years before the nationwide deadline of 2009. In 2013, the city opened Europe's most modern biogas facility, turning food waste (from Oslo and other municipalities) into biogas to power some 150 buses and into fertilizer to supply some 100 local farmers. Meanwhile, landfill gas is captured and generates electricity for local schools, and CO_2 emissions have been reduced.[15]

Still, reducing the overall amount of waste generated in Oslo appears to be an elusive goal. Between 2006 and 2011, the city's waste volume increased nearly 20 percent—from just over 200,000 tons to 240,000 tons—while the population grew only 11.6 percent. Like many European cities, Oslo relies heavily on waste incineration, in part to generate district heat and power; however, experience suggests an inherent contradiction between incineration and waste reduction, creating a sustainability dilemma for the city. (See Box 13–1.) Oslo's waste-to-energy capacity is 410,000 tons per year—far above the current waste volume—giving the city a strong incentive to import waste in addition to using its own flows.[16]

One of the problems with landfills is that anaerobic digestion produces methane, which may be released into the atmosphere. Methane is over 20 times more potent as a greenhouse gas than CO_2. In the past, wasteful practices such as flaring the methane abounded. More recently, cities have been capturing landfill gases and turning them into heat or electricity as a waste-to-energy strategy. In the United States, where landfills are the third-largest source of methane emissions, the Environmental Protection Agency counts 645 operational landfill energy projects and another 440 candidate sites. In Europe, methane recovery rates vary from as high as 72 percent in Ireland and 62 percent in the United Kingdom to as low as 11 percent in Denmark and Austria. In South Africa, the city of Johannesburg uses methane from five landfill sites to generate electricity for some 12,500 households. Dar es Salaam in Tanzania and Addis Ababa in Ethiopia hope to implement similar policies.[17]

Linköping, a city with about 150,000 inhabitants in southern Sweden, has pioneered the use of landfill gas in public transportation. With the aim of

---

Box 13–1. What a Waste! Incineration versus Waste Reduction

Relying on waste incineration for heating or electrical energy creates a sustainability dilemma that Oslo shares with many other European cities. Because incinerators use waste feedstocks to produce energy, they contribute to a city's energy supply. However, to earn a profit from the waste-to-energy plants, which are costly to build, operators need to have a guaranteed stream of waste. As a consequence, policies to support incinerators potentially are at odds with the goals of waste reduction and recycling. Cities typically sign contracts for more than 20–30 years with the facilities to supply them with trash. Such long-term investments can divert funds from recycling and waste reduction efforts.

The countries of Northern Europe together produce some 150 million tons of burnable trash annually, but the region's current incineration capacity is well above that, at 700 million tons. This gap between capacity and supply creates an incentive to import waste from elsewhere, especially if cities and countries in the region dramatically reduce their domestic waste streams. (Since 1995, Norway has reduced its generation of MSW, but Denmark, Finland, and Sweden have increased theirs.)

Many other European countries—particularly Germany and the Netherlands—also rely on incineration and imported waste. As of 2013, the continent was home to some 459 waste incinerators. Between 1995 and 2013, total MSW in the EU-27 grew by about 8 percent; the volume going to landfills was halved (to 73 million tons), while the amount incinerated nearly doubled (to 62 million tons). Recycling grew by an impressive 163 percent (to 66 million tons), and composting by 153 percent (to 36 million tons), but these results could have been much higher in the absence of large-scale incineration (as occurred in Belgium's Flanders region, which has not increased incineration capacity since the early 1990s).

By comparison, Eastern and Southern European cities have few incinerators, recycle little, and rely on landfilling for three-quarters or more of their waste. In the United States, more than half of all municipal waste ended up in landfills in 2013; of the remainder, about 25 percent was recycled, 13 percent burned, and 9 percent composted.

Source: See endnote 16.

---

reducing air pollution, the city decided in the early 1990s to switch its bus fleet from diesel to methane gas obtained from wastewater treatment plants and landfills as well as from a local slaughterhouse, crop residues, and manure. The city inaugurated a methane-manufacturing facility in 1996, and farmers use the by-product of the process as a substitute for fossil fuel-based fertilizer. By 2002, Linköping's entire bus fleet had been converted to bio-methane, and the city introduced the world's first biogas train in 2005. Overall, the venture cut gasoline and diesel use by 5.5 million liters per year and reduced CO_2 emissions by more than 9,000 tons annually. The scheme also benefits the local economy by replacing imported energy. Sweden as a whole generates 60 percent of its biogas at sewage treatment plants and 30 percent from landfills, while the rest comes from codigestion plants.[18]

In France, the Lille Métropole Communauté Urbaine (which groups 87 local authorities with a combined population of more than 1 million people) launched a project in 1990 to use biogas from a sewage plant to power urban transport buses, with the first buses running on biogas in 1994. Half of the metropolitan area's biodegradable wastes are turned into methane, and its 400 buses run on a mix of biogas and natural gas. As in Linköping, the by-product of the methane manufacturing serves as compost for local agriculture. The European Commission has calculated that converting all of the EU's organic wastes into methane could supply a third of the region's current demand for transport fuels.[19]

Cities elsewhere are beginning to follow a similar route. In North America, Pierce Transit, near Seattle, became the first U.S. transit agency to use landfill gas for public transportation purposes. And in July 2015, Santa Monica's Big Blue Bus, a municipal bus operator in California, announced that it had converted its fleet from compressed natural gas (CNG) to methane gas harvested from organic waste in landfills.[20]

Landfill gas collection can be a useful way to address problems arising from traditional waste management. However, a better strategy than dumping organic material in landfills and then extracting the resulting methane is to divert these materials to a biogas facility directly. Even so, biogas policies, as with waste incineration, rely on a steady flow of waste volumes, thus providing a de facto incentive for cities to keep generating unabated waste flows.

Empowering Waste Pickers

Waste management and recycling policies vary greatly around the world. Unlike the formalized systems found in industrialized countries, many cities in the developing world have inadequate and overburdened waste collection services, or none at all. In many developing regions, urban growth is rapid and unplanned, and vast urban areas are slums and other informal settlements. Informal waste picking is the dominant way of collecting and sorting wastes, carried out by an estimated 15–20 million people worldwide, typically from impoverished and marginalized groups.[21]

Waste picking can be an extremely hazardous means of eking out a living, especially for those who scavenge landfills. A 2012 report from the Global Alliance for Incinerator Alternatives described the harsh realities at sites in the city of Pune in western India:

> Forced to use bare hands to rummage through putrefying garbage containing glass shards, medical waste, dead animals, toxic chemicals, and heavy metals, waste pickers collected bits of reusable, repairable, and marketable materials. Many sustained repeated injuries, illnesses, and diseases as a result of their work. Tuberculosis, scabies, asthma, respiratory infections, cuts, animal bites, and other injuries were common.[22]

Much greater quantities of recyclable materials are recovered by informal waste pickers than by formal waste management companies. Formal operations typically focus on collection and disposal, and the experience in cities like New Delhi or Cairo suggests that contracts with municipal authorities generally require only very low recycling rates. In contrast, the livelihoods of informal waste pickers depend on extracting and selling valuable materials from waste streams, so pickers routinely reach material recovery rates of 80 percent or more.[23]

In Pune, a union of waste pickers established in 1993 has successfully organized door-to-door waste collection for about half the city's population. The union pushes the households that it serves toward greater source separation and treats organic materials separately. Formalization of the operation resulted in improvements in the pickers' working conditions and livelihoods. In 2003, the Pune municipality decided to pay health insurance premiums for the pickers, in recognition of their contribution to the city's financial and environmental well-being. The pickers' cooperative is trying to reduce disposal rates further by pioneering a zero-waste program. (See City View: Ahmedabad and Pune, page 231.)[24]

In Ho Chi Minh City, Vietnam, a lack of financial and human resources limits the formal waste collection system to no more than one-third of households. In the poorest districts, 9 out of 10 households have no access to formal waste collection. Instead, thousands of informal collectors provide door-to-door collection for up to 90 percent of households in areas that cannot be accessed by vehicles because of the predominance of narrow alleyways. Ho Chi Minh City and several other Vietnamese cities are promoting programs in which community groups, cooperatives, and syndicates of individual collectors are responsible for collection activities. But there is no legislation that mandates waste collection or that addresses occupational health and safety concerns among the waste pickers.[25]

City administrations in many countries have tended to be hostile to waste pickers, sometimes attempting to sideline or even criminalize them. Yet efforts to promote the formalization and organization of such workers—and to grant official recognition to their organizations—can be highly successful, as has been demonstrated in several Latin American countries. In Colombia's capital of Bogotá, the Asociación de Recicladores de Bogotá, created in the early 1990s, brings together 24 waste-picker cooperatives that provide services to 10 percent of the city under a contract with the municipality. Brazil has perhaps the most extensive experience, given longstanding policies that support picker cooperatives at both the municipal and national levels. (See Box 13–2.)[26]

In the 1980s, to address ever-expanding waste flows resulting from both population and consumption growth, Curitiba became the first large Brazilian city to launch a recycling program, called Lixoquenão é lixo ("Garbage That

---

Box 13–2. Supporting Waste-picker Cooperatives in Brazil

Since the 1980s, legislation in a growing number of Brazilian cities has enabled the creation of municipal partnerships that recognize the role of waste pickers. The Movimento Nacional dos Catadores de Materiais Recicláveis, founded in 2001, is the world's largest national waste-pickers movement, with more than 500 affiliated cooperatives representing some 60,000 pickers. The Brazilian government has put in place an effective mix of policies, including legal recognition, local- and national-level organization, municipal government contracts and facilities, skills training, and occupational safety and health instructions, as well as measures to prevent and discourage child labor.

Highlights of these measures have included:
- 2001: Federal legislation recognizes waste picking as a legitimate occupation.
- 2007: Legislation is enacted to allow municipalities to hire waste-picker organizations.
- 2009: The Cata-Ação project is launched in five Brazilian cities, offering professional training and socio-economic integration assistance to waste pickers.
- 2010: The National Policy of Solid Waste law is approved, mandating that informal recyclers be included in municipal recycling programs and promoting cooperatives.
- 2011: Brasil Sem Miséria, Brazil's national poverty eradication plan, establishes a goal to integrate 250,000 pickers into municipal recycling programs and to improve their working conditions. It provides for training and infrastructure support.

Source: See endnote 26.

Is Not Garbage"). The city also created a garbage purchase program to involve neighborhood associations in narrow alleyways that could not be accessed by garbage trucks. As an incentive, families receive a free bus ticket for every 8–10 kilogram bag of collected recyclables. The city's efforts divert 2,400 cubic meters of recyclable materials from landfills each day, or about one-quarter of the total MSW volume.[27]

Although Curitiba's waste pickers initially feared that the city's plans would threaten their livelihoods, the pickers continue to account for the bulk of recyclables recovery—nearly 10 times the amount collected by regular garbage trucks. Since the 1990s, the Câmbio Verde ("Green Exchange") program also has allowed residents to receive 1 kilogram of local produce for every 4 kilograms of recyclable materials traded in, providing a further incentive for recycling and a means to improve access to healthy food for the urban poor.[28]

The experience of Buenos Aires, Argentina, illustrates some of the difficulties and contradictions in moving to effective new policies. The work of the city's waste pickers (*cartoneros*)—whose ranks swelled enormously after the country's economic meltdown in 2001—was illegal until 2002. In 2004, a plan to establish new landfills was met with mass popular opposition. This led to the Zero Waste Act, passed in 2005, which established the goal of reducing the amount of MSW going to landfills by 50 percent by 2012, 75 percent by 2017, and 100 percent by 2020 (all relative to 2004). Implementation proved difficult, however, in part because the city administration dragged its feet. In 2010, only about 51,000 tons of MSW was recycled, compared with 2 million tons of material entering landfills.[29]

Cartonero cooperatives, representing nearly half of Buenos Aires's waste pickers, were instrumental in changing perceptions about recycling in the city. The cooperatives gained government recognition and, in 2010, for the first time, were given exclusive responsibility for sorting dry waste. Although support from the city has not always been consistent, and landfilling still received higher budget allocations, the authorities began to provide collection trucks, child care facilities, health and accident insurance, uniforms, and safety equipment to the pickers. Working conditions improved markedly.[30]

By 2012, a landfill crisis in Buenos Aires—prompted when the provincial government decided that it would no longer accept trash from the city—provided a fresh impetus for the city to ramp up its recycling efforts. Buenos Aires now has a goal to reduce overall waste sent to landfills 83 percent by 2017, and it expects the recycling rate to reach 68 percent. The amount of waste entering landfills declined 44 percent in 2013 alone. Today, the 10,000 or so *cartoneros* recycle about one-sixth of Buenos Aires's trash. Some 4,500 of the pickers have been given formal jobs, with 2,000 more expected to join them.[31]

Conclusion

Global waste flows show no signs of abating, indicating the tremendous challenge that urban leaders, as well as national and state governments, still face. Cities are pursuing diverse strategies to deal with waste and are undertaking commendable efforts to divert materials from landfills. Yet many urban policies still assume that large-scale waste flows are a given, and the prevailing concern is only what share of the waste is dumped, burned, converted to energy, or recycled.

Waste-to-energy initiatives, although touted as a form of renewable energy, are predicated on an unabated continuation of waste flows, and they result in the perception that waste avoidance strategies could endanger the production of heat or electricity for communities. The environmentally preferable options—recycling and reuse, as well as the overall reduction of waste through better design and the avoidance of unnecessary packaging—typically account for only a small share of the solutions being pursued. Even some of the most pioneering cities seem hard pressed to reduce overall waste volumes, indicating just how difficult a challenge it is to align waste management practices with the demands of long-term sustainability.

CITY VIEW

Ahmedabad and Pune, India

Kartikeya Sarabhai, Madhavi Joshi, and Sanskriti Menon

Ahmedabad Basics

Municipal population: 5.6 million
Land area: 466 square kilometers
Population density: 11,950 inhabitants per km^2
Source: See endnote 1.

Pune Basics

Municipal population: 3.1 million
Land area: 244 square kilometers
Population density: 12,746 inhabitants per km^2
Source: See endnote 1.

Kite flyers on the rooftops of Ahmedabad.

Shweta Kaushik

231

Leapfrogging to Sustainable Cities: Challenges and Opportunities for India

In 2011, India's 377 million urban residents accounted for about 32 percent of the country's population; however, the number of city dwellers could surge to 900 million by 2050. It is critical that this future urban growth not follow the traditional path of carbon-intensive development. With many of India's cities still undergoing modernization, the country has an opportunity to leapfrog toward sustainability, integrating the traditional with the new, and domestic with international best practices.[2]

Most of the infrastructure needed to accommodate this enlarged urban population, as well as to meet the needs of today's urban poor who still lack minimum facilities, is yet to be built. India currently averages an estimated 13 cars per 1,000 people nationwide, and 100 cars per 1,000 people in major cities, compared with some 450 cars per 1,000 people in developed countries. As income levels rise, and if current trends continue, the pressure on already congested roads will be enormous. Housing also is required for the estimated 17.4 percent of urban households living in slums, as well as for new migrants and the growing population overall. These pressures are significant, but they can be seen as opportunities to build sustainable cities, rather than retrofit cities that already have most of their infrastructure in place.[3]

The leapfrogging challenge is to be able to build cities of the future by learning from existing cities and not imitating models that essentially are unsustainable. It is to realize that the urban habitats that are being built now are in a very different technological era than those built earlier. Internet-based applications are rapidly changing the way people shop, learn, do business, transact money, interact socially, and access music and entertainment. In searching for sustainable solutions, looking at traditional solutions is also important. Solutions are context-specific and culture-specific, so what might be good in one place may not be suitable or work elsewhere. The challenge of leapfrogging is to be able to analyze a problem, to search for or innovate with alternatives, and to choose the right solution for a particular situation.

In 2015, the Indian government launched three key schemes to improve the physical, institutional, social, and economic infrastructure of cities and to enhance quality of life. The Smart Cities Mission seeks to develop 100 smart cities as satellite towns of larger cities and to modernize existing mid-sized cities by 2020. It complements the Atal Mission for Rejuvenation and Urban Transformation, which focuses on projects in the areas of water and sewerage, greenery and open space, and non-polluting transportation (transit, walking, and cycling). The third scheme, the Housing for All (Urban) initiative, seeks to address a housing shortage of 20 million units during 2015–22 through the rehabilitation of slums, credit and subsidy schemes for building or improving housing units, and the development of affordable housing.[4]

The cities of Ahmedabad and Pune, in western India, offer a flavor of the context, efforts, and challenges in ensuring that urban India's transformation is socially and environmentally sound. Located close to Mumbai, India's most populous city, both cities are—and historically have been—important urban centers. They also are among the first 20 cities that will receive funds under the Smart Cities Mission.

Sustainability Lessons from an Historic Urban Form

Ahmedabad and Pune both have an old core city area, a modern city that has evolved around it, and emerging suburban and peri-urban neighborhoods. The core city areas typically are densely populated places, characterized by the integration of workplaces and residences, narrow streets, houses or buildings with inner courtyards, spaces and systems for water recharge, the use of local building materials, and climate-appropriate building designs. The streets, roofs, and courtyards are designed to facilitate community interactions and celebrations. As such, many old cities are good examples of urban ecosystems that integrate social, cultural, economic, and environmental aspects—offering lessons that are relevant to sustainability in the modern context.

These historic, cultural, aesthetic, economic, and community elements are enduring qualities that make cities livable and that will be important for the cities of the future. Unfortunately, this urban form often gets destroyed through efforts to "modernize" core cities, such as attempts to widen roads to address traffic congestion in pre-motorization neighborhoods, or to raze old buildings, pool land parcels, and "revitalize" economic activities, often in a quest for profits from the high land values.

The challenge is not only to preserve the identity and livability of historic core cities, but also to recognize their relevance in a "leapfrog" context. Innovations such as public bicycling schemes or auto aggregation (such as on-call cabs and auto-rickshaw services), combined with congestion charging, can help address traffic issues while retaining walkable/cyclable streets. A mix of conservation, renewal, and replacement of buildings can help retain the built form.

City governments and urban planners need to recognize the new urbanism and sustainability principles that core cities embody and to evaluate them in the modern context, with the goal of extending them to other parts of the city. Leapfrogging will involve revisiting the concepts of urban planning and zoning.

Bus Rapid Transit

Since the advent of bus rapid transit (BRT) in India in 2004–05, at least 13 cities nationwide—including Ahmedabad and Pune—have developed or are developing BRT systems. Pune was the first city in the country to operationalize a pilot BRT project, covering a 16.2 kilometer stretch, with support from the Ministry of Urban Development. Using what was

Nitin Warrier

Buses of the Rainbow BRT traveling in their dedicated lanes, Pune.

learned from the pilot project, a new Rainbow BRT system was launched in Pune and Pimpri Chinchwad in 2015, with three corridors totaling 30 kilometers of dedicated bus lanes serving more than 100,000 people daily. The first few months of operation saw a 12 percent increase in bus ridership. Three more corridors are under development, and the future proposed network would extend some 147 kilometers across the two cities. Rainbow BRT received recognition from the Volvo Sustainable Mobility Awards 2015 for its "Outstanding Contribution to Sustainable Mobility."[5]

Ahmedabad's Janmarg system, which started in 2008 with a 12-kilometer pilot corridor, today boasts a ring radial network of 82 kilometers, with daily ridership of 130,000. It has won national and international awards, including the Government of India's Best Mass Transit project in 2009, the Sustainable Transport Award in 2010, and the United Nations (UN) Momentum Change award in 2012.[6]

Bazaars and Streets for People

Ahmedabad and Pune celebrate the color, vibrancy, and multi-use nature of streets and public spaces. Like many developing-country cities, both cities have a large informal sector. Much of the informal, self-employed workforce works at home or in open public spaces (such as hawkers, vendors, and waste pickers).

Ahmedabad's Gujribazaar, a unique, centuries-old market, provides space to some 1,200 traders (many of them women), and more than 20,000 people rely on the market for making, transporting, and selling goods. When a redevelopment project for the area considered relocating or removing the bazaar's street vendors, a more sustainable solution emerged and the vendors were incorporated into the new design proposal, helping to improve their organization and awareness, and protecting their livelihood.

Neighborhood streets are multi-use areas, meant not only for transportation, but also for activities such as walking and cycling, vending and small business arrangements, processions and festivals, street cricket, and meeting over a cup of tea at a roadside vendor. As the density of uses—especially motorized traffic—increases, stresses and conflicts emerge around the use of precious street space.

One way to deal with these different needs is through locale-specific participatory design processes. In both Ahmedabad and Pune, architecture colleges, the Centre for Environment Education (CEE), the Sustainable Urban Mobility Network, and other partners are carrying out "Streets for People" processes, seeking to create democratic discussion spaces as well as designed physical spaces that meet the needs of diverse stakeholders. Street vendors are getting online as well: the Street Saathi application for mobile devices, for example, points users to the nearest vendors of delicious food or other products.[7]

View of street life from the Law Garden, Ahmedabad.

The vitality and vibrancy of neighborhood spaces can remain intact if their spirit can be captured through formal city development processes that integrate the informal sector. The recently passed Street Vendors Act of 2014 recognizes street vending as a source of livelihood and requires municipal governments to create vending zones. Cities need to consciously protect livelihood rights and to promote a safe and secure working environment for all, especially women, as recognized in the UN's Sustainable Development Goals.

Participatory Governance

As a democracy for over 65 years, India has a wealth of experience with the struggles and advocacy of people and organizations in areas such as housing, livelihood, waste, and transport, as well as in the creation of positive legislation and initiatives such as the Right to Information Act. However, given the complexity of civic issues, the unmet needs of marginalized groups, and the varying scales and scope of decisions needed, new methods of participatory governance need to be developed that are culturally appropriate and yet meet challenges such as illiteracy and a highly diverse society. Combining online platforms with citizens' assemblies and other deliberative forums are some innovations being used elsewhere.

Since 2006, the Pune Municipal Corporation (PMC) has implemented an annual participatory budgeting process in which citizens can submit suggestions online or by using a paper form. Over the years, at least 1–2 percent of the city's total capital expenditure has been allocated in this manner, which includes suggestions from the poor. More than 800 neighborhood improvement projects were included in the Citizens' Budget section of the PMC Budget for 2015–16, including footpath repairs, drainage work, and the installation of benches, toilets, signage, and vendor platforms. Still, there is room for improvement in enhancing the scale and quality of participation, tracking citizens' suggestions, and providing information about projects at all stages.[8]

In early 2016, the PMC made it easier for people to access budgetary information by making the budget statement simpler to understand and by placing it online. The WISE (Ward Infrastructure, Services and Environment) Information Base and the WISE Index, developed by the PMC and CEE, categorize and rank city wards using 26 indicators related to municipal services, population, and geographical area. The aim is to allocate proportions of city funds to the wards based on the Index, with less-developed areas getting more funds. This information then can be used in public deliberations organized both by the municipality and by civil society, helping to increase transparency and accountability.[9]

Waste and Recycling

In India, municipal corporations are mandated to provide solid waste management services in cities. However, solid waste management in both Pune and Ahmedabad has been enhanced due to the formation of associations of waste pickers. The Self Employed Women's Association (SEWA) in Ahmedabad and the Kagad Kach Patra Kashtakari Panchayat (KKPKP) in Pune have worked to establish and assert waste pickers' status as workers, their crucial role in urban solid waste management, and their contribution to the environment.

The KKPKP has successfully negotiated with the Pune government to integrate waste collectors in doorstep collection of household waste. SWaCH (Solid Waste Collection and Handling), a formal institution formed by the city government, is a wholly owned workers' cooperative of self-employed waste pickers and other urban poor operating as a "pro-poor public-private partnership," providing front-end waste management services to residents. All stakeholders benefit: the city is cleaner, households and businesses get a waste collection service, materials are recycled or processed (with lower environmental impacts and economic costs because they are decentralized), and workers' conditions are improved.

In Ahmedabad, the municipal government (AMC) has developed both a Solid Waste Management Plan and a Zero Waste Strategy. The city's concerted waste collection efforts include privatization of collection from households, and disposal. SEWA has organized waste pickers in waste collection and the recycling of paper waste into products, helping to lend dignity to their work.

Water

In 1865, Ahmedabad had over 200 lakes, a number that decreased to 113 in 1975 and to only 62 today. During the 1980s and 1990s, the city's large-scale growth led to considerable building activity and infrastructure development. As a result, the few remaining water bodies no longer could function as water collectors during the monsoon season, leading to frequent flooding. In 2004, the Ahmedabad Urban Development Authority initiated a watershed planning approach to link seven small and big ponds in the city in order to rejuvenate them as water recharge sites and to address the flooding to a large extent.[10]

Historically, Ahmedabad's wealthier homes and traditional housing complexes had storage facilities for rainwater harvesting, but these fell into disuse when the AMC introduced piped water to the city. The Heritage Cell of the AMC undertook an initiative to revive 10 *tankas* in one of the city's pols (a historic housing form in Ahmedabad). Of the 10,000 houses in the area that have *tankas*, about 1,500 are still used, many of them over 150 years old.[11]

Green Spaces

Many of Ahmedabad's institutions have green, biodiversity-rich campuses, but the city's only non-park, forested public green space is the Manekbaug Nature Park, located in a residential area. CEE transformed a barren plot into the green space, which was handed over to the AMC after 18 years. The park now attracts a variety of birds and smaller animals and is being managed by residents and visiting nature enthusiasts.

A path through the Manekbaug Nature Park.

Mudita Vidrohi

Pune, in comparison, has always been well-endowed in green space. However, residents had to protest to save the hills from development, a successful effort that remains a vigilant grassroots movement.

Both cities are fighting to maintain a balance between the rapid development of areas for residential and commercial purposes, and the availability of green spaces for recreation.

Energy

Gandhinagar Rooftop Solar Programme, started by Gujarat Energy Research and Management Institute in Gujarat's capital city, just north of Ahmedabad, aims to install 5 megawatts of solar photovoltaic systems on the rooftops and terraces of private homes and commercial, institutional, and government buildings. The state government will select a number of project developers who will install, own, and maintain the systems and sell the electricity to the grid. Property owners will receive "green incentive" payments based on the electricity they generate.[12]

India's Energy Conservation Building Code (ECBC), launched in 2007, was the first step toward promoting energy efficiency in the building sector. The ECBC provides design norms for building envelopes; lighting, HVAC, and electrical systems; and water heating and pumping systems, with a focus on energy conservation. Energy simulation analysis has indicated that ECBC-compliant buildings may use 40–60 percent less energy than similar buildings.

Environmental Education

The Indradhanushya Centre for Citizenship and Environment Education is a public facility of the PMC that aims "to develop Pune as a city of responsible citizens and environmental stewards." The purpose is to help citizens improve their environmental literacy and hence the environmental performance of the city. Activities include deliberations on the city's annual *Environmental Status Report* as well as events and workshops on issues such as urban farming, biodiversity, energy efficiency, climate change, and transportation. Educators at the center also facilitate a structured learning experience for schools.[13]

Lessons Learned: Making Leapfrogging Happen

Leapfrogging is a process of choice. It involves picking appropriately from a range of options, from the traditional to the most innovative and contemporary. It involves adapting lessons from elsewhere and developing context-specific solutions. It requires continuous feedback and an ability to change when things do not work, as well as finding the optimal solutions in a cultural context. For example, residents of both Ahmedabad and Pune have a long tradition of eating street food. Designing public spaces without accommodating this practice resulted in the emergence of street food facilities in an unregulated way, without meeting the necessary waste-management or hygiene conditions and creating blockage of urban spaces. Over the coming decades, India's drive for urban transformation through the Smart Cities Mission needs to recognize these unique local conditions and to incorporate a diversity of leapfrogging ideas.

Kartikeya Sarabhai is Director of the Centre for Environment Education (CEE) in Ahmedabad, India. **Madhavi Joshi** and **Sanskriti Menon** are Program Directors at CEE.

CHAPTER 14

Solid Waste and Climate Change

Perinaz Bhada-Tata and Daniel Hoornweg

Urbanization and the growth of cities is driven largely by the ability of cities to use materials more efficiently, to bring people together, and to provide better access to health care, education, and employment. Accompanying that urbanization and growth, however, is an increasing stream of waste. As more people move from the countryside to urban areas, their per capita waste levels are rising, commensurate with the higher-consumption lifestyles associated with cities.

Municipal solid waste (MSW) is linked inextricably to urbanization, economic development, and climate change, and local waste managers are at the forefront of dealing with these global trends—or being overwhelmed by them. As countries develop, populations tend to shift from rural to urban areas, where they can find better employment opportunities, lifestyle choices, and education. This urbanization fuels economic growth, savings, and improved standards of living. But with improved living standards come increased consumption and more trash.

Inadequate waste collection and uncontrolled dumping—realities that are common in low-income and even some middle-income countries—contribute to local environmental and public health problems. One key impact is the pollution of ground and surface water, which is caused by several waste-related factors: the lack of containment of leachate (contaminated liquid that is generated when water, usually from rainfall, passes through the waste); high levels of biochemical oxygen demand in local watercourses, resulting from the presence of

Perinaz Bhada-Tata is a solid waste consultant living in Dubai, United Arab Emirates. **Daniel Hoornweg** is an associate professor and holds the Richard Marceau Chair at the University of Ontario Institute of Technology, Canada.

food waste and other organic material in the waste; and the presence of microbial contaminants, such as fecal coliform bacteria. All of these factors contaminate drinking water, adversely affect aquatic life, and cause soil pollution.

A second significant challenge comes from indiscriminate disposal of waste, characterized by a lack of physical boundaries around waste sites or the absence of daily cover on top of the garbage. Open waste sites attract vermin and scavenging animals and provide food and habitat for disease vectors such as rats and mosquitoes, which can lead to the spread of ailments such as dengue fever, plague, and other infectious diseases. Surat, India, a city of roughly 3 million people, was hit by the plague in 1994. The main causes were determined to be the lack of waste collection coupled with flooding due to blocked stormwater drains (blocked mostly by uncollected waste). Since this crisis, Surat Municipality has transformed the city into one of the cleanest in India, with effective solid waste facilities and wastewater treatment plants.

A third problem, as in the case of Surat, is improper collection of waste, which can result in increased local flooding from blocked storm drains and lead to the spread of water-borne and communicable diseases such as malaria and cholera. In many cases, medical waste (such as syringes and bandages) and household hazardous waste (such as paint, electronic waste, and batteries) also are mixed with municipal waste. Waste pickers, sorting through waste at informal dumpsites, can become infected or injured. In some cities, such as Jakarta in Indonesia, financial losses from flooding often far exceed the costs that would be required to properly manage solid waste.

As a final challenge, improper waste disposal also leads to significant air pollution. A 2014 study in Bangalore, India, found that the air around garbage dumps contained elevated levels of pathogens and drug-resistant bacteria, associated with the presence of medical waste. Waste managers sometimes unofficially practice open, uncontrolled burning of waste in order to reduce odors as well as the volume of waste. In many cases, waste pickers start fires to be able to easily identify and collect recyclables. Open burning of waste results in thick smoke that contains carbon monoxide, soot, and harmful organics that degrade air quality and that can compromise human health by causing respiratory diseases and increased cancer risk for waste pickers and nearby communities.[1]

Besides these local impacts, solid waste is one of the largest sources of pollution in the oceans. Plastics make up the largest type of waste, but sewage, oil, and other wastes contribute to the poisoning and mortality of numerous marine species as well. A 2016 study by the World Economic Forum concluded that, by 2050, there will be more plastics than fish (by weight) in the oceans.[2]

On a broader scale, solid waste contributes to climate change in several important ways. Indirectly, before a product becomes waste, it goes through a process of raw material extraction, manufacturing, and transportation to market. Each step requires energy, which is generated mostly by fossil fuels. Another contribution is the decomposition of organic (essentially, food and horticultural) waste under anaerobic (without the presence of air) conditions, such as those experienced when waste is thrown in local waterways. This results in the production of methane, a greenhouse gas that is many times more potent than carbon dioxide (CO_2). Methane is a particularly important greenhouse gas because it has a much higher short-term global warming potential (GWP) than CO_2. Over the typically used 100-year time horizon, methane has a 25 times higher GWP, but over the shorter time frame of 20 years, methane has a 72 times higher GWP than CO_2.[3]

Overall, the waste sector is estimated to account for some 3–5 percent of global greenhouse gas emissions. This estimate is probably low, because uncollected waste in local watercourses usually decays anaerobically, likely generating large amounts of methane that have not yet been quantified. Waste collection vehicles also contribute greenhouse gas emissions, and the combustion of solid waste—especially informal burning of garbage at low temperatures—generates significant black carbon (soot), which is an important short-term contributor to climate change.[4]

Trends in Waste Generation

Global solid waste generation is increasing rapidly. In 1900, 13 percent of the global population lived in cities and generated less than 300,000 metric tons of trash per day. By 2000, some 2.9 billion urban residents (49 percent of the world's population) were generating more than 3 million tons of waste a day. By 2025, the waste volume will be twice that, and, by 2100, it is projected to reach 11 million tons per day—enough to fill a line of garbage trucks stretching from Tokyo to Denver every single day.[5]

High-income countries generate more garbage than low-income countries, and, historically, the focus on reducing waste generation, improving recycling rates, and recovering materials and energy from waste started in high-income countries. However, waste generation in high-income countries is likely to plateau and even start to decline by the end of the century. The focus now needs to shift to low- and middle-income countries, where waste generation is rising markedly. Urbanization growth rates in the second half of this century are expected to peak in South Asia and sub-Saharan Africa, and the bulk of waste generation is expected to come from these rapidly emerging cities. As a result, it is unlikely that "peak waste" will be reached before the end of the century, unless significant advancements are made in lowering overall raw material usage.[6]

Different countries—and even different cities within the same country—face different challenges related to solid waste management, depending on their income level and geographic setting. (See Table 14–1.) Cities in low-income countries, for example, are focused primarily on improving waste collection rates, with most of a typical city's waste budget going for garbage collection. High-income countries, in contrast, spend only a small fraction of their solid-waste management budget on collection, with most going toward disposal.[7]

Efforts to move toward circular economies are gaining support, particularly in Europe. This concept involves a more aggressive re-introduction of "waste" or secondary materials as feedstocks for other related processes, a much greater emphasis on waste minimization, and more emphasis on use of materials that can be readily re-introduced into the economy. The more-typical material flow-through economy is replaced by a circular approach where materials at the end of their useful lives in one process are feed material for another.

The Solid Waste System and Climate Change

Solid waste management systems can be generalized into four components: generation, collection, transport and transfer, and disposal and treatment. Table 14–2 summarizes the sources of greenhouse gas emissions originating from the solid waste management system and where potential savings could be achieved.

Estimating greenhouse gas emissions from the generation of MSW is challenging because of the complex life cycles of products, the heterogeneous nature of waste, and innumerable production processes and standards. The total upstream greenhouse gas emissions from the production of a product would always be higher than the downstream emissions from proper waste management of that product. Hence, waste prevention always should be given the highest priority.[8]

Waste collection and transport typically contribute less than 5 percent of greenhouse gas emissions in the solid waste sector and thus are minor contributors to climate change. (This figure is for the European Union but is common across most regions.) The main driver of emissions from the sector—and

Table 14–1. Comparison of Solid Waste Management Practices, by Country Income Level

Activity	Low-Income Country	Middle-Income Country	High-Income Country
Source Reduction	• No organized programs, but reuse and low per capita waste generation rates are common	• Growing awareness of the three Rs (reduce, reuse, and recycle), but rarely incorporated into an organized program	• Organized education programs emphasize the three Rs • More producer responsibility and focus on product design
Collection	• Sporadic and inefficient • Service is limited to high-visibility areas, the wealthy, and businesses willing to pay • High fraction of inert and compostable materials affects collection • Overall collection below 50 percent	• Improved service and increased collection from residential areas • Larger vehicle fleet and more mechanization • Collection rate in range of 50–80 percent • Transfer stations slowly incorporated into the solid waste management system	• Collection rate greater than 90 percent • Compactor trucks and highly mechanized vehicles and transfer stations are common • Waste volume is a key consideration • Aging collection workers are often a consideration in system design
Recycling	• Done mostly through the informal sector and waste picking; however, rates tend to be high both for local and international markets and for imports of materials for recycling, including hazardous goods such as e-waste and ship-breaking • Markets are unregulated and include a number of middlemen • Large price fluctuations	• Informal sector still involved; some high-tech sorting and processing facilities • Rates are still relatively high • Materials often are imported for recycling • Markets are somewhat more regulated • Material prices fluctuate considerably	• Recyclable materials collection services and high-tech sorting and processing facilities are common and regulated • Increasing attention to long-term markets • Overall rates are higher than in low- and middle-income countries • Informal recycling still exists (e.g., aluminum can collection) • Extended product responsibility is common
Composting	• Rarely undertaken formally, even though the waste stream has a high percentage of organic material • Markets for, and awareness of, compost are lacking	• Large composting plants often are unsuccessful due to contamination and operating costs (little waste separation); some small-scale projects at the community/neighborhood level are more sustainable • Eligible for Clean Development Mechanism (CDM) projects, but this is not widespread • Increasing use of anaerobic digestion	• Becoming more popular at both backyard and large-scale facilities • Waste stream has a smaller portion of compostables than in low- and middle-income countries • More source segregation makes composting easier • Anaerobic digestion increasing in popularity • Odor control is critical

continued on next page

	Table 14–1. continued		
Activity	**Low-Income Country**	**Middle-Income Country**	**High-Income Country**
Incineration	• Not common, and generally not successful because of high capital, technical, and operation costs; high moisture content in the waste; and high percentage of non-combustibles	• Some incinerators are used, but many experience financial and operational difficulties • Air pollution control equipment is not advanced and is often bypassed • Little or no stack emissions monitoring • Governments include incineration as a possible waste disposal option, but costs are prohibitive • Facilities often are driven by subsidies from wealthier countries on behalf of equipment suppliers	• Prevalent in areas with high land costs and low availability of land (e.g., islands) • Most incinerators have environmental controls and some type of energy recovery system • Governments regulate and monitor emissions • Treatment of bottom and fly ash varies by jurisdiction
Landfill/ Dumping	• Low-tech sites, usually open dumping of wastes • High polluting to nearby aquifers, water bodies, settlements • Often receive medical waste • Waste regularly burned • Significant health impacts on local residents and workers • Informal waste pickers often conflict with site management • Waste can be disposed at low-lying areas	• Some controlled and sanitary landfills with some environmental controls • Open dumping is still common • Clean Development Mechanism projects for landfill gas are more common	• Sanitary landfills with a combination of liners, leak detection, leachate collection systems, and gas collection and treatment systems • Often problematic to open new landfills due to concerns of neighboring residents, "NIMBY"ism, and costs • High landfill tipping fees are increasingly common • Post-closure use of sites is increasingly important, e.g., golf courses and parks
Costs	• Collection costs represent 80–90 percent of the MSW management budget • Waste fees are regulated by some local governments, but the fee collection system is inefficient • Only a small proportion of budget is allocated toward disposal • No funds set aside for formal recycling	• Collection costs represent 50–80 percent of the MSW management budget • Some local and national governments regulate waste fees • More innovation in fee collection, e.g., included in electricity or water bills • Expenditures on more-mechanized collection fleets and disposal are higher than in low-income countries	• Collection costs can represent less than 10 percent of the budget • Large budget allocations made to intermediate waste treatment facilities • Upfront community participation reduces costs and increases options available to waste planners (e.g., recycling and composting) • Quality of secondary materials important, e.g., recyclables, compost

Source: See endnote 7.

the area most amenable to mitigation efforts—is disposal. Methods to reduce the volume of waste generated (e.g., incineration) or conversion of waste (e.g., composting, recycling) sometimes are confused with disposal. Yet there is a distinction between technologies that enable the waste to be converted to new products or energy and those processes that simply provide long-term storage without benefiting from the inherent resource or energy-recovery potential in that waste.[9]

Landfills are a source of methane, which (discounting water vapor) is the second most abundant greenhouse gas after CO_2. That, as well as its much higher potency, result in methane accounting for more than one-third of all

Table 14–2. Greenhouse Gas Emissions and Savings from the Solid Waste Management System

Solid Waste Management Component	Source of Greenhouse Gas Emissions	Opportunities for Greenhouse Gas Savings
Waste Generation	• Use of virgin materials in the manufacture of goods that are consumed by households and institutions • Wasteful production processes in the manufacture of goods	• Use of recycled materials in the manufacturing process • More-efficient manufacturing design to prevent wastage • Reducing the amount of goods used, reusing products in order to prevent or delay their disposal, and recycling waste that cannot be reused
Waste Collection	• Inefficient or infrequent waste collection practices, such as lack of segregation of wastes at source	• Requiring segregation of recyclable products from organic waste and inert materials • Improving the frequency of waste collection, especially for organic waste
Waste Transport and Transfer	• CO_2, nitrous oxide, and black carbon emissions from fossil-fueled vehicles used for collecting and transporting waste to transfer points or final disposal or treatment facilities • Inefficiency at transfer stations in terms of vehicle size for final transfer and frequency of transfer, especially for organic waste	• Optimizing route frequency and collection • Investing in fuel-efficient fleets • Decreasing transportation requirements by having local, community-based waste facilities and treatment options
Disposal and Treatment	• Decomposition of organic waste in landfills results in the generation of methane • The majority of greenhouse gas emissions in the waste sector comes from landfilling of waste	• Enabling the capture of landfill gas through piping systems • Using landfill gas for energy production as a substitute for fossil fuel-based sources • Eliminating open dumping and burning of waste by investing in sanitary landfills

human-caused emissions. Globally, landfills account for 11 percent of methane emissions. (However, most well-managed landfills collect and use the methane as a fuel source or, at minimum, flare the gas—converting it to CO_2—to prevent it from accumulating and posing a risk of explosion. Landfill gas recovery is discussed further below.)[10]

A European Union study estimated the net emissions from treating and disposing organic waste and found that the priority of options, in terms of lowest net emissions, is, first, incineration (with energy recovery), then anaerobic digestion, then composting, and, finally, landfilling. Incineration with energy recovery actually can result in negative emissions, presumably by displacing fossil fuels as energy sources. Composting, to a very small extent, produces net greenhouse gas emissions (presumably due to the transportation involved), whereas landfilling without methane recovery does so to a larger degree.[11]

Waste management practices can be rank ordered according to environmental preferability. (See Figure 14–1.) This waste management hierarchy is an important policy tool that encompasses the "three Rs"—reduce, reuse, and recycle—as well as waste treatment and disposal. The aim is to minimize waste generation, conserve natural resources, reduce final waste disposal, and maximize potential benefits such as energy recovery and secondary materials. In some jurisdictions, additional "Rs" include (resource) recovery, reclamation, replacement, and repair. However, the three Rs broadly encompass overall waste principles.

The hierarchy of waste management provides an aspirational or prioritized

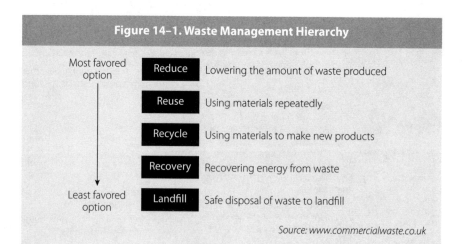

Figure 14–1. Waste Management Hierarchy

Most favored option

Reduce	Lowering the amount of waste produced
Reuse	Using materials repeatedly
Recycle	Using materials to make new products
Recovery	Recovering energy from waste
Landfill	Safe disposal of waste to landfill

Least favored option

Source: www.commercialwaste.co.uk

approach, but day-to-day implementation can be challenging. A city's immediate priority is to dispose collected waste properly, which alone can often overwhelm a city's capacity. Working with globally disparate manufacturers to redesign or reduce waste is difficult, as is seeking to move toward an "industrial ecology" system in which waste from one manufacturer becomes another's material source. More-achievable options can include bottle return systems, bans on plastic bags, and sending specific materials to special waste facilities, such as tire recycling centers and electronic waste collection facilities. Two key areas of waste reduction and recovery for a city to address—almost always the responsibility of local governments—are organic waste (horticultural and food) and construction and demolition waste (as well as waste generated during emergencies such as earthquakes, flooding, and windstorms).

Briefly, the three Rs are as follows:

Reduce

The most effective way to deal with waste is to not generate it in the first place, a practice known as source reduction. This includes minimizing the use of raw materials in the production process and purchasing or consuming only the necessary amount. Reduction can be achieved through careful selection of durable materials and through incentivizing changes in consumer behavior. Opportunities for waste reduction include renting rarely used items instead of buying them; two-sided printing; avoiding disposable goods (especially those that have extra packaging); and using electronic documents and correspondence instead of printed copies.

Policies that affect behavioral changes and that are successful in various cities include bans on plastic bag usage and variable pricing for trash collection (commonly called "pay-as-you-throw" schemes). In Seoul, South Korea, the amount of waste disposed decreased by almost 18 percent in the first year after the introduction of a volume-based waste fee system, and the amount of recyclables collected increased 18 times within 20 years. Such policies have multiple cobenefits beyond the obvious ones of extending the life of landfills and conserving resources: they decrease costs of collection and disposal, provide employment opportunities, and mitigate climate change.[12]

Variable waste-pricing policies are best implemented at the city or metropolitan level. They are not without drawbacks, however, as they can encourage illegal dumping, backyard burning, and over-compacting waste. A well-documented example of the latter from the mid-1990s is the "Seattle stomp," in which residents and businesses increased the amount of garbage in their trash containers by manually compacting it, in response to the introduction of a per container unit-pricing scheme. City officials countered by increasing the cost of trash collection but kept recycling fees steady. Volume-based pricing is effective only if the fees charged are high enough to trigger behavioral change.[13]

Importantly, although the focus of waste reduction policies is almost always on residential areas, the bulk of waste generation comes from industrial, commercial, and institutional sources. These waste sources tend to be influenced by financial drivers in the solid waste sector and often cross-subsidize residential solid waste services.

Reuse

Reuse refers to the selection of more-durable materials that can be used several times. Examples include rechargeable batteries, retaining scrap paper and glass containers for other uses, and using durable shopping bags instead of plastic bags. Fees or penalties for specific types of waste can meet multiple goals. When Washington, D.C., introduced a five-cent tax on disposable plastic and paper bags in 2010, use of the bags plunged by 85 percent. Presumably, people either reused plastic bags or bought durable ones that could be used over and over again. The city has raised $10 million in revenue from the "bag tax" since the policy was introduced. Even small businesses found that the policy increased cost savings. The policy reduced plastic bag consumption, encouraged residents to reuse, and raised revenue that was used to fund cleanups in the local Anacostia River.[14]

Policies based on fees or financial incentives can be implemented at either the local or state level, so that the revenues can be fed back into programs that can directly benefit or be monitored by residents. The concept of reuse also is gaining traction in the "sharing economy," in which peers participate in various types of digital technology-facilitated exchanges such as trading, lending, recycling, and upcycling. Lawn mowers, power tools, and camping equipment, for example, lend themselves well to sharing.

Recycle

Recycling is the processing of materials to turn them into new commodity materials with economic value, through which resources are conserved and overall waste is minimized. Recycling saves energy, preserves landfill space, and reduces greenhouse gas emissions. A policy that helps address the growing problem of waste is "extended producer responsibility" (EPR), which requires the manufacturer of a product to maintain responsibility for it throughout its full life cycle. EPR encourages manufacturers, which have the most control over product design, production, and packaging, to reduce toxicity and simplify product recycling. EPR can be established in both low- as well as high-income countries at the local, state, or national level. Although EPR for electronic waste and automotive parts may be more suited to high-income countries (where electronic products and vehicles are replaced frequently), low- and middle-income countries can begin implementation of EPR policies with container deposits for glass bottles or aluminum cans.

In addition to reducing greenhouse gas emissions, the use of virgin materials, and energy inputs, recycling plays an important role for local employment, particularly in low- and middle-income countries. The International Labour Organization estimates that 15–20 million people earn a living from informal recycling activities: collecting, sorting, cleaning, and recycling. In this way, waste pickers play a pivotal but often overlooked role in the waste economy. Recycling rates in many low-income countries tend to be high only because of these recycling activities, as more-formal recycling policies and centers do not yet exist.[15]

Reducing the Impact of Solid Waste on Climate Change

A number of policy options are available to help local and national governments directly mitigate greenhouse gas emissions from the solid waste sector. Many policies can be implemented in countries regardless of their income level, whereas others might be more applicable to one income level or the other.

Composting and Anaerobic Digestion

These processes control the decomposition of organic waste in order to prevent the generation or release of methane into the atmosphere. Composting involves the breakdown of organic waste in the presence of oxygen into a nutrient-rich soil conditioner over a period of weeks to months. Composting reduces the volume of organic waste by as much as 90 percent. Although CO_2 is also released during the decomposition process, composting is considered to be renewable because the amount of CO_2 released is equal to the amount that was absorbed from the atmosphere by the organic matter (green waste, vegetables, fruits, etc.) while it was growing.

In addition to reducing greenhouse gas emissions, composting offers various cobenefits. Using compost decreases the need for fertilizers by at least 20 percent, thus avoiding significant emissions of CO_2 and nitrous oxide in the manufacture, transport, and use of synthetic fertilizers. Using compost also improves water retention and decreases water usage, facilitates reforestation, increases soil health, helps to sequester carbon, and promotes higher yields of crops.[16]

Composting is practical at a local, community, city, or regional scale and can be conducted in the open or in purpose-designed vessels. Various composting methods can suit different scales of operation and availability of capital funding. As a result, composting is a practical technology for low- and high-income countries.

Anaerobic digestion is the biological process of converting organic waste to two useful products—digestate and biogas—in the absence of oxygen. Anaerobic digestion is a proven technology that increasingly is being used to treat the organic fraction of MSW, yielding a liquid digestate useful as fertilizer and biogas (primarily methane) that can be used to generate electricity or to augment existing natural gas supplies. The process itself also produces heat that can be used for heating and cooling buildings. Like composting, anaerobic digestion can be carried out at small or large scales. This technology has been used commonly with animal waste in low- and middle-income countries, such as India and China, but it is becoming common in Europe to treat MSW.

The main advantages of anaerobic digestion are to divert waste from landfills, thus prolonging their life spans, and to mitigate climate change, because it is a renewable technology. However, energy production is considered to be only a secondary benefit, because anaerobic digestion can only augment, but not displace, traditional sources of energy. One study calculated that if all U.S. MSW produced in 2006 had been diverted from landfills to anaerobic digestion, it would have generated enough energy to power 1.3 million households, or some 1 percent of the country's total. It also would have resulted in an almost 2 percent reduction in total U.S. 2006 greenhouse gas emissions. By diverting organic material for composting or digestion, the quality of subsequent recycling material increases and the remaining material is more suitable for incineration (as the moisture-laden organics are removed).[17]

Waste-to-Energy

Waste-to-energy encompasses incineration of waste, refuse-derived fuel, pyrolysis, and gasification. Although all of these options are considered to be thermal treatment technologies, there are important differences that result in varying costs, efficiencies, and suitable scales of application. All methods significantly reduce the volume of waste, destroy harmful pathogens, and reduce nuisances such as odor and attraction to vermin. Waste-to-energy plants reduce greenhouse gas emissions by avoiding methane emissions from landfills, generating electricity that otherwise may have come from conventional fossil fuel energy sources, and recovering metals from recycling, which saves energy compared to mining virgin materials. In most cases, waste-to-energy plants emit more biogenic carbon than non-biogenic (i.e., fossil fuel-based products, such as plastics). For all of these reasons, waste-to-energy is not considered to add net greenhouse gas emissions to the atmosphere.[18]

The U.S. Environmental Protection Agency's Waste Reduction Model (WARM) estimates greenhouse gas emissions from various waste management options. Table 14–3 shows the estimated direct and avoided emissions per metric ton of MSW from waste-to-energy. Since waste going to a waste-to-energy facility otherwise would have been destined for a landfill, the greenhouse gas emissions for landfilling also are presented. Net greenhouse gas emissions from waste-to-energy are similar to landfills only if landfill gas is collected and used for electricity generation. In other cases, such as when no landfill gas is collected or if the gas is flared, waste-to-energy is preferable to

Table 14–3. Comparison of Estimated Direct and Avoided Greenhouse Gas Emissions for Waste-to-Energy and Landfilling	
Method of Disposal	**Emissions**
	Million tons of CO_2-equivalent per ton of MSW
Waste-to-Energy (Incineration)	
Direct emissions (CO_2, transportation)	0.40
Avoided emissions (utilities)	-0.39
Avoided emissions (metals recovery)	-0.05
Net greenhouse gas emissions	-0.04
Landfill	
No landfill gas recovery	3.10
Landfill gas recovered and flared	0.31
Landfill gas recovery with electricity generation	-0.03

Source: See endnote 19.

landfilling in terms of greenhouse gas emissions avoided. These estimates are particularly important for infrastructure planning, as waste treatment and disposal options tend to have long life spans (about 20 years for a waste-to-energy facility and 30 years for a landfill), and cities generally will be locked in for those periods.[19]

Incineration is the combustion of waste under controlled conditions and with advanced air pollution control (APC) technology. It has developed into a mature technology and is found primarily in high-income countries because it is still expensive to install and requires advanced technological training. Incineration almost always includes energy recovery, meaning that the energy and heat generated are captured for electricity generation and/or district heating. Suitable feedstocks for these facilities can generate 500–600 kilowatt-hours (kWh) per metric ton of MSW. In the European Union, it is common to find cogeneration of electricity (500 kWh per ton) and district heating (1,000 kWh per ton). Incineration also requires sophisticated regulatory capacities and air-quality monitoring facilities.[20]

Refuse-derived fuel (RDF) refers to the process of modifying MSW by one or more of the following processes: sorting, screening, shredding, drying, and/or converting to bricks or pellets, to make a more homogeneous product. RDF has a relatively higher energy content than unprocessed MSW and can be used as a supplementary fuel in conventional boilers; it therefore has more flexibility for use in industrial processes. It also is transported more easily than other feedstocks. However, RDF incurs additional costs of waste pre-processing and usually is not recommended for direct combustion in common waste facilities. There is a growing demand, in both low- and high-income countries, to use RDF as a fuel source in coal-fired power plants or cement kilns. In Poland, for example, an average of 36 percent of coal has been replaced with RDF across all cement kilns.[21]

The incineration of waste and RDF generates harmful air pollutants, such as dioxins and toxic fly ash (fine particles emitted from smokestacks). Increasingly strict government regulations can require the use of highly advanced APC equipment to capture the air pollutants for treatment at the facility. Emissions are monitored continuously and generally are significantly lower than those required by environmental regulations. The drawback is the high cost of the APC equipment and the fact that operators can bypass it. Although the environmental considerations of these technologies can be addressed, the costs can be prohibitive for low-income countries.

Pyrolysis is the process of heating waste to high temperatures in the absence of oxygen, thus producing synthetic gas (syngas), tar, and char (the burned waste remainder). Syngas can be converted to a fuel to produce electricity or heat, or used as a chemical feedstock. The outputs from pyrolysis can be adjusted by varying the reactor temperature: higher temperatures (greater than 760 degrees Celsius) result in more gas production, while lower temperatures (450–730°C) produce both liquid and gas. Liquids from pyrolysis can be refined for use as chemicals, motor fuels, and other products.

Finally, gasification involves the heating of waste in the presence of oxygen, but not as much as in incineration. The waste is heated to temperatures of 1,000–1,500°C, resulting in the production of syngas. Liquids, such as tar and oils, and solids in the form of char and ash also may be produced. The syngas can be processed into fuels such as chemicals and fertilizers, which would reduce the greenhouse gas emissions associated with those processes. In one type of gasification, called plasma arc, extremely high temperatures are used to create an inert material (slag) that looks like glass and can be disposed safely in landfills or used as construction material. The high temperatures also inhibit the production of toxic air pollutants, such as dioxins.[22]

Since emissions are significantly lower in pyrolysis and gasification compared to incineration and RDF, the APC costs for the former are low. However, the energy requirements to run these technologies are high because they require external heating; further energy also may be required to clean the syngas. Pyrolysis and gasification are emerging technologies in the treatment of MSW, with only a few full-scale facilities in operation, mostly in Japan. The costs of these technologies for waste disposal presently are very high, even for high-income countries.[23]

Regardless of the technology employed, landfilling cannot be avoided completely. Landfills are necessary in order to receive byproducts from treatment methods, to collect overflow waste from other options, and to function as a backup when waste-to-energy and other facilities are shut down. When landfills are planned, landfill gas recovery should be included in the design. This greatly affects the reduction of greenhouse gas emissions and enhances safety. Landfills should never be built without gas recovery because of the risk of explosions. Thought also should be given to integrating the landfill into local land uses—for example, golf courses and parks—upon eventual closure.

Landfill Gas Recovery

Globally, the dominant form of waste disposal is by means of landfills or open dumps, as historically they have been the cheapest method of disposal.

However, there is a growing awareness of the harmful environmental effects of improper waste disposal, and the overall costs of landfilling often do not take into consideration environmental and social costs, potential revenue from the generation of electricity (for alternative options such as incineration), and site closure and maintenance costs.

Capturing landfill gas (primarily methane) reduces the contribution of the waste sector to climate change. Landfills need to be constructed with pipe networks to collect landfill gas generated during waste decomposition. (In some cases, pipes can be added to existing landfills to recover the gas.) The methane fraction of landfill gas can be separated and either flared or combusted to generate electricity. If the latter, this provides two important climate benefits: methane is combusted instead of escaping to the atmosphere, thereby converting it to CO_2, which is much less harmful as a greenhouse gas than methane; and it displaces other energy sources (e.g., oil, coal) that otherwise might have been used to generate electricity. Obviously, landfills or open dumps that do not have gas collection systems contribute to the climate change problem.

Methane gas from the Los Reales landfill is piped from this pumping station to Tucson Electric Power for electricity generation.

Landfill gas collection systems are not airtight: around 4 to 10 percent of the gas escapes a typical collection system. Approximately 120 kilograms of methane are generated from every metric ton of MSW landfilled. But reducing methane emissions from landfills is considered to be relatively cost-effective and efficient compared to other methane mitigation technologies: an investment of up to $60 per ton of CO_2-equivalent can reduce methane emissions by 76 percent. A key priority should be to reduce the organic fraction of the waste prior to disposal (by composting, perhaps).[24]

Conclusion

The waste sector can contribute substantially to greenhouse gas reductions through various policies and technologies that already exist and are proven. Still, a number of challenges remain. Waste generation, especially in low-income countries, is expected to increase continuously until the end of this century as a result of increased urbanization and improved living standards. Technology transfer, higher public awareness, and thoughtfully implemented policies can encourage the shift toward greater sustainability in the solid waste sector.

Solid waste managers make locally driven decisions. Their decisions require knowledge of local waste quantities, management capabilities, technical know-how, marketability of outputs (compost, sale of energy in the form of electricity/heat), municipal finance options, and planning for long-term waste infrastructure. One principle that can guide decision makers in both high- and low-income countries is a focus on integrated solid waste management. This encompasses the waste hierarchy, public awareness and outreach, financing, planning, human resources (including integrating waste pickers), and technical and environmental factors. An integrated approach can benefit local communities and—because local decisions have global ramifications—the planet at large.

CITY VIEW

Barcelona, Spain

Martí Boada Juncà, Roser Maneja Zaragoza, and Pablo Knobel Guelar

Barcelona Basics

City population: 1.6 million
City area: 102 square kilometers
Population density: 15,824 inhabitants per square kilometer
Source: See endnote 1.

The Montjuïc cliffs, a biodiversity hotspot inside the city of Barcelona.

Biodiverse City Between the Forests and the Sea

Barcelona, a city renowned for its historical avant-garde urbanism, is located in a flatland along the Mediterranean coast of the Iberian Peninsula. It has natural boundaries on all sides: the Llobregat and Besòs rivers to the southwest and northeast, the Mediterranean Sea to the southeast, and Collserola Natural Park to the northwest. A large, forested section of the park extends into Barcelona itself and is an important asset for biological diversity. Bio-diversity has become a key element in the city's urban governance, with a focus on greater harmony between humans and nature.[2]

Continuous Urban Transformation

By the end of the nineteenth century, the city of Barcelona was growing enough to connect directly to smaller surrounding towns. The City Council decided to rationalize and unify the city's growth, leading to the approval of the Jaussely Project in 1907. Influenced by Ebenezer Howard, the British founder of the garden city movement, the project outlined the creation of an extensive park system to supplement Ciutadella Park, Barcelona's only green park at that time.

Between 1917 and 1937, architect and urbanist Nicolau Maria Rubió i Tudurí had great influence over the design of Barcelona's urban parks. His main ideas, inspired by the humanistic visions of Howard and of French architect Léon Jaussely, included simplicity in park design, the rational distribution of parks over the city, and the reclamation of as much open space as possible. Today, Barcelona's green area totals 36 square kilometers, compared to a built-up area of 66 square kilometers. Of the green area, 18 square kilometers is forest, 11 square kilometers is public, and some 7 square kilometers is private.

Rubió i Tudurí also proposed the creation of a semi-circular green ring around the city, closed by the Collserola reservoir and with two lateral axes that coincide with the Besòs and Llobregat rivers. He imagined concentric inner rings featuring different kinds of green sites: small urban gardens (of 8–10 hectares, comprising the inner circle), outer parks (comprising a second, larger circle), and landscape reservoirs (large forest areas forming the last circle and shared with adjacent municipalities). In the late 1990s, ecologist Ramon Margalef reformulated this theory as the Metropolitan Green Belt of Barcelona, a 3,200 square kilometer system of green spaces and protected areas that affects more than 4 million inhabitants.[3]

Modern Barcelona also reflects the revolutionary urban expansion ideas of engineer Il-defons Cerdà. His original plan, dating to 1859, included 20-meter-wide streets with buildings on only the ends of each block, allowing for large green spaces in the middle. The plan focused on street geometry, with streets set at strict right angles except for a few large diagonals. Due to political and demographic pressures, the project was redesigned with more (and higher) buildings in the spots initially reserved for open space, and with most industrial activity moved outside the city.[4]

Barcelona is still undergoing transformation as it creates new nodes of activity (in 2014, the Plaça de les Glòries, a strategic city square, was completely remodeled with the aim of decentralizing the city and improving urban sites). The city also is rethinking basic urban functioning through ambitious efforts such as the Superblocks project, a series of urban areas that reflect a new model for transportation and public spaces. Going beyond traditional demographic and geographic criteria for urban redesign, the Superblocks allow for improvements in many aspects of the city: mobility, space revitalization, biodiversity and green improvement, social cohesion, energy self-sufficiency, and citizen participation.[5]

Biodiversity Overview

The city of Barcelona is estimated to be home to more than 2,000 plant species, more than 200 tree species, and more than 235,000 urban trees, which translates to 0.15 trees per inhabitant. It also hosts 28 mammal species, 184 bird species, 16 reptile species, 10 amphibian species, 57 butterfly species, and 4 fish species. On the Shannon-Weaver Index, a conventional indicator of biodiversity, Barcelona scores 2.96 out of 5, or at the high end of medium.[6]

Marti Boada

An egret flying over an urban park in Barcelona's Poblenou neighborhood.

Understanding the current state of urban biodiversity in Barcelona requires appreciation of the major role played by the city's two biggest and most biodiverse natural sites: Montjuïc Mountain and Collserola Natural Park. With a kaleidoscope of habitats, these sites hold a large number of species (1,711 taxa and 1,500 taxa, respectively). They also stand inside the ecotone line—the species-rich zone between the city and forest—making them even more biologically diverse.[7]

On the other end of the habitat spectrum are the small, isolated green sites of the city that lack fluent ecological connection (57 percent of these sites are less than 1,500 square meters in area). Consequently, urban green corridors, represented by streets and avenues, play a critical role in connecting "recharge nodules," such as Collserola Natural Park and Montjuïc Mountain, with smaller green areas scattered across the city. The corridor system creates a green infrastructure that facilitates the propagation of wildlife.[8]

Forest area represents half of Barcelona's total green area. But whether the portion of Collserola Natural Park that extends within the municipality is included in the calculation of green area per person is important. Including the park, the green area per inhabitant is 18.1 square meters; excluding it, the green area decreases to 6.6 square meters per inhabitant. Other indicators, such as the share of tree canopy and the number of trees per hectare, also are affected by the inclusion of the park. In terms of forest area, Barcelona can be compared to Boston if Collserola Natural Park is included, but is more like San Francisco or Chicago if it is not.[9]

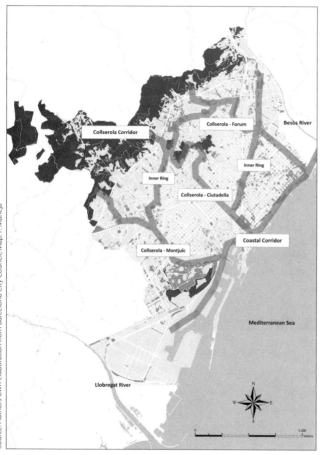

Source: Authors' own elaboration from Barcelona City Council. Map: F. Maneja

Barcelona's urban green corridors connect the recharge nodules of Collserola Natural Park and Montjuïc Mountain.

Urban Green Governance

Urban green governance is not new for Barcelona. In the last two decades, the city has committed to several initiatives to promote the enhancement of nature. The city's approach to urban biodiversity is understood as an ongoing commitment to global sustainability through appreciation of the local environment. This implies an increasing connection between urban biodiversity and citizens while ensuring the provision of ecosystem services, with particular attention to promoting the well-being of city dwellers.[10]

In 2004, after four years of civic discussion among residents and representatives from different city entities, the Citizen Commitment for Sustainability (Agenda 21) agreed to 10 overall objectives and 10 action lines for a more sustainable city. A key theme was city biodiversity. In 2008, Barcelona joined LAB (Local Action for Biodiversity), a global program that commits the city to a biodiversity action plan and to implementing several major biodiversity initiatives.[11]

In 2013, Barcelona City Council launched the Green and Biodiversity Barcelona Plan 2020, which sets the main thrusts of action for the coming years and outlines the challenges, objectives, and commitments inherent in the city's effort to enhance its natural spaces and biodiversity. The Plan treats nature and biodiversity as a whole, taking into account the differences between natural sites and the relationships among them. It assesses the current state of nature and biodiversity in Barcelona and offers short- and long-term actions to improve it.[12]

The City of Barcelona looks forward to creating a more resilient and fertile environment that prepares the city infrastructure and metabolism for global changes that are under way or are coming. Among the planned actions are a green-corridors project throughout the city and several projects promoting green roofs and walls as well as urban vegetable gardens. All of these are designed to increase biomass in the city, to enrich the existing green infrastructure, and to enhance its habitat function.

Connecting Citizens with Urban Biodiversity

A variety of civic initiatives aim to promote nature and biodiversity in Barcelona. They include the following:

BioBlitzBCN Project. Launched in 2010 by the Natural Science Museum of Barcelona in collaboration with the city and municipal governments and other institutions, the project aims to develop an inventory of Barcelona's urban biodiversity through collaboration in species censusing between biodiversity experts and citizens. It is based on the BioBlitz experience, first held in Washington, D.C., in 1996 and organized in 13 other countries since then. The project has documented a sharp increase in animal and plant species in Barcelona, including the presence of 13 different ant species in Ciutadella Park (in 2010) and the first appearance in Catalonia of the flying insect *Dolichopeza hispanica* in Laberint d'Horta Park (in 2013).[13]

SOCC (Monitoring of Common Birds in Catalonia). Launched in 2002 by the Catalonian Ornithology Institute in conjunction with the Catalan government, the project aims to improve environmental quality through the tracking of common bird species in the region. The census is undertaken by committed volunteers (both professional ornithologists and amateurs) and is included in the European Common Bird Monitoring Scheme, which gathers data from 25 countries. SOCC's flagship project, the *Atlas of Nesting Birds of Barcelona*, started in 2011 and describes the distribution of all the nesting species in the city.[14]

Aula Ambiental Bosc Turull. This municipal facility aims to promote environmental education, mainly to primary and secondary schools. Through participatory guided activities in different parts of the city (for example, the installation of nest boxes in urban parks and the celebration of the World Day of Birds), the facility aims to raise citizen awareness of the importance of conservation and the vital role of urban biodiversity.[15]

Falcon chicks discovered in Sagrada Família basilica, one of the most visited landmarks in Barcelona.

Amics del Jardí Botànic de Barcelona. Founded in 1993, this association aims to promote and preserve the city's botanic garden. Over the last several years, the association has organized weekend guided tours that help citizens appreciate biodiversity and allow them to enjoy several interests related to the botanic garden.[16]

Peregrine Project. For many years, the peregrine falcon nested in the tallest buildings in Barcelona. In 1973, with the change of the hunting law, the species disappeared from the city. In 1999, a reintroduction program supported by the City Council began to facilitate the falcon's return. The Peregrine Project introduced three falcon pairs to the city in 2005, and a fourth pair in the metropolitan area. Ten years later, a pair of peregrines had bred in two different parts of the city: Montjuïc's cliffs (resulting in two chicks) and the Calatrava Tower (resulting in a minimum of two chicks).[17]

La Fàbrica del Sol. This facility, supported mainly by the City Council's Department of Urban Ecology, aims to promote environmental education through various approaches that help motivate primary school students to consider the importance of urban biodiversity in a global context, among other goals.[18]

Martí Boada Juncà is a senior researcher and a professor in the Department of Geography at the Institute of Environmental Science and Technology – Autonomous University of Barcelona (UAB). **Roser Maneja Zaragoza** is a senior researcher at the Institute of Environmental Science and Technology at UAB. **Pablo Knobel Guelar** is a junior researcher at the Institute of Environmental Science and Technology at UAB.

CHAPTER 15

Rural-Urban Migration, Lifestyles, and Deforestation

Tom Prugh

The means of addressing greenhouse gas emissions related to building and transport energy use, urban form, and waste are technically straightforward, even if socially and politically challenging. Tackling these areas successfully would go a long way toward mitigating the emissions load that cities impose on the global ecosystem. However, cities also lie at the root of an additional important source of emissions—deforestation and changes in land use—and these drivers are both under-appreciated and perhaps more problematic to address.

According to researchers at Winrock International and the Woods Hole Research Center, tropical deforestation accounts for an estimated 3 billion tons of carbon dioxide (CO_2) per year, or about 10 percent of all heat-trapping emissions—equivalent to the emissions of some 600 million cars. (There are many ways to calculate both deforestation and total greenhouse gases, and other estimates suggest that deforestation's share may be even higher.) Urban growth is one of two main drivers of deforestation, along with land clearing for the export of agricultural products.[1]

Urban growth drives deforestation in at least two ways. First, as rural migrants to cities adopt city-based lifestyles, they tend to use more resources as their incomes rise and as their diets shift from starchy staples to a greater share of animal products and processed foods. (There is a strong relationship between income and meat consumption; see Figure 15–1.) This, in turn, drives land clearance for livestock grazing and fodder, either in the migrants' own countries or in other countries that export such products or their inputs.

Tom Prugh is a senior researcher at the Worldwatch Institute and Codirector of the *State of the World* 2016 project.

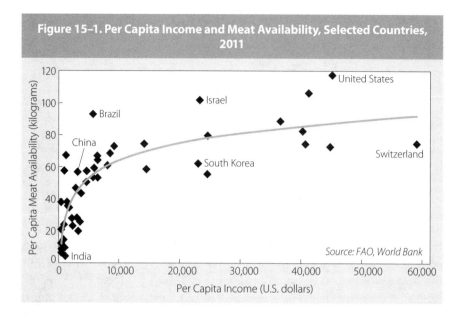

Figure 15–1. Per Capita Income and Meat Availability, Selected Countries, 2011

Source: FAO, World Bank

(Population growth in rural areas does not appear to be closely associated with deforestation, indicating that urban and international demand for agricultural products is the primary culprit.) This clearance has been extensive: from 1980 through 2000, nearly 80 percent of new agricultural land in the tropics was converted from forests (more than half of it from intact forests and 28 percent from disturbed forests).[2]

The effect on land use of adopting richer diets is striking. The hidden multipliers lie mainly in the efficiency with which plants are converted into food. It is far more efficient for humans to consume plant calories directly (via vegetarian diets) than to require livestock to consume plants and then to consume the meat in the livestock, because this second step entails serious additional energy losses. By one estimate, the amount of food produced relative to total plant production—the efficiency of production—is 78 percent for cereal grains but only 20 percent for poultry, 18 percent for pork, and 2 percent for beef. Even in relatively highly productive European agriculture, it takes an estimated 0.3 square meters to produce an edible kilogram of vegetables, 0.5 to produce a kilogram of fruit or beer, and 1.4 for cereals—but 1.2 square meters for a kilogram of milk, 3.5 for eggs, 7.3 for chicken, 8.9 for pork, 10.2 for cheese, and 20.9 for beef. On European Union farms, on average, it takes nine hectares of pasture plus three hectares of cropland to produce 1 ton of beef, compared with only one hectare of cropland per ton of poultry or pork.[3]

A second (and likely lesser) factor linking urban growth to deforestation is that, with the influx of migrants, cities often expand into areas of natural habitat, including forests. Cities worldwide are growing by 1.4 million new inhabitants every week, and urban land area is expanding, on average, twice as fast as urban populations. Under current trends in population and urban growth, the number of people in cities is projected to rise by nearly 3 billion by 2050, and the area covered by urban zones is projected to expand by more than 1.2 million square kilometers between 2000 and 2030. This expansion destroys wildlife habitat and threatens biodiversity (especially in "hotspots" that are particularly rich in species); it also releases carbon stored in biomass and thereby increases atmospheric carbon concentrations. Ironically, even as urban expansion drives forest clearance for agriculture, it simultaneously consumes existing farmland; by one estimate, urbanization may cause the loss of up to 3.3 million hectares of prime agricultural land each year.[4]

Deforestation is believed to have additional, local climate effects with global implications. Forests draw huge amounts of water vapor from soils and transpire it into the atmosphere—billions of tons per day—creating clouds and aerial rivers of water vapor that travel long distances. The theory of "biotic pumping" holds that this process of vapor transpiration creates low-pressure weather systems that draw moisture inland from the oceans. Through that mechanism, deforestation in the Amazon—some 5,000 square kilometers were lost in 2014—is thought to be a major cause of Brazil's current extreme drought. Simulations using climate models interactively linked to vegetation models suggest that reducing the Amazon rainforest by 40 percent from its original extent could precipitate a vast ecological transformation to much drier savannah and ultimately displace much or all of the remaining forest. About 20 percent of the original forest is already gone.[5]

Although the rate of deforestation globally has been reduced by half in the last 25 years, net forest cover continues to decline. About 129 million hectares has been lost over the period, an area roughly equal to South Africa, while global carbon stocks in forest biomass declined by over 17 gigatons. Deforestation and land-use change are now severe enough that the Stockholm Environment Institute, developers of the "planetary boundaries" system for gauging the sustainability of the global economy, has concluded that the category of deforestation has shifted out of the "safe operating zone," joining the boundary categories of climate change, biodiversity loss, and eutrophication—thus, four of the nine identified categories—in that alarming state.[6]

Tele-deforestation?

To a considerable extent, current rates of deforestation are the outcome of globalization and expanding international trade. A team of researchers led by Karen Seto at Yale University has proposed the term "urban land teleconnections" to conceptualize the ways that people, goods, and services flow around the global economy, as well as the changes in land use that drive and respond to urbanization. The researchers point to the strengthening connectivity between distant locations and the growing separation between places of consumption and production as central to the idea of teleconnections.[7]

The separation of consumption and production has exploded over the last 300 years, and especially so in recent decades. Cross-border trade in food commodities increased by a factor of five between 1961 and 2001, while the international trade in wood products expanded sevenfold. Dependence on such trade can be extreme. In 2001, for example, Switzerland imported agricultural products that would have required more than 150 percent of the country's arable land if grown domestically.[8]

In Brazil, a surge of deforestation in the Amazon in the early 2000s has been attributed to the expansion of pasture and soybean croplands in response to international market demand, particularly from China, where economic growth and diets richer in meat products have boosted soy imports from Brazil to feed pork and poultry. The Brazilian government has long encouraged frontier farmers to clear forest. Although the legally defined latitude for clearance has been tightened recently, a variety of factors—including corruption and a lack of compliance incentives and enforcement capacity—has reduced the effectiveness of the restrictions.[9]

Dependence on trade in food commodities seems likely to increase the already substantial pressures on forests. Meeting the food needs of a rising and urbanizing global population could require an additional 2.7–4.9 million hectares of cropland per year. This amount could be higher or lower depending on trends in global diets, food waste, and the efficiency of converting feed to animal biomass (which varies among livestock species, as noted previously). It also could be driven up by the anticipated degradation of existing farmlands due to ongoing climate change under business-as-usual scenarios, especially in the medium to long term (after 2050).[10]

Teleconnections between developed and developing countries express themselves in other ways as well. Countries that seek to conserve lands of value for habitat, carbon storage, or both may simply displace clearance elsewhere, to one degree or another. One modeling study suggested that protecting 20 hectares of forest in North America or Europe induces clearance of roughly 1 hectare of primary forest in the tropics or in Russia. Likewise with food commodities: between 1990 and 2004, developed countries that introduced conservation set-aside policies increased their per capita cereal grain imports by more than 40 percent, compared with average increases of under 4 percent in countries without such policies.[11]

Roosewelt Pinheiro/Abr

These Brazilian soybeans have been harvested for export.

These factors combine with various biofuel mandates and/or market demand, erosion and other kinds of degradation of arable land, demand for plantation products, and climate change (which opens some lands to cultivation and closes others) to create something of a perfect storm of stresses on forests that boosts pressures for additional forest clearance. Most or all of these stresses are artefacts of the wealth and lifestyles of urbanites. One frequently heard response is to intensify agriculture, applying the industrial model of mechanization and higher inputs to raise outputs, reduce cultivated area, and spare more land for nature. But apart from the other factors dimming the long-term prospects of this more "efficient" mode of agriculture, it may backfire with respect to forests because it may be more profitable, depending on the "demand elasticity" of the crops grown. Crops that consumers readily buy more of when prices fall will be in higher demand, a trend that can lead to the expansion of cultivated land.[12]

Forests are a prime example of ecologist Garrett Hardin's adage that "we can never do merely one thing." In complex systems, any action has multiple, often unintended, consequences. As researchers Eric Lambin and Patrick Meyfroidt put it:

In a more interconnected world, agricultural intensification may cause more rather than less cropland expansion. Land use regulations to protect natural ecosystems may merely displace land use elsewhere by increasing imports. Mitigating climate change by mandating the use of biofuels in one place may increase global greenhouse gas emissions due to indirect land use changes in remote locations. A decrease in rural population due to outmigration may increase land conversion through remittances being invested in land use.[13]

Mitigation

The impacts may be difficult to quantify, but it appears that there is some urgency in attempting to mitigate urbanization's direct and indirect destruction of forests. The impact of urban expansion can, in principle, be attenuated by focusing on proven methods of shaping urban form to emphasize compact development and higher densities. (See Chapter 7.)

The so-called wealth effect, on the other hand, presents a thornier issue. A major factor in urban growth is rural-urban migration, a widespread and long-term trend that has accelerated in recent decades. Rural dwellers may move involuntarily to cities for a variety of reasons, including displacement by natural disasters, such as the Dust Bowl events in the United States in the 1930s or the recurrent floods, droughts, and cyclones in Bangladesh. Political processes also can drive farmers off their lands, as in the "enclosure movement" in Britain. But a key driver of migration is the hope and expectation of higher incomes.[14]

According to the standard economic model of rural-urban migration (the Harris-Todaro model), the perception of wage differentials between rural and urban settings draws workers to cities. Although many urbanites live in impoverished circumstances, the perception often turns out to be true, and workers moving to cities in search of jobs often achieve greater prosperity than they might have by remaining in the countryside. This, in turn, results in higher consumption and its impacts on forests. (The migrants do not necessarily sever their country roots, however. Many would prefer to remain in their rural communities and near their families. Millions send money home in the form of remittance payments, which were estimated to reach $586 billion for all migrants globally in 2015.)[15]

Suppose policy makers sought to retard urban expansion by reducing or eliminating this income differential? There is evidence that the types of smallholder farming that are so prevalent in developing countries can be at least as productive, or perhaps more so, than the globalized and industrialized model that is spreading throughout the world, undermining smallholder incomes and contributing to the displacement of so many farmers. The adjustment or abolition of certain international trade policies could help reduce the undue advantage that these policies confer on large farming enterprises and make smallholder farming more profitable. (See Box 15–1.) It is by no means clear that this would reduce consumption overall, however, as consumption tends to increase with economic activity, regardless of location.[16]

There is a more subtle factor to consider as well. Urbanization and rising incomes are "dynamically codependent" upon each other, with "strong positive interlinkages," in the words of Felix Creutzig of the Mercator Research Institute on Global Commons and Climate Change, in Berlin. Cities tend to be more economically productive environments than rural communities—which may well be the main reason that cities exist. The reasons for this higher productivity have to do with "agglomeration economies," a long-noted and studied phenomenon in which high densities of firms and workers drive productivity in a synergistic manner. Economists have developed three general theories about the process: 1) high densities of firms reduce the costs of moving goods around, 2) high densities of people produce compact pools of labor that can be shifted easily to where they are optimally needed, and 3) the concentrations of both people and firms promote the ready exchange of ideas, boosting human capital and innovation. Although all three notions have intuitive appeal, Harvard economists Edmund Glaeser and Joshua Gottlieb argue that attention has focused increasingly on the third:

> The largest body of evidence supports the view that cities succeed by spurring the transfer of information. Skilled industries are more likely to locate in urban areas, and skills predict urban success. Workers have steeper age-earnings profiles in cities, and city-level human capital strongly predicts income.[17]

It appears that greater wealth generation (and thus higher incomes and consumption) are more or less inherent in cities that have benefited from agglomeration—not for all cities or all urban dwellers, but in the aggregate. Even though cities may be more economically efficient in supporting that higher consumption, the dynamics that encourage higher consumption with increasing wealth are not absent in cities.

Box 15–1. Is Urbanization Really a Green Anti-Poverty Strategy?

Rural dwellers may be drawn to cities in the hope of improving their economic lot, but they do not always succeed. There is a substantial debate over which strategies are best in combating rural poverty, and it is not clear that urbanization is superior. It is equally unclear, according to some analysts, that this sort of demographic shift moves the world closer to sustainability.

Chris Smaje, a U.K.-based farmer and social scientist, cautions against assuming that urbanization is a green anti-poverty program. He points out that the simple model of poor farmers moving to cities and beginning the climb into the middle class masks a much more complex reality:

> . . . [R]ural-urban migration is rarely a final, one-way thing. That's just too risky for poor rural families. Rather, it's a cyclical way of spreading risk and increasing the rural household's wealth by sending young adults to seek wage labour in the city: for them, living on the street or under bridges for a few months is bearable in return for a decent wage, but it's not a viable strategy of household improvement.

Smaje notes that both the city and the countryside are complex places, in ways that muddle the simple distinction between urban and rural. Cities also vary widely from one to another, as do different areas within any given city. Some city dwellers may have lower environmental impacts than those living in the country, but not all of them do. Perhaps most important, cities and rural areas are not separate realms so much as different parts of a coupled system. Many studies of environmental impact, Smaje argues, consider only the direct impacts of city living and fail to account for the indirect effects: "Much of the [greenhouse gas] emissions from farm traction, agro-chemicals, tillage, livestock, the anthropogenic nitrogen cycle, and the many other environmental impacts of industrial, export-oriented agriculture must . . . be attributed to cities" but instead are ignored or assigned to the countryside. Yet those impacts are incurred in support of city-dwellers' lifestyles.

Smaje's bottom-line argument is that urbanization is supported and driven by the system of "industrial cash crop agriculture" and that much rural-urban migration is an artefact of displaced small farmers being driven off their lands by national and international policies and trade agreements that "undermin[e] the possibility of rural livelihood." Apart from the direct and severe impacts of such agricultural practices, this does not necessarily create space for nature, but rather tends to create incentives for more-intensive cultivation of existing lands, thus raising output and making such cultivation more profitable—thereby encouraging expansion. Industrial agriculture displaces small farmers to cities and "replaces resource-frugal peasant smallholding, generally diminishes genetic diversity, pollutes air and watercourses, and exhausts soils. In short, mass urbanisation is a social and environmental disaster. . . ." "Ultimately," Smaje concludes, "the debate about whether urban or rural life is 'greener' is fruitless. We do not yet have enough data."

Source: See endnote 16.

Various approaches exist for addressing these impacts on deforestation-related greenhouse gas emissions. The first and most obvious option is to increase the efficiency of economies at delivering human well-being per every unit of resource input. A wide range of technical means and supporting policies would help achieve gains in the energy efficiency of buildings, in transport, in urban form, and in waste handling. (See Chapters 7 through 14.)

The impact of the dietary share of higher consumption could be reduced sharply by doing two things: reducing food waste and creating incentives for much lower meat consumption. About one-third of all food produced globally is lost or wasted as it moves through the stages of growth, harvest, processing, shipping, retailing, and consumption. Such losses can, and must, be reduced at every stage. Projections of the much greater supplies of food necessary for future populations generally assume that this wastage is not successfully reduced, that land is available to produce the required food quantities without addressing waste (when, in fact, land clearance—and deforestation—would inevitably result), and that everyone now living primarily on a vegetarian diet will undergo the "nutrition transition" that demands more meat and related products.

None of these assumptions should be allowed to go unchallenged, and cities have a role to play in questioning these assumptions as well as in shaping food handling practices and dietary habits. One approach might involve "delegitimizing" meat consumption, using techniques similar to those deployed to create social support for leaving fossil fuels in the ground rather than burning them. The Meatless Monday campaign is an example. New—and more convincing—meat substitutes, such as the soy and pea protein-based products of Beyond Meat, could contribute significantly to reduced meat consumption and its related impacts.[18]

Cities also may have a role in determining broader agricultural policies. Some experts argue that, in addition to reducing meat consumption, it is possible to reduce the impacts of meat production by de-emphasizing ruminant animals (cattle, goats, and sheep) in favor of poultry and pigs (which are more efficient ecologically), integrated aquaculture, and other more-efficient protein sources, and by shifting from intensive, fossil fuel-based livestock systems to more-diverse, coupled systems that emulate the structure and functions of ecosystems and thereby conserve energy and nutrients. Cities could marshal their political influence and clout behind policies to support these transitions among their own citizens as well as at the national and international levels.[19]

Conclusion

Deforestation and its effects on greenhouse gas emissions and concentrations are outcomes of economic activity, whether by rural smallholding farmers or by their distant city customers who seek to tap the opportunities for affluence that cities offer many urbanites. In a world where the productive capacities of tightly coupled ecosystems are increasingly under strain, it is hard to find ways to alleviate poverty that do not have far-reaching effects. Those living in the countryside in material poverty rightly wish to improve their lives, and many uproot themselves and their families, temporarily or permanently, in search of greater comfort, affluence, and security in cities.

But if limits to economic growth do exist—a proposition supported by the planetary boundary transgressions identified by the Stockholm Environment Institute and much other evidence—and are to be acknowledged and respected, it would seem inevitable that aggregate consumption must still decline. This may be accomplished by lifestyle "downshifting" among wealthy urbanites and/or by wringing much greater efficiencies out of resource use without aggregate declines in prices (to avoid rebound effects). In either case, deep and widening wealth gaps that have long histories of intractability will have to be confronted.

Politics, Equity, and Livability

CHAPTER 16

Remunicipalization, the Low-Carbon Transition, and Energy Democracy

Andrew Cumbers

The term "remunicipalization" has become associated with a global trend to reverse the privatization wave that swept many countries—both industrialized and developing—in the 1980s and 1990s. Outside of Germany, the trend is associated primarily with the water sector; however, the push to take back formerly privatized resources and services into local forms of public ownership and control is happening in other sectors as well, including transport, waste management, energy, housing, and cleaning.[1]

What is behind these developments? A simple answer is generalized dissatisfaction with the consequences of global privatization initiatives, which on the whole have not delivered the cost efficiencies, performance improvements, and infrastructure investment and modernization that their advocates had promised. At a time when local governments around the world face deteriorating public finances, and in a context of continuing recession and broader global economic austerity, bringing vital utilities, sectors, and revenue streams back under public ownership and control is increasingly popular.[2]

In addition to these more pragmatic considerations, advocates of remunicipalization (and of public ownership more generally) suggest that it holds the possibility for renewing public engagement and democratic accountability in the economy. Linked to this are claims that the processes of decentralization that are inherent in remunicipalization can challenge the power of vested interests (such as large private corporations) and provide local actors with the tools to effect more progressive forms of public policy. This is especially relevant for efforts to develop integrated local strategies to

Andrew Cumbers is a professor of urban and regional political economy at the University of Glasgow in Scotland.

tackle climate change, encourage energy efficiency, and advance renewable energy solutions.[3]

In particular, remunicipalization processes in the energy sector have the potential to create significant momentum in combating climate change. If the climate challenge is to be tackled seriously, the evidence suggests that forms of public ownership—at the national level and increasingly at the local scale—are likely to be critical. Across the energy sector, privatization not only is creating security-of-supply problems for many locales, but it fails to address the long-term investment needs required to convert energy infrastructure to low-carbon systems. In Germany, remunicipalization (*Rekommunalisierung*) has played a key role in facilitating the country's energy transition (*Energiewende*), with implications for emerging remunicipalization processes around the world.

Remunicipalization as a Global Push-Back Against Privatization

The concept of "remunicipalization" describes a growing trend by local and regional governments to take back utility sectors into public ownership, following growing resistance to the global privatization agenda that international bodies such as the International Monetary Fund, the World Bank, and the European Commission have promoted as part of their conditions for loans or other external support. Although the term has been used broadly to define any reversals of privatization (even where assets and resources remain partly in the private sector), remunicipalization is best understood as a process describing "the passage of services from privatization in any of its various forms—including private ownership of assets, outsourcing of services, and public-private partnerships (PPPs)—to full public ownership, management, and democratic control."[4]

So far, the remunicipalization debate has focused mainly on the water and sanitation sectors, where, since 2000, 235 towns and cities worldwide have taken back services and assets into local public ownership. Although the trend is global, it has been especially prominent in France (94 cases) and the United States (58 cases), with some of the early iconic examples happening in Latin America. In Bolivia, after privatization efforts in the late 1990s resulted in the raising of water rates, the outbreak of "water wars" in Cochabamba and La Paz spurred local remunicipalization efforts, which then spread to other cities and areas throughout the region.[5]

Worldwide, cities as diverse as Berlin (Germany), Bordeaux (France), Dar es Salaam (Tanzania), Houston (Texas, United States), Paris (France), and various municipalities in Malaysia have followed suit. In Mali and Uruguay, national water services have been returned to public hands after failed privatization experiments. In many Latin American cities, new and more-participatory models of local public ownership have been developed with the support of hybrid organizations that combine local government, trade unions, and sometimes residents' cooperatives. (See Box 16–1.)[6]

Remunicipalization has been on the increase in other sectors as well, although so far it has become a major trend only in Germany's energy sector. Worldwide, in sectors as diverse as electricity, transport, waste management, cleaning, and housing, local government authorities have been able to bring services back "in-house," typically following the expiration of an existing contract. (See Table 16–1.)[7]

Perhaps the most celebrated example of remunicipalization is the successful 2013 referendum in Hamburg, Germany, where a grassroots citizens' campaign won the vote to take back the city's electricity grid from the Swedish utility Vattenfall. The initiative also will buy back the city's natural gas and district heating system by 2019. Other examples include the cancellation of two public-private partnerships to run parts of London's public transport system, and the buying back of the local public transport system in Kiel,

---

Box 16–1. Hybrid Public Ownership in Buenos Aires Province, Argentina

In 1999, the province of Greater Buenos Aires in Argentina, home to some 10 million people, signed a concession agreement to hand over its water services to Azurix, a company owned by the U.S. energy utility Enron. Opposition to the privatization scheme developed quickly as water rates to customers increased, water quality suffered, and promised investments in infrastructure failed to materialize.

When Enron went bankrupt in 2001 and Azurix pulled out of the concession, the provincial government decided to establish a new public company with support from the water employees' cooperative (5 de Septiembre), which already held 10 percent of the organization's shares. The cooperative, through the support of the Water and Sanitation Workers' Trade Union, was able to provide valuable technical expertise that had been lost through the privatization debacle.

The hybrid organization has been highly successful, reducing technical costs by 75 percent compared to the privatization regime. The cooperative also has provided expertise and consulting support to other local public water authorities in Latin America that are seeking alternatives to privatization.

Source: See endnote 6.

Germany. In the United Kingdom, a study in 2011 found that 80 out of 140 local authorities surveyed had taken back in-house private contracts for services as diverse as housing management, waste management, street cleaning, and information technology.[8]

Beyond the local level, renationalizations have included oil and gas in Argentina, Bolivia, and Venezuela; the energy system as a whole in Lithuania; and the Finnish government's decision to buy back 53 percent of the country's national grid.[9]

A leading driver of remunicipalization has been the failure of privatization to deliver the improvements in performance that its adherents promised. Research shows that many cities and regions that pursue privatization are faced with deteriorating services and receive none of the investment, modernization, or "know-how" that they expect from private ownership; meanwhile, users frequently face rising rates and become aware that their local taxes are subsidizing private profits, often at great public expense. A 2013 analysis of the

Table 16–1. Examples of Remunicipalization Campaigns in Various Sectors		
City	**Sector**	**Details**
Paris, France	Water	After the city's contract with the private companies Veolia and Suez was not renewed in 2010, the public entity Eau de Paris was able to reap €35 million ($37 million) in savings in its first year and to reduce water tariffs by 8 percent in its second year.
Berlin, Germany	Energy	Following a successful campaign to remunicipalize the water sector, citizens launched a grassroots initiative to take back the city's privatized electricity grid concession from the Swedish company Vattenfall. Although a 2013 referendum in favor of the takeback garnered 83 percent of the vote, the attempt failed because voter turnout was below 25 percent. The ruling political coalition has responded by creating its own public energy company with a view to taking control of the grid.
Boulder, Colorado, United States	Energy	In 2013, the city voted by a two-thirds majority to take back the city's electricity supply into public hands from the private utility Xcel Energy. However, continuing legal battles over pricing of assets and compensation have frustrated the process from taking any further shape so far.
Islington Council (London borough), U.K.	Cleaning services	Following the expiration of a private cleaning contract in 2010, the Council made the decision to take back services in-house, including paying workers a "living wage."

Source: See endnote 7.

effects of rail privatization in the United Kingdom could be applied easily to the privatization experience worldwide, describing a situation whereby "risk- and investment-averse private companies positioned themselves as value extractors, thanks to high public subsidies."[10]

In a time of fiscal austerity, many cash-strapped city councils and other local authorities that have the power to operate and manage utilities and public services are bringing back critical assets and revenue streams under public control. Some of the most significant developments in countries that have decentralized political systems have been in the United States and France (in the case of water, where governance remains largely the preserve of towns and cities) and Germany (in the case of energy). But even in the United Kingdom, where privatization has resulted in highly centralized forms of utility regulation and provision, new forms of local public and community ownership are emerging in some sectors.

The Climate Change Imperative: Remunicipalization in the Energy Sector

In addition to the problems related to performance and value capture, privatization has been associated with growing concerns about energy security and supply, as electricity blackouts from Auckland (New Zealand) to California have demonstrated. In the United States, Superstorm Sandy in 2012 reminded people about the perils of relying on older, centralized utility models after more than 8 million people in 21 states lost their electricity, whereas many local and decentralized power sources stayed on during the storm.[11]

In the European Union, the push to introduce competition in the energy sector was a strong driver for privatization in the 1990s, with internal market directives for energy and gas aiming to achieve "a complete opening of the markets while at the same time guaranteeing high public service standards and maintaining universal service obligations." However, privatization failed to deliver on its promises, as large privatized utilities were able to use their dominant market positions to enhance profitability and short-term efficiencies at the expense of investing in new capacity and infrastructure. The result has been a series of crises in supply, culminating in grid overload and power cuts across large parts of Europe in November 2006.[12]

Remunicipalization in the energy sector also has been driven by obligations to tackle climate change. As powerful government and corporate interests create obstacles to the low-carbon transition at a national scale, new urban and regional coalitions are forming to drive the environmental agenda within regions and municipalities. This is leading not only to the taking back of privatized energy companies into public hands, but also to the setting up of new public companies and community-owned enterprises, as well as to the revitalization of many existing public energy companies.

In the United States, many municipal initiatives involve new forms of public and community ownership related to clean energy and climate change mitigation. The city of Minneapolis, Minnesota, considered remunicipalizing its power system, although it opted for a partnership agreement with the private operator Xcel Energy rather than taking full control. There also have been two failed attempts at remunicipalization: in South Daytona, Florida, and Thurston County, Washington. The Center for Social Inclusion has highlighted public- and community-owned energy schemes to encourage renewable power and energy efficiency in marginalized urban areas—ranging from the City of St. Paul's district heating system to Delaware's Sustainable Energy Utility.[13]

National Capital Planning Commission

Vision for the SW Ecodistrict Initiative in Washington, D.C., which aims to transform an isolated federal precinct into a highly sustainable workplace and livable neighborhood.

Other notable U.S. initiatives include local public-private partnerships such as the EcoDistricts concept of multi-modal, low-carbon neighborhoods that was pioneered by Portland, Oregon, but that is spreading to cities such as Austin, Boston, San Francisco, and Seattle. These partnerships offer an important contrast with top-down public-private finance initiatives by attempting to involve local community actors rather than outside corporations, an approach described as "rooted in authentic collaboration that honors and respects a community's collective wisdom."[14]

The United Kingdom, too, has seen a growth in municipal energy companies in cities as diverse as Aberdeen, Nottingham, Woking, and, most recently,

Bristol, the largest British city (population of 440,000) to establish its own public energy company. Although privatized utilities still dominate the country's energy market, the emergence of decentralized forms of energy, often linked to new district heat and power schemes, is offering the possibility for cities to both meet climate change obligations and tackle fuel poverty (a situation in which residents cannot afford to adequately heat their homes) through more-efficient heating systems. However, the most significant developments in remunicipalization worldwide have occurred in the energy sector in Germany, where they are linked to the country's ongoing energy transition.

Germany's *Rekommunalisierung* Wave and the *Energiewende*

In recent years, Germany has seen a massive process of remunicipalization in the energy sector. In various parts of the sector, authorities have reversed local and regional privatization contracts and reinstated public ownership. Additionally, a new generation of local energy companies is taking advantage of the opportunities provided by renewable energy to develop more decentralized and locally autonomous forms of power. These two elements sometimes overlap: in some cases, new companies are created to take over contracts as part of remunicipalization, whereas in other cases the services revert to existing public utilities.

Reversing Privatization

Since 2000, more than 100 contracts for energy distribution networks or service delivery in Germany have returned to the public sector. As elsewhere, dissatisfaction with the consequences of privatization has accounted for most of the return of local utility companies to public hands. Many German towns and cities had privatized in response to deteriorating public finances and rising debt levels in the 1990s, only to find that the privatized services were even more expensive on a rented-back basis.[15]

An equally important driver has been Germany's strong environmental agenda. The country's retreat from nuclear power and the setting of strong national renewable energy objectives as part of the energy transition, or *Energiewende*, has led many policy makers and activists at the local level to challenge the power of the "big four" private utilities (E.ON, Vattenfall, RWE, and EnBW) and their links to carbon-based energy, particularly natural gas and coal. Fulfilling Germany's climate change obligations requires estimated investments of €25–€42 billion ($27–$45 billion) in infrastructure improvements alone. The more-progressive local politicians are realizing that only a renewal of public ownership and investment will achieve this. The country's remunicipalization efforts range from big-city campaigns, to small town and rural district initiatives, to the takeover of larger regional concerns.[16]

As mentioned earlier, the most celebrated example of a big-city campaign is in Hamburg, Germany's second largest city, with a population of 1.7 million. Hamburg was the first large city to take back its electricity grid into public ownership, following privatization by Vattenfall in 2002. After a local referendum vote in September 2013, the city council was forced to buy back the grid and to set up a new public utility to manage it. The remunicipalization campaign was particularly impressive because it was opposed by the two main political parties, the CDU and SPD, and originated from a grassroots coalition of green and left activists (Unser Netz, or "Our Network").[17]

Campaigns for remunicipalization in other major cities—notably Berlin, Bremen, Dresden, and Stuttgart—have had mixed results. In the case of Berlin, the referendum to return the grid to public ownership failed to secure enough votes. Yet these efforts all have propelled momentum for local authorities and citizens' groups to extend existing public and community-owned energy companies, or to establish new ones.[18]

Equally significant have been remunicipalization campaigns in smaller German towns and cities (see Box 16–2), with the first return to public ownership occurring in Nürnbrecht in North-Rhine Westphalia in 1996. Researchers Oliver Wagner and Kurt Berlo with the Wuppertal Institute have identified strong clusters of remunicipalization in many rural areas, such as in the countryside close to Munich, the Bodensee and Black Forest regions in the south, the Rhineland in the west, and East Westphalia-Münsterland in the north. Often, these are grassroots campaigns led by local residents that push local governments into taking action.[19]

A final form of remunicipalization involves the taking back of regional or pan-regional entities by German states. One of the most high-profile and controversial examples was the decision by the Christian Democrat (CDU)-led administration in Baden-Württemberg to buy back EnBW (Energie Baden-Württemberg) from the French utility Électricité de France, which had owned a 45 percent controlling share. The decision likely reflected narrow political self-interest rather than any broader ideological or environmental agenda, but it did not stop the CDU from being ejected from office after nearly 60 years in power and being replaced by a Green-led administration.[20]

Two other large utilities, Steag and Thüga, have been remunicipalized

Box 16–2. The Pioneering Remunicipalization Town of Wolfshagen, Germany

An oft-quoted German example of remunicipalization is the town of Wolfshagen (population of 14,000) in the state of Hessen, which has won a federal government award as an "energy-efficient town." The local town council took back the grid from the private utility E.ON Mitte in 2006. Although the original contracts were for 20 years, a break clause in the contract allowed the town to bring the network back into public ownership after 10 years. As in many parts of Germany, Wolfshagen retained a small energy-producing public company, which gave it the technical expertise both to strike a tough bargain with E.ON and to devise a new strategy to promote renewables.

The town initially had a contract with an Austrian hydroelectric supplier to produce 100 percent of its electricity from renewables, but Wolfshagen's aim is to be self-sufficient in renewable energy by the end of 2015, realized through the construction of five wind turbines and a 42,000-panel solar park, completed in 2012. Two-thirds of the town's energy now comes from wind, with the remainder from solar and biomass. The form of public ownership—part local council and part cooperative (with a community cooperative created to give local residents a 25 percent stake)—is also typical of the demand to share revenues and encourage greater civic engagement.

Source: See endnote 19.

through trans-municipal alliances—in the case of Steag, the Rhine-Ruhr region involving Dortmund, Duisburg, Bochum, Essen, Oberhausen, and Dinslaken, and, in the case of Thüga, a consortium of municipal utilities led by the cities of Frankfurt, Hannover, and Nuremberg. Another example of regional remunicipalization is the purchase of E.ON Mitte by the states of Hessen, Niedersachsen, and Rheinland-Pfalz for €617 million ($656 million). Such examples are not without their problems and seem less aligned to energy transition objectives than to the need for hard-pressed local and regional governments to recapture lost revenue streams.[21]

A New Generation of Local, Collectively Owned Energy Companies?

Alongside the return of privatized assets to local public sector control, Wagner and Berlo identified 72 new public energy companies that have been established in Germany since 2005. As impressive as the scale of new enterprises is the diversity and innovation in forms of collective ownership. These range from the creation of new local state-run entities, such as Hamburg Energie, set up in 2009 by the Green Party in coalition with the CDU in the city government; to Stuttgart's new municipal utility, created in 2011; to smaller-scale rural cooperatives (*Genossenschaft*) throughout the former West Germany. Fewer examples exist in the former East Germany, where a lack of a history of civic engagement and local mutualism seems to be an important barrier to the emergence of more-grassroots energy initiatives.[22]

Hamburg Energie (HE) offers a good example of how state action can supplement effective grassroots mobilization to facilitate a low-carbon transition locally. In the six years since HE was established, the utility has grown its electricity supply business to more than 100,000 customers and is now operating at a small profit. Envisaged as the vehicle to shift Hamburg toward a 100-percent-renewable electricity and heat supply, HE has begun to invest in its own power sources, including six wind farms within the city's boundaries and 10 megawatts of solar photovoltaic capacity. However, HE also represents an example of the continuing political and economic interests that can block energy transition. Despite the successful referendum campaign to take back the grid, the ruling Social Democrats (SPD) and energy trade unions still have strong vested interests in coal-based power plants and strong ties with Vattenfall. The new company that the SPD established to operate the grid is not integrated with HE, posing potential problems for creating more-holistic energy policies that also address energy efficiency, carbon reduction, and fuel poverty issues.[23]

Similar blockages are frustrating efforts elsewhere in Germany, where many local and even state governments, especially in the most populous state of North Rhine-Westphalia, have strong interests in carbon and nuclear industries. The trade unions and SPD in particular are viewed as playing regressive roles in frustrating both remunicipalization and the energy transition. As one activist involved in the Berlin remunicipalization initiative remarked, "The big energy firms are basically in bed with the Social Democrats in much of the established energy sector in Germany and are very cosy with the main unions."[24]

More promising are a number of cross-communal initiatives that have created new utilities at the regional level. These include Hochsauerland Energie GmbH, created in 2009 and involving four smaller towns in North Rhine-Westphalia, and the Regionalwerk Bodensee, created in 2008 by seven municipalities along Germany's southern border with Switzerland. Rather different, but equally interesting, are ongoing discussions—involving medium-sized towns such as Marburg and Göttingen—about creating new hybrid municipal and cooperative energy companies by breaking up the newly remunicipalized E.ON Mitte.[25]

Below the municipal level, more than 800 smaller community cooperatives have been created in recent years, investing some €1.3 billion ($1.4 billion) in new renewable energy projects. The German government's feed-in tariff has provided a major boost to individual ownership—which, in 2012, represented 35 percent of the country's installed renewable power capacity, of which cooperatives represented 21 percent—whereas the big four privatized utilities accounted for a much smaller share (around 5 percent). (See Figure 16–1.)[26]

A final key ingredient of the remunicipalization process has been Germany's decentralized and largely socially owned banking sector, with funding for renewable energy projects coming primarily from local state-owned banks (Sparkassen and Landesbanken) and cooperative banks (Volksbanken). Cooperative banks in particular have been natural supporters of community energy schemes, sharing many of the same values

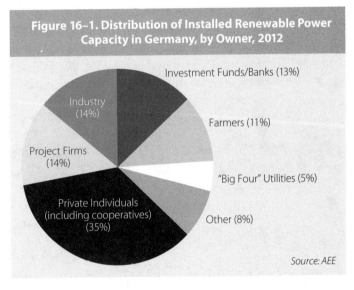

Figure 16–1. Distribution of Installed Renewable Power Capacity in Germany, by Owner, 2012

Investment Funds/Banks (13%)

Industry (14%)

Farmers (11%)

Project Firms (14%)

"Big Four" Utilities (5%)

Private Individuals (including cooperatives) (35%)

Other (8%)

Source: AEE

and ethics to promote local and community-centered forms of development and providing up to €30 million ($32 million) for local cooperative renewable energy schemes.[27]

Regional state banks also have played an important role in providing investment funding for larger-scale environmental and renewable energy projects. In Frankfurt, the city's municipal utility, Mainova, received a loan of €100 million ($106 million) from the state Landesbank for a project to "couple" and integrate the city's varied power sources to improve efficiency and storage. The ability to borrow at interest rates of less than 2 percent means a level of "patient capital" and long-term stability for investment planning that is not available to companies that trade on stock exchanges or that are reliant on private bank loans. Similarly, the impressive expansion of the Munich municipal utility's renewable energy portfolio to become self-sufficient in renewables by 2025, including its participation in offshore wind consortia in the North Sea, has been underpinned by local and regional state bank support, which has allowed it to embark on a €9 billion ($9.6 billion) investment strategy.[28]

The German Experience and Broader Lessons

It is important not to over-romanticize Germany's process of energy remunicipalization or to take for granted its ability to continue to deliver important renewable energy objectives in the fight against climate change. Although the share of renewable energy has risen through national energy policies and the emergence and revival of local individual and cooperative ownership in recent years, there are still some major obstacles in the way of achieving transition. New municipal companies are playing an important part in sustaining and deepening the energy transition, but coal generation is still the largest contributor to the country's electricity supply, and the retreat from nuclear energy has resulted in the big four utilities increasing their production of coal. (See Figure 16–2).[29]

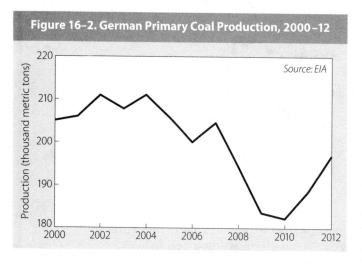

Figure 16–2. German Primary Coal Production, 2000–12

Source: EIA

The *Energiewende* has been described as "coming to a standstill," in the sense that a recent federal government bill to impose a carbon levy on the oldest coal-fired power stations was blocked by the main coal-producing regions of North Rhine-Westphalia, Brandenburg, and Sachsen, and a policy u-turn resulted in a subsidy being negotiated. At the same time, not all municipal utilities are playing progressive and enabling roles in facilitating the energy transition. Many of them, particularly in old industrial regions such as North Rhine-Westphalia, have shareholdings and therefore considerable financial interests in the four big utility companies, complicating the simple binary between public and private ownership. One also should not over-emphasize the importance of the environmental agenda: for many city and local governments, remunicipalization is aimed first and foremost at regaining control of revenue-producing assets during a period of heightened fiscal austerity and welfare retrenchment.[30]

Nevertheless, decentralized and locally owned energy systems have played a positive role in facilitating the growth of renewable energy in Germany. Unlike in the United Kingdom, where there is a very centralized grid structure and a concentrated industry structure with an effective private oligopoly throughout the energy supply chain, Germany's historically decentralized energy system based around municipal utilities and rural cooperatives has meant that local initiatives have the space to mobilize both civic support and infrastructure to enable the shift toward a low-carbon model.[31]

At the same time, the growth in local collective ownership, through both new municipal utilities and smaller cooperatives, has had a positive effect in facilitating public engagement and participation in the transition process. Again, it is important not to overstate the degree of commitment of the average citizen to a radical environmental politics in a country where the large BMW or Mercedes is still a strong symbol of consumer identity and where the income-generating aspects of renewable energy ownership have been an important draw for private households. Nevertheless, the growth of massive campaigns and movements attests to the degree to which individual and collective ownership do draw citizens into the movement for low-carbon transition and the battle against climate change, forging both a personal commitment as well as important socialized and collective learning processes around environmental goals. This, again, can be contrasted with more centralized and privatized systems, notably in the United Kingdom, where, as one commentator describes it, "the depoliticisation of energy policy has resulted in embedded corporate power, a widening disjuncture between experts and majoritarian institutions and limited knowledge structures."[32]

Although many of the political and economic features of Germany's experience reflect a distinctive national flavor, there are broader themes that can be deployed elsewhere. Most countries have cooperative traditions—both in the banking sector and in other aspects of society, particularly in rural and agricultural areas—that can be enrolled in transition policies, give the right state levers and regulatory support. In Denmark, for example, massive advances in renewable energy have been achieved through the combination of national government subsidies, market incentives, and legislation to promote both small-scale and localized production and ownership. Even the United States has strong cooperative and mutualist traditions that can build momentum—with the right government support.[33]

Conclusion

The global remunicipalization wave has important potential for enhancing public engagement with climate change and achieving key targets in the form of low-carbon transition and renewable energy production. At its most progressive, it offers more democratic and participatory forms of public and collective ownership of essential resources such as water, energy, and transport that can challenge marketized and commodified values and provide more humane and environmentally driven agendas around social need and the common good. As the damaging effects of privatization become evident, beyond attempts by hard-pressed local governments to regain control of key services and assets, there also is the potential for the emergence of new and innovative models of public organization and ownership. The global extent of remunicipalization processes indicates the potential of the phenomenon across geographical boundaries and among very different national political-economic cultures and trajectories.

The example of Germany's *Rekommunalisierung* process in the energy sector provides inspiration for what can be achieved with the right institutional structures, support mechanisms, and political mobilization. But it also highlights the continuing blockages that exist at the national level, where—as in many large and advanced countries—strong vested interests in privatized and corporatized carbon-based economic sectors can thwart the progress toward lower-carbon and clean energy transitions. On the positive side, remunicipalization highlights the potential for local actors to initiate important energy transition projects, as well as the trend toward growing trans-local collaboration among cities and towns to, in part, sidestep—but also push—national and state actors along more progressive pathways.

Notwithstanding such developments, important battles remain to be fought in the years ahead, not least in tackling the continuing neo-liberal marketization and competition agenda that dominates national and supranational government policy agendas within the European Union, the United Nations, and other key institutions. It is important to develop new and decentralized forms of public ownership that engage citizens and social movements in the battle against climate change from the bottom up, rather than allowing public agendas to be captured by vested interests. Older forms of centralized and top-down state ownership—such as those developed in France and the United Kingdom—effectively closed down public debate and set in train devastating nuclear and carbon-based solutions to the problem of energy supply. Although remunicipalization and other forms of local collective ownership, such as cooperatives, do not hold all the answers, and higher-level state strategic planning is still required, they do at least encourage public engagement, collective learning, and vibrant discussion about the future of the utility sectors and how they might be reorganized to tackle climate change.[34]

CITY VIEW

Portland, Oregon, United States

Brian Holland and Juan Wei

Portland Basics

City population: 609,456

Metropolitan area: 376 square kilometers

Population density: 1,621 inhabitants per square kilometer

Source: See endnote 1.

A Portland light rail train decorated for the MAX Orange Line grand opening in September 2015.

Sam Churchill

U.S. Leader in LEED-Certified Buildings and Biking Infrastructure

The City of Portland has created and implemented strategies to reduce greenhouse gas emissions for more than 20 years. In the early 1990s, it became the first city in the United States to adopt a comprehensive carbon dioxide reduction strategy. In 2001, Multnomah County (the most populous of Oregon's 36 counties) and the City of Portland (which is the seat of Multnomah County and Oregon's largest city) passed their joint Local Action Plan on Global Warming.

In 2009, Multnomah County and Portland adopted an updated climate action plan (CAP) with expanded categories for actions and more-rigorous reduction targets. The plan identifies 93 action steps in 8 categories to reach its emissions reduction goals, ranging from curbside pickup of residential food scraps to expanding the city's streetcar and light rail system.[2]

Thanks to strong government leadership, science-informed policy making has long been practiced in Portland. To avoid the catastrophic consequences of climate change, the city set its latest emissions reduction target by referring to current science from the Intergovernmental Panel on Climate Change (IPCC). Portland adopted an 80 percent emissions reduction target by 2050, with an interim goal of 40 percent by 2030. In line with IPCC recommendations, 1990 was set as the baseline year for the reduction target.[3]

Expanding Transit and Biking Options

Portland has developed a broad set of policies and programs to achieve its ambitious emissions reduction targets. Some measures far predate the concern about the changing climate but offer important tools in this fight. As early as the 1970s, Oregon adopted a statewide land-use policy to prevent urban sprawl by establishing urban growth boundaries. Guided by this policy, cities were encouraged to develop more-dense urban neighborhoods while preserving farmland and wilderness. This successful policy set the stage for a series of effective greenhouse gas emissions reduction programs in Portland.[4]

With a focus on development that aims to provide accessible transportation options to people within its city limits, Portland has made the expansion of streetcar and light rail systems a priority in the past several decades. Since 1990, Portland has added four major light rail lines (with a fifth under construction) and the Portland Streetcar. Construction is near-

The content for this City View is adapted from Mike Steinhoff et al., *Measuring Up 2015* (Washington, DC: WWF-US and ICLEI USA, 2013). The report analyzes data from 116 local governments in the United States and uses in-depth profiles to highlight four U.S. cities (Atlanta, Cincinnati, Minneapolis, and Portland) that have set particularly ambitious targets.

ing completion on the nation's first multi-modal bridge that is off-limits to private automobiles, which will carry bikes, pedestrians, and public vehicles over the Willamette River.[5]

In addition, Portland now has 513 kilometers of bikeways, including 95 kilometers of neighborhood greenways; 291 kilometers of bike lanes, cycle tracks, and buffered bike lanes; and 127 kilometers of dedicated bike paths. Portland received the League of American Bicyclists' highest rating for being a bicycle-friendly community. In addition, *Bicycling* magazine designated Portland as the number-one bike-friendly city in the United States.[6]

This cable-stayed span across the Willamette River, named Tilikum Crossing, has the distinction of being the only bridge in the United States dedicated to light rail, buses, bicycles, and pedestrians only.

As a result of these efforts, Portland drivers travel fewer vehicle miles than those in most other similarly sized cities. Transit ridership has more than doubled in the past 20 years (totaling 100 million rides in 2013), and, today, at least 12,000 more people bike to work daily in Portland than in 1990. Six percent of Portlanders commute to work by bike, nine times the national average. Although the population of Portland has increased 31 percent, gasoline sales have decreased 7 percent compared to 1990.[7]

Building Greener and Smarter

In addition to providing more transportation options, Portland has implemented a series of clean energy and energy efficiency programs. A strong focus on green buildings has led to more than 180 certified green buildings. Data for 2012 show that Portland had more LEED Platinum-certified buildings than any other city in the United States. The city also is expanding the use of solar energy in its facilities and neighborhoods; the number of solar energy systems has increased from only 1 in 2002 to 2,775 today.[8]

Portland's energy efficiency program, Clean Energy Works (CEW), was started in 2009 with 500 pilot homes. Aimed at reducing energy consumption by 10–30 percent, CEW provides long-term, low-interest financing to homeowners for whole-home energy upgrades, with on-bill utility repayment of the loan. Because of its innovation and success, CEW attracted $20 million from the U.S. Department of Energy to scale up the pilot into a statewide effort.[9]

Mike Boucher

Stormwater drainage system on the campus of Portland State University.

The program has realized multifaceted benefits. As of April 2014, more than 3,700 homes in Oregon had been upgraded for energy efficiency. These upgrades help avoid more than 5,000 tons of greenhouse gas emissions each year, equal to powering nearly 500 homes for one year. Meanwhile, the program has generated $70 million in economic activity and created some 428 jobs.[10]

Stormwater, the runoff created by rainfall, is another challenge faced by modern cities. Like many older cities, Portland has a combined stormwater and wastewater system, which has resulted in the pollution of local rivers and streams when high storm volume causes the system to overflow. To protect rivers and natural systems, Portland voted to enforce a series of policies that promote green infrastructure, including requiring all new construction to manage 100 percent of stormwater on-site through structures such as green streets and green roofs.[11]

Thanks to these new policies and the city's ongoing promotion of green roofs, a number of buildings and structures in Portland now have living, vegetated roof systems that decrease runoff and offer aesthetic, air quality, habitat, and energy benefits. Portland is now home to more than 390 green roofs, covering nearly 8 hectares of rooftops. The city also has invested heavily in green infrastructure, such as rain gardens and bioswales (landscaping elements designed to remove silt and pollution from surface runoff water), with more than 1,200 such facilities in the public right-of-way. Portland uses green infrastructure to manage millions of liters of stormwater each year.[12]

Recycling, Saving Energy, and Creating Jobs

Portland also is a national leader in recycling efforts. It has a 70 percent overall recycling rate for residential and commercial waste. Due to the addition of a weekly food scrap composting service and a shift to every-other-week garbage collection in 2011, residential garbage taken to the landfill has decreased by more than 35 percent, and collection of compostable materials has more than doubled.[13]

Leading by example, Portland also has been setting more-aggressive emissions reduction targets for its own operations. Through efficiency improvements, including traffic lights, water and sewer pumps, and building lighting systems, the city has realized energy savings of more than $6.5 million a year, which adds up to around 30 percent savings in Portland's annual electricity costs.[14]

Contrary to the widely held assumption that pursuing emissions reduction goals will likely slow down the local economy, the experience in Portland shows that climate actions have reduced the cost of doing business and created more-equitable, healthier, and livable neighborhoods. The number of green jobs is growing in Portland. More than 12,000 jobs in the city can be attributed to the clean technology sector, including green building, energy efficiency, and clean energy. Portland also is a national leader in innovative bicycling product manufacturing and services.[15]

Portland's emissions reduction programs have been successful. Local greenhouse gas emissions in 2013 were 11 percent below 1990 levels (equal to a 32 percent per capita reduction), and Portland homes now use 11 percent less energy per person than in 1990. With all of these efforts and achievements, the City of Portland became one of the 16 local jurisdictions across the United States to receive recognition as a Climate Action Champion from the White House in 2014. In the same year, Portland was among 10 cities worldwide to receive the C40 Cities Climate Leadership Award for its Healthy Connected City strategy. The award honors cities all over the world for excellence in urban sustainability and leadership in the fight against climate change.[16]

Moving Forward

Multnomah County and the City of Portland are in the process of reviewing and revising their 2009 climate action plan. Building on previous successes and lessons learned, the 2015 update incorporates recommendations for action and social equity into the development process.

For the energy program, the city is planning to advance net-zero energy buildings and to require energy disclosure for large commercial buildings. The focus on solar and low-carbon fuel sources will remain, and efforts to encourage the adoption of electric vehicles will be enhanced.

Portland has adopted a set of Sustainable City Principles to guide daily operations by city agencies, officials, and staff. In addition to promoting greener choices in city procurement, these principles seek to balance environmental quality, economic prosperity, and social equity, and to encourage thinking beyond first costs and consideration of the long-term, cumulative impacts of policy and financial decisions. They encourage innovation and cross-bureau collaboration; engage residents and businesses in the promotion of more-sustainable practices; and include measures in favor of a diverse city workforce and ensuring equitable services to communities of color and other underserved communities.[17]

The city now is seeking reductions in global lifecycle emissions from consumption. Lifecycle emissions are those created by the production and use of products, from furniture to computers to appliances. For this, Portland has taken the innovative step of measuring lifecycle emissions generated through consumption by households, public agencies, and businesses. The consumption-based inventory revealed that Portland's global greenhouse gas emissions are double the in-boundary emissions traditionally measured.

Portland is planning to increase its efforts in this area and to find an effective way to communicate these findings to the local community. There also is a need to help businesses and residents better understand that their consumption choices contribute significantly to global emissions.[18]

Portland recognizes that cities around the country and the world need to collaborate more in order to succeed in their efforts to reduce urban climate impacts. In June 2014, Portland was one of 17 cities worldwide to launch the Carbon Neutral Cities Alliance, which is committed to achieving aggressive long-term carbon reduction goals. The Alliance aims to strategize how leading cities can work together to attain emissions reductions more effectively and efficiently.[19]

Brian Holland is Director of Climate Programs at ICLEI–Local Governments for Sustainability USA, and **Juan Wei** is a former research fellow at ICLEI USA.

CHAPTER 17

The Vital Role of Biodiversity in Urban Sustainability

Martí Boada Juncà, Roser Maneja Zaragoza, and Pablo Knobel Guelar

Cities often are perceived as the antithesis of nature, as places where plants, animals, insects, and their homes are relegated to the margins of human activity. Composed mainly of non-living materials and showcasing non-biological systems, such as subways and skyscrapers, they have long stood almost as monuments of human disregard for the natural world. But this is an increasingly dated concept. Many citizens and city planners now understand the importance of nature—the ecological component of urban life—as an aesthetic amenity and as an important environmental and economic asset.

In recent decades, advances in the environmental sciences have created new educational and conceptual toolboxes for understanding the place of nature in cities. Since the 1992 Rio Earth Summit, for example, urban biodiversity has become a sustainability indicator, and the importance of urban "green governance" is increasingly apparent. In the coming years, cities will be challenged to operationalize these advances by giving nature a thriving space that favors biodiversity and that weaves natural functions throughout city life. In a sustainable city, not only wildlife and ecosystems, but also humans and their well-being, can flourish.[1]

In considering the role of urban biodiversity, the concepts of "naturation" and "naturalizing cities" can be used to describe initiatives aimed at blending nature more broadly and deeply into urban life. Using biodiversity as one of the central measures of urban sustainability makes it possible to outline ways

Martí Boada Juncà is a senior researcher and a professor in the Department of Geography at the Institute of Environmental Science and Technology (ICTA) at the Autonomous University of Barcelona (UAB). **Roser Maneja Zaragoza** is a senior researcher at ICTA, and **Pablo Knobel Guelar** is a junior researcher at ICTA.

to promote and assess urban biodiversity, the ecosystem services that it provides, and broader biological functions in cities. Cities in the Mediterranean region, such as Barcelona (Spain) and Jerusalem (Israel), offer examples of this approach (see City Views, pages 257 and 311), and a newly devised "urban green governance index" can serve as a toolbox and guideline for managers of urban biodiversity both in the region and beyond.

Cities as Ecosystems

An ecosystem is a functional unit of the physical environment that serves a community of organisms, featuring the flows and exchanges of matter and energy among them. Ecosystem diversity is one of three key dimensions of biological diversity, along with genetic diversity and species diversity. Healthy ecosystems enhance the robustness of both genetic and species diversity, allowing for a broad mix of plants, animals, and other organisms.[2]

An urban ecosystem extends across an entire city and includes biological as well as built elements. It is characterized by flows and exchanges of matter and energy, and its biological components provide important "ecosystem services" to the city as a whole. (See Box 17–1.) But urban ecosystems differ from natural ecosystems in an important way: whereas a natural ecosystem is largely self-sufficient in materials and energy, an urban ecosystem depends heavily on outside sources of energy and matter. In Mediterranean cities, for example, energy comes mainly from oil wells in the Middle East (although many cities are increasingly embracing renewable energy), and materials such as food and wood are imported from throughout the region and the world.[3]

In an urban context, it makes sense to think about biodiversity not merely as an isolated sector of urban activity, but as a sector that is present throughout the entire city. It is more accurate to say that cities *are* ecosystems than that cities *have* ecosystems. In large part, urban ecosystems are shaped by the built environment: urban structures influence physical parameters such as temperature, wind flows, greenhouse gas concentrations, and pollutants, among others—which, in turn, determine the kind of urban biodiversity present. In the face of change, urban ecosystems can be more complex and more vulnerable than global ecosystems.

Urban ecosystems consist of three subsystems: green (all living matter in natural soil), gray (built-up areas), and blue (coastal zones, rivers, standing water, and fountains). Each can be divided further into specific biotopes— living spaces that provide suitable conditions for the development of certain

Box 17–1. Let Nature Do the Work

Cities are home to more populations of animal and plant species than we might think. They provide habitat for about 20 percent of the world's bird species and 5 percent of vascular plant species. The *Cities and Biodiversity Outlook* estimates that 34 of the world's biodiversity "hotspots"—areas with exceptionally high biodiversity that have lost at least 70 percent of their original habitat area—are located in cities, including Brussels (Belgium), Curitiba (Brazil), New York City, and Singapore (see City View: Singapore, page 211).

Biodiversity is the foundation of ecosystem services that we often take for granted. Ecosystem services are important not only in forest and rural areas, but also in urban areas, where we are highly dependent on them. In cities, we need trees for shade and for their aesthetic value, and we need parks for stormwater management, physical activity, and stress recovery. The advantage of ecosystem services is that they provide several benefits simultaneously, compared to technical solutions that often focus on solving one aspect at a time.

For example, unlike traditional stormwater systems, which serve the single purpose of handling and storing water when it rains, nature can effectively handle stormwater while simultaneously providing many additional benefits. On a rainy day, a park with 10,000 square meters of lawn has the potential to delay 75,000 cubic meters of stormwater. On a sunny day, when the technical stormwater system is not functioning, the park serves as a recreational area for playing soccer, picnicking, or recovering from a stressful day at work. Numerous studies show that people who spend time in parks or live in neighborhoods with trees have lower perceived stress levels than people who do not. Spacious and serene areas with natural characteristics also seem to encourage more physical activity, prevent mental disorders, and promote children's development.

Large parks and natural environments in cities not only are great places for humans, but also serve as important habitats and feeding areas for plants, birds, and pollinators. A single oak tree can function as the habitat for several hundred species. These species, in turn, provide the city with pollination, pest control, noise and air pollution reduction, and carbon sequestration.

One study estimates the value of the services provided by Eurasian jays in Stockholm National Urban Park in Sweden at $4,300 annually. Each year, the birds bury and abandon acorns in the park, which leads to a continuous regeneration of oak trees. Meanwhile, Stockholm's goshawk population kills an estimated 2,500 pigeons per year, contributing to efficient pest control. Without the free service provided by the goshawks, the City of Stockholm likely would need to increase the number of hunters employed to deal with the pigeons.

Trees, bushes, and hedges are an important aesthetic component in the urban landscape, but they also have the function of decreasing the temperature in cities. Cities

continued on next page

Box 17–1. continued

often are warmer than surrounding areas due to the heat-island effect, especially at night. Buildings and pavement absorb and store large amounts of heat that is produced in the city. During the summer, high temperatures lead to greater cooling needs and to increased levels of heat stress, which is especially severe for the elderly, children, and other sensitive individuals. Higher volumes of vegetation provide canopy cover and retain water, contributing to a cooler microclimate in built areas. The shading effects of broadleaf trees also are an efficient measure for decreasing the indoor temperature in buildings.

Trees and hedges serve as noise barriers as well, with the potential to reduce noise levels by 2–10 decibels. In cities with risk for erosion, the root systems of trees and other vegetation provide a cost-efficient and effective means of preventing landslides.

These examples highlight only a small number of the important roles that parks, vegetation, and trees play in the city. Parks and trees are part of a larger green network, and, as such, they can provide 10-fold as many benefits as traditional built infrastructure. Nature not only provides more services, but it often does so at a lower price tag. Green infrastructure, in many cases, provides a good return on investments.

—*Anna Larsson and Peter Wrenfelt, U&We Business Sustainability Consultancy, Stockholm*
Source: See endnote 3.

living organisms—or localized elements, such as trees (in the green system), sidewalks (gray), and ponds (blue). (See Figure 17–1.) In this approach, walls and buildings are as much a part of the urban ecosystem as is forested area: the green, gray, and blue systems are given equal importance, in contrast to other conceptualizations of ecosystems in cities that regard the green system as superior to the gray and blue ones.[4]

In the Mediterranean region, many everyday examples illustrate the importance of gray (or built-up) biotopes. The caper bush, a perennial plant that has rounded, fleshy leaves and large white-to-pinkish flowers, grows wild in stone walls throughout the cities of Rome, Italy, and Amman, Jordan. Common geckos can be spotted in urban infrastructure across Amman and Barcelona. And the Wailing Wall in Jerusalem is one of the world's most important nesting sites for swifts, a family of small birds with slender bodies and long, curved wings. (See City View: Jerusalem, page 311.)

The idea of cities as ecosystems dates to at least 1925, when the Chicago School of Sociology sought to translate ecological terminology to urban sociology. But it wasn't until the 1970s that ecologists started to see cities as ecosystems. The influence and acceptance of this new line of thinking, known

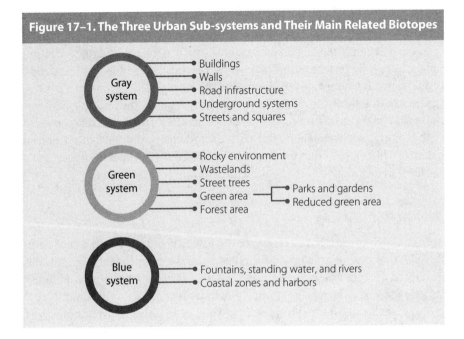

Figure 17–1. The Three Urban Sub-systems and Their Main Related Biotopes

as urban ecology, led disciplines other than ecology—such as urbanism, environmentalism, and geography—to become involved in the study of cities using a holistic and socioecological approach.[5]

The city is not a marginal environment for biodiversity: in certain geographic areas, some species find their refuge in highly urbanized settings, such as Alpine swifts that have found conditions in city buildings to be equal to or better than those on mountain cliffs. Peri-urban spaces, located in the transition zone between rural and urban environments, also allow for the existence of biodiversity, and some even become unique ecosystems within the biogeographical areas where they are located.[6]

Species that are capable of adapting to human interaction tend to be more successful in urban areas than species that are suited primarily to natural environments. Like many cities, Barcelona has devoted resources to reintroducing peregrine falcons by promoting a participatory project that enables citizens to observe nesting sites remotely and to post information on local sightings and behaviors. (See City View: Barcelona, page 257.) The falcon chicks born in the city are living evidence of the influence that this kind of initiative can have in boosting biodiversity in urban settings. The reintroduction's success also is an excellent example of the role that city landmarks can play in providing breeding and sheltering sites.[7]

A city's dependence on outside resources, such as energy and materials, not only affects the production of these resources, but also influences the design of the urban landscape, given that resources enter the city from varying distances. Getting these resources to cities quickly requires the building of roads, ports, and other infrastructure, a process that often results in fragmentation of the natural environment. Cities are thermodynamically isolated systems of balance, which means that they are self-organized at the expense of increasing the level of disorder or entropy in the surrounding environment.[8]

"Naturalizing" a City

To increase the presence and resilience of a diversity of species, cities can "naturalize"—that is, support a broad variety of natural elements—through specific, grounded actions. These "naturation" projects, aimed at attracting wild biodiversity (especially beneficial animals) include the creation of feeding, breeding, and sheltering sites within the city, whether in green, gray, or blue areas. Establishing urban infrastructure, such as parks and gardens, is a common naturation tactic, but activities also can include the creation of green roofs, walls, façades, and balconies. The result is an expansion in the number or area of ecosystems within a city that can function autonomously, without human input.[9]

The Mediterranean region, because of its biological richness and other unique characteristics, is home to many possibilities for naturation. (See Box 17–2.) Examples of strategies that bring wild animals into the region's urban settings include Jerusalem's Gazelle Valley, a large nature site near the city center where gazelles and other

Marti Boada

Tree of Heaven in Collserola Natural Park, an example of an Asian species that has become invasive throughout Barcelona.

Although the Mediterranean Sea represents less than 1 percent of the world's ocean surface, the region surrounding it is home to more than 10 percent of known species, including many found nowhere else on Earth. The huge numbers of endemic plants (around 13,000 native plants), animals (46 percent of reptiles are endemic, as are 25 percent of mammals, 3 percent of birds, and 2 out of 3 amphibians), and freshwater fish (some 250 species) represent a remarkable diversity of life. However, rising human pressure from overfishing, pollution, coastal development, unsustainable tourism, and increased sea traffic poses a serious threat to the region's biological health.

Located at the continental intersection of the southern coastline of Europe, North Africa, and western Asia, the Mediterranean region is very diverse biogeographically. It includes marine areas and terrestrial zones that traditionally have been recognized as important stopovers for migratory species. The region can be described using three primary characteristics: its history, its diverse landscapes, and its climate.

Human history: The Mediterranean region is an example of environmental and cultural hybridity, combining diverse landscapes with a mixture of cultures and religions. Considered one of the main cradles of civilization, the region illustrates complexity at varying levels. Historical examples include the remains of the Roman city of Barcino in Barcelona and the villas (such as Villa Medici and Villa Borghese) scattered throughout Rome. Still-living displays include Gethsemane Gardens and its ancient olive trees in Jerusalem's Mount of Olives, and the Royal Parks in Amman.

Diverse landscapes: The region's predominant habitats are woodland and scrub, dominated by low-water species that have a high tolerance to stress. The best example of Mediterranean adaptation is the holm oak forest, present in many of the existing ecosystems. Among the region's diversity of landscapes, three habitats have particular socio-ecological significance: cork oak forests in Tunisia, which represent the region's highest evolutionary expression of forest; the arid region of maquis shrubland in the south of France, which is a clear example of climatic hardiness; and vineyards in the Priorat Region of Catalonia, which exemplify a quality landscape at the human scale.

Climate: The Mediterranean climate is characterized by two main elements: strong seasonality in the distribution of temperatures and highly unpredictable rainfall. Summers, which generally are warm and dry, bring conditions of water and heat stress to ecosystems.

In contrast with the urban sprawl found in many countries in the world, especially in the United States, Mediterranean cities usually are characterized by complexity and compactness (as exemplified by Rome). The proximity of uses and functions facilitates a more sustainable mobility model by reducing the need for cars and easing pedestrian movement.

Source: See endnote 10.

The Jardí Tarradellas, a green building at an important intersection in Barcelona.

species are being reintroduced, and Barcelona's Jardí Tarradellas, a "green wall" structure in the city's Eixample district that is home to extensive bird life.[10]

Naturation also includes creating natural connectors that criss-cross a city and that link to natural areas outside the city, all in support of habitats. The naturation process adapts the classic models of "corridors" and "patches" used in the discipline of landscape ecology to the urban environment, with streets and avenues being the corridors and parks being the patches. By promoting a resilient network of habitats and of feeding and breeding sites, naturation initiatives stimulate the entry of biodiversity from so-called recharge nodules, or areas near the city that have a high level of naturalness that nourishes the city's biodiversity. By connecting the city with these areas, the naturation process essentially blurs the line between city and nature.[11]

It also is necessary to strengthen ecological resilience in urban green areas, incorporating natural cycles as much as possible without reducing the aesthetic quality of these spaces. This can be achieved by promoting strategies and actions aimed at connecting urban biodiversity with citizens and the entry of biodiversity in the city.

In sum, a naturation process is a key tool for promoting a city's naturalization objective, relying on urban green built-up areas as an entryway for biodiversity (mainly animals) from outside of the city system. These species find in urban biotopes (green, gray, and blue) suitable life conditions not only for their survival, but also for the establishment of consolidated populations that are highly adapted to the urban ecosystem. When naturalized, a city makes valuable green areas available for citizens while also providing and promoting urban biodiversity services.

Birds are among the most relevant indicators of these processes. Due to their high mobility, birds use street trees and avenues as corridors that connect recharge nodules with urban and peri-urban areas by providing permeability to the urban system. Those elements also support the richness of bird populations, providing feeding, breeding, and sheltering sites.[12]

Rome's Villa Borghese, a public park and garden spanning 80 hectares, serves as a remarkable recharge nodule within the city, containing majestic stone pine trees (also called umbrella pines due to the shape of their tops) that serve an important function as sheltering and breeding sites and that contribute to the park's high biodiversity. The wide range of habitats inside the park promotes the presence of many species of animals, including squirrels, hedgehogs, and frogs.

Rome's Villa Borghese, an exceptional recharge nodule inside the city with a clear predominance of stone, or umbrella, pines.

Animals living in urban ecosystems face less pressure from natural predators than animals living in peri-urban and surrounding natural areas, where rates of predation are higher. As a result, urban animals show reduced stress levels and a decrease in "alert distance," or the point at which an animal begins to exhibit alert behaviors in response to an approaching human. The animals' decreased stress also results in a shortened "escape distance," or the distance between an approaching human and the point at which the animal flees.[13]

Based on its presence and origin, urban biodiversity can be classified into three typologies: captive, induced, and drawn. Captive fauna are those animal populations that continue to live in longstanding, pre-urban habitats that are predominantly green, with examples being certain birds, amphibians, and squirrels found in ancient gardens, vestigial forests, courtyards, and private gardens. Captive fauna are a qualitative indicator of sustainability, as they demonstrate a historical sustainability that has enabled a species to maintain its presence in an urban area despite a city's growth over time.[14]

Induced fauna, in contrast, are animals that exist as a result of human activities and installations that have favored the presence of certain species that originally are from other habitats (and even other continents). An example in the Mediterranean region is parrots that have escaped from captivity. Induced fauna challenge the resilience of the urban system in the face of a new living organism.

The third urban biodiversity type, drawn biodiversity, are those people-loving species, such as sparrows, that are linked symbiotically to human activities, taking advantage of available resources and materials flows without causing either negative or positive effects. They demonstrate the non-aggressive, tensionless relation between today's urban culture and the existence of spontaneous biodiversity.[15]

Urban Biodiversity Services

Urban biodiversity is an important indicator of quality of life. An increase in biodiversity improves the quality of the environment and enhances the quality of life for humans. Many studies have established a positive relationship between urban naturalization processes and citizens' well-being. Others note that merely viewing nature results in a more relaxed physiological state, reduced stress levels, increased satisfaction and personal well-being, decreased mental fatigue, and a changed state of mind.[16]

A healthy state of biodiversity has direct effects on human well-being by generating ecosystem services that are obtained directly from natural assets (soil, biodiversity, air, and water) and bring beneficial effects for people.

Marti Boada

A statue disfigured by rock-dove feces is a good example of an urban "disservice."

Three general categories of ecosystem services include: *regulating,* for example purifying air and water or mitigating floods; *provisioning,* which includes the supply of food, water, or medicines; and *cultural,* which covers aesthetic, spiritual, recreational, and intellectual benefits. (See Table 17–1.)[17]

Urban biodiversity does not always produce beneficial effects, however. Some urban species have a negative impact on human well-being. Authors have coined the term "urban disservices" to describe the negative effects derived from biodiversity. For example, just as people perceive some plants and animals as providing urban services ("beneficial species"), others—animals such as rats, pigeons, flies, cockroaches, and mosquitoes, and plants such as stinging nettles—often are perceived as disservices, in part because they are scary or unpleasant, cause domestic disturbances, or, in some cases, may carry disease.[18]

The case of wild boars in the upper areas of Barcelona, which are closer to

Table 17–1. Ecosystem Services Provided by Urban Biodiversity		
Regulating	Air filtering	Vegetation can help to reduce air pollution and related environmental and public health problems caused by transportation and the heating of buildings.
	Micro-climate regulation, at the street and city level	Cities can affect the local climate and even weather. A single large tree can transpire 450 liters of water per day.
	Noise reduction	Vegetation and open space can help to increase the distance and reduce the volume of noise from traffic and other sources that may create health problems for people in urban areas.
	Rainwater drainage	The soft ground of vegetated areas allows water to seep through, and the vegetation takes up water and releases it into the air through evapotranspiration.
	Sewage treatment	Wetland plants and animals can assimilate large amounts of nutrients and slow the flow of sewage water, allowing particles to settle out on the bottom.
Provisioning	Food supply	Urban gardens can be an important source of local vegetables, and peri-urban areas can provide food for both humans and animals.
	Medicines	Some vegetation species produce medicines.
	Shade	Trees and other urban vegetation provide shade, create humidity, and block wind.
	Smell	Some flora species, such as *Tilia* (linden), *Buxus* (boxwood), and *Lonicera* (honeysuckle), can produce pleasant smells.
Cultural	Educational/Scientific	Urban biodiversity provides environmental education, helping to connect people with nature, seasonality, and the notion of the living world and our natural origins.
	Aesthetic/Art	Urban biodiversity can be a source of artistic inspiration.
	Social	The appearance of animals, such as birds and fish, should be accounted for in recreational values. Green spaces are very important psychologically.

Source: See endnote 17.

forest areas, illustrates how a species can be perceived as pernicious. Wild boars are progressively invading the city, causing traffic disturbances, damaging urban infrastructure, laying waste to gardens and bird nesting sites, and (in a less

Raquel Baranow

Urban coyote crosses a street in Tucson, Arizona.

visible trend) carrying diseases of animal origin toward the city. Parallel cases exist around the world, such as the raccoons and coyotes in some U.S. cities, the monkeys in New Delhi, India, and the baboons in Durban, South Africa.

Urban Green Governance

In the urban context, and given the importance of the interaction between citizens and biodiversity, it is possible to rethink the functions of biodiversity using criteria relevant to human motivations. In this sense, "urban green governance" refers to naturation strategies that take place in a city, with the aim of naturalization. This governance model is complemented by a wide range of participatory processes that connect urban biodiversity with citizens. The three main motivations that cities have for supporting and cultivating urban biodiversity are naturalization, urban-dwelling biodiversity, and ecosystem services. (See Table 17–2.)[19]

In the context of today's environmental crisis, many sustainability-minded cities are choosing to support greater tree diversity, which can offer adaptability and resilience in the face of change. In some cities, modern urban planning includes specific diversification criteria for trees and their distribution. The urban forest plan of Melbourne, Australia, sets a goal for 2040 of limiting the species prevalence of trees to 5 percent, genus prevalence to 10 percent, and family prevalence to 20 percent. (See City View: Melbourne, page 155.) Similarly, Barcelona proposes having no more than a 15 percent prevalence of any single tree species. Strategic policies such as these may help cities become both more resilient and more sustainable in a global change scenario.[20]

Urban Green Governance Index

Urban biodiversity is a strong indicator of human well-being. It serves as a tool for monitoring global change and as a benchmark against which to measure ongoing city efforts to harmonize city activities with nature. In recent decades, numerous urban biodiversity indicators have been developed and used, among

Table 17–2. Motivations and Functions of Urban Biodiversity

Motivation	Functions	Main Objective(s)
Naturalization	Conservation	To preserve local biodiversity in an urban environment and protect important populations or rare species
	Connection with recharge nodules	To create natural corridors or connectors inside the city and promote feeding, sheltering, and breeding sites
Urban-dwelling biodiversity	Urban resilience	To promote adaptive responses to global change To have self-sufficient food reservoirs (e.g., urban vegetable gardens) To serve as a bioindicator of urban quality
Ecosystem services	Well-being	To improve human well-being and quality of life To provide resounding and silent landscapes through active and passive perception To contribute to high social cohesion
	Urban biodiversity services	To regulate: air filtering, micro-climate regulation, noise reduction, rainwater drainage, and sewage treatment To provision: food supplies, medicines, shade, and smell To provide cultural benefits: educational/scientific (reconnecting and reconciliating citizens with nature), aesthetic/art, and social

them the Shannon Index, the Simpson Index, and the Singapore Index on Cities' Biodiversity, also known as the City Biodiversity Index, or CBI.[21]

A more recent indicator, the Urban Green Governance Index (UGI), has been created as a tool to help policy makers and urban planners assess and manage urban green space according to the three urban biodiversity motivations (naturalization, urban-dwelling biodiversity, and ecosystem services). Naturalization includes indicators related to feeding, breeding, and sheltering (for example, referring to the state and seasonality of fruits and flowers, the production of edible fruits, pollinator-attraction capacity, species types and heights, the kind of pruning used, crown density, cavity formation, and leaf retention); urban-dwelling biodiversity includes indicators such as water requirements, susceptibility to plagues and illness, invasive species, and adaptation to climate change; and ecosystem services are divided into two main blocks: health and well-being, and urban ecosystem services.

The UGI is especially useful for cities that are home to certain urban species found commonly in the gardens of Mediterranean cities. It also is a suitable tool for helping urban managers confront global change (such as the effects of climate change) and for providing guidance for developing urban master plans. The UGI was developed based on experience gained in Barcelona, with the goal of becoming a replicable management tool in other cities in the Mediterranean region, such as Amman, Jerusalem, Rome, and Tunis, where it will be tested on a pilot basis.[22]

The UGI can be used to make cities aware of important gaps in information about their biodiversity. For this purpose, it considers two main kinds of indicators: urban model indicators, which contain relevant data on the city's socioeconomic parameters, and an urban biodiversity review, which evaluates the city's biodiversity status.

Within cities, the "relative abundance" of species—or how common or rare a species is relative to other species within the area—is important because it affects many other elements of urban ecosystems in a cross-cutting way. Cities that are more species-diverse also are more resilient toward climate change, have better offerings for feeding and breeding sites, and offer certain enhanced ecosystem services, such as enabling city dwellers to enjoy the beauty of different seasonal effects. Overall, a deeper understanding of the importance of urban biodiversity can lead to improvements in the relationship between humans and the planet, giving sustainable cities hope for the future.

The authors would like to thank the following individuals for their support with this chapter: Hakam Al Alami, Adrià Costa, Raed Daoud, Sabina Giovenale, Munther Haddadin, Francesc Maneja, Benedetto Proietti Mercuri, Franco Paolinelli, Beti Piotto, Maen Smadi, and Mohammed Zaarour.

CITY VIEW

Jerusalem, Israel

Martí Boada Juncà, Roser Maneja Zaragoza, and Pablo Knobel Guelar

Jerusalem Basics

City population: 829,900

City area: 125 square kilometers

Population density: 6,587 inhabitants per square kilometer

Source: See endnote 1.

The Old City walls of Jerusalem host a colony of rock doves.

A Shrine for Urban Biodiversity

Despite its complex biogeographical location and sociocultural dimensions, the city of Jerusalem has been successful in nurturing its urban biodiversity in line with sustainability. At between 650 and 850 meters above sea level and midway between the Mediterranean and Dead seas, the municipality of Jerusalem occupies 125 square kilometers of a plateau in the Judean Hills, 57 percent of which is built up and 43 percent of which represents non-built-up areas in and around the city.[2]

A delicate mix of religions and cultures shares the city. Jewish communities comprise the majority (61 percent), followed by Muslims (36 percent) and, with much smaller shares, Christians (2 percent) and other faiths (1.1 percent). The city is divided into the East side, occupied by the Arabs, and the West side, occupied by the Jews, where, simultaneously, the orthodox Jews live in their own neighborhoods. Some 46 percent of the city population is considered to be poor, but the share is considerably higher for orthodox Jews (59 percent) and Arabs (76 percent).[3]

The gap between East and West Jerusalem also reveals differences in biodiversity, geology, and even climate. The old city developed in a narrow strip around the watershed line, leaving the semi-arid rim of the Judean desert to the east and the fertile Mediterranean plain to the west. The modern city, requiring a wider area, grew toward the hills that surround the plateau. On the east side of the city, following the streams that spill into the Dead Sea, the slopes are arid, with limited precipitation, a rocky surface, and desert-like biodiversity. Westward of the city, the scene is very different, with streams, ridges, and Mediterranean-like biodiversity.[4]

Jerusalem's climate is affected heavily by the difference between slopes, and precipitation varies greatly between the East and West zones. Sun exposure is very sensitive to slope orientation and differs across the city, especially in areas oriented to the south and north. All of these elements have a direct influence on land use in the Jerusalem hills.[5]

Biodiversity Overview

Due to its unique biogeography, Jerusalem is considered an important site for biological diversity. The city is home to some 1,000 species, including 738 plant species, 176 bird species, 16 mammal species, 18 reptile species, and 3 amphibian species. Because Jerusalem is located at the confluence of the Mediterranean and Judean bioregions, species from both regions coexist. The differences in rainfall and sunlight allow species with different needs to find a place in the city that suits them. The most common cultivated plants are species with low water requirements, although other important plant groups also are present, such as Mediterranean tree species (pines, olives, and cypress), hardy shrubs, northern-origin plants (*Quercus pedunculiflora, Ulmus* species) and desert plants (*Agave americana*).[6]

Jerusalem's location is also important for bird diversity. Israel is located at the confluence of three continents, representing a major migration path that funnels birds into the territory. The city's green infrastructure and its wide array of habitats offer good feeding and breeding sites, making Jerusalem an attractive stopover site for migrating birds. During the migration season, more than 500 million birds can be seen in Jerusalem's skies, and the Old Testament even makes reference to historical bird migrations over Judea.[7]

Local conditions are not the only reason for the city's rich biodiversity. As elsewhere in the Mediterranean region, human interaction shapes the biodiversity mix. Two important characteristics that have important implications for Jerusalem's biodiversity are the city's age (the second temple was completed around 485 BCE) and the fact that the city commands great cultural and religious respect. Many sites that would have been repurposed if they had been located elsewhere were preserved over time, and some are important for biodiversity. The nooks and crannies of the Old City walls, for example, house a wide variety of animal and plant species and serve as a refuge for birds, mollusks, reptiles, and rodents.[8]

Some of the older olive trees in the Garden of Gethsemane.

Similarly, the Garden of Gethsemane, where Jesus is said to have prayed after the Last Supper, is surrounded by many houses of worship, including, prominently, the Church of All Nations. A recent study revealed that the eight olive trees growing in a nearby enclosed garden were planted during the twelfth century, making them among the oldest broad-leafed trees in the world, at between 800 and 900 years old. Another study assessing the age of Jerusalem's trees found that some 12 percent were more than 80 years old and 75 percent were around 50–80 years old.[9]

This biodiversity is sustained by a network of natural areas, including rocky ground, Batha (the Israeli equivalent of soft-leaved scrubland), semi-steppe Batha, Mediterranean groves, *Pistacia atlantica* forest, and wetland habitat, as well as rock and wall flora, orchards and vineyards, planted forests, traditional and contemporary agricultural areas, and roadside vegetation. A large diversity of cultivated plants complements these natural areas.[10]

Urban Green Governance

Jerusalem's total green area covers 36.5 square kilometers, of which 25.2 square kilometers is forest and 4.1 square kilometers is public open space. The approach to urban biodiversity in the city reflects two main elements: 1) the importance of bottom-up initiatives in building local understanding of urban biodiversity, and 2) a comprehensive approach to existing urban biodiversity and commitment from a wide range of international organizations.

In the last decade, to manage the city's human and urban tapestry, Jerusalem has developed the New City Urban Master Plan. Sustainability has a cross-cutting role in the plan, which guides the establishment of a sustainable transport system and the protection of significant urban nature sites. Nature protection is realized through the Urban Nature Master Plan (LBSAP), which defines urban nature as a distinct infrastructure system. The main objectives of the LBSAP are to integrate natural spaces into the city fabric, to rehabilitate ecological corridors, and to restore rare or threatened habitats.

Nature Sites in Jerusalem

The *Urban Nature Infrastructure Survey* divides Jerusalem's nature sites into four groups: open nature sites, open agricultural sites, parkland, and nature sites in built-up areas. Fourteen typologies are defined to cover all of the nature sites in the city:

Blossoming plant sites	Unique landscape formations
Mature tree sites	Mediterranean groves
Vineyards and olive groves	Orchards
Green roofs	Planted forests
Wet habitats	Sites with concentrations of insects
Additional flora sites	Sites with concentrations of reptiles
Bird sites	Sites with concentrations of mammals

Source: See endnote 11.

As one of Jerusalem's key biodiversity initiatives, a team of experts gathered vital information about 151 nature sites across the city to include in the municipal data system. The *Urban Nature Infrastructure Survey*, published in 2010, provides comprehensive written descriptions and images, as well as a species inventory. Incorporating these data into the city's administrative system and keeping the survey up-to-date allows developers, planners, and decision makers to consider nature in their activities and increases the possibilities of synergies between city development and environmental plans, as well as with education and tourism.[11]

To strengthen sustainability planning, Jerusalem has committed to a variety of international agreements, including the Kyoto Protocol, the Rio Declaration, the Durban Commitment to Biodiversity Protection, and the Rio+20 Declaration, among others. It also participates in diverse environmental networks such as ICLEI–Local Governments for Sustainability, the International Union for the Conservation of Nature (IUCN), the Urban Biosphere Initiative (URBIS), the United Nations Educational, Scientific and Cultural Organization (UNESCO), the Green Pilgrimage Network (GPN), and Local Action for Biodiversity (LAB).

LAB has played a major role in the city. Jerusalem joined the network in 2010 and already has put in place many important mechanisms and actions, including creating a stakeholder forum with representatives from the government, nature-protection organizations, and public-interest groups. LAB also has helped with a Legacy Project program focused on the Gazelle Valley, approval of the Municipal Urban Nature Master Plan, and publication of the city's biodiversity report for 2013.[12]

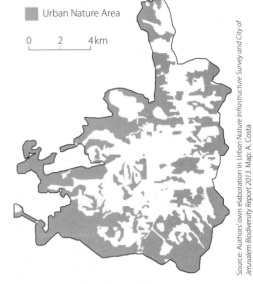

☐ Urban Nature Area

0   2   4 km

Source: Authors' own elaboration in Urban Nature Infrastructure Survey *and City of* Jerusalem Biodiversity Report 2013. *Map: A. Costa*

Of a total area of 125 square kilometers, 52 percent of Jerusalem is considered urban nature areas.

Connecting Citizens with Urban Biodiversity

Promoting nature and biodiversity in Jerusalem is undertaken through a series of civic initiatives. These include:

Jerusalem Bird Observatory. The Jerusalem Bird Observatory (JBO), located between the Israeli Parliament and the Supreme Court, was launched by community activists in 1994 as a community-based urban wildlife site. Today, it is maintained by the Society for the Protection of Nature in Israel (SPNI) through a collaboration of volunteers, educators, and researchers. As a research facility that houses the Israel National Bird Banding Center, the JBO is an ideal place to conduct conservation studies and population and migration monitoring. It also is an educational center, where city residents can participate in activities created by educators, interact with wildlife, and volunteer, heightening their participation.[13]

Wildflower Sanctuary. Once an illegal dumping site between two residential neighborhoods, the Wildflower Sanctuary is now a natural open space managed jointly by the municipality and surrounding residents. Its wide diversity of species makes it an important site for school outings and scientific research, as well as a popular recreational site.[14]

Jerusalem Green Map. The Jerusalem Green Map is an interactive online map that inventories environmentally friendly elements throughout the city. Through a participatory process, city mapping teams chart parks, gardens, green businesses and accommodations, and other environmentally friendly locales. The Green Map includes the city's 151 surveyed nature sites and allows tourists and residents to enjoy the city in a more-sustainable way.[15]

Railway Park. The park is a restoration of an abandoned railway strip that once was a dividing element between neighborhoods. When the railway stopped operating, the land became an illegal dumping site and was regarded as an annoyance. The SPNI, among other organizations, planned the park with the aim of rehabilitating the space and bringing together the communities on both sides of the track. The long, narrow park works as a green connector, rich in biodiversity and hosting separate bike and pedestrian lanes.[16]

Green Pilgrimage City. The Green Pilgrimage Network (GPN) brings together religious and secular authorities from pilgrimage sites around the world with the aim of introducing good environmental practices in order to transform these sites into models of sustainability. Jerusalem, as a pilgrimage destination for people of different faiths from around the world, is a GPN member and has undertaken initiatives to "green" the pilgrim experience. Among the city's plans are opening a new light rail system, creating a green pilgrimage map, and restoring the Kidron Valley.[17]

Pablo Knobel

A mountain gazelle in southern Jerusalem's Gazelle Valley.

Gazelle Valley. An area that was once planned for urban development in southern Jerusalem now houses the Gazelle Valley, a green lung within the city composed of many different habitats. Management of this urban wildlife park is shared between the city and community administrators, whose objective is to protect and restore the site's natural biodiversity. As the first of its kind, the park holds major importance for Israel, as its success could ease the way for similar initiatives around the country.[18]

Martí Boada Juncà is a senior researcher and a professor in the Department of Geography at the Institute of Environmental Science and Technology (ICTA) at the Autonomous University of Barcelona (UAB). **Roser Maneja Zaragoza** is a senior researcher at ICTA, and **Pablo Knobel Guelar** is a junior researcher at ICTA.

CHAPTER 18

The Inclusive City: Urban Planning for Diversity and Social Cohesion

Franziska Schreiber and Alexander Carius

Every week, about 3 million people move to cities worldwide. Over the coming decades, such migration will contribute to an increase in the urban share of the global population from 54 percent in 2014 to 66 percent in 2050. Although migration is not a new phenomenon, the current pace of rural-urban migration, both within and between countries, is unprecedented. In developing and emerging economies, this has led to the mushrooming of megacities such as Cairo, Jakarta, Lagos, Manila, and Mumbai. However, cities are not only growing in population, but also becoming increasingly diverse and ethnically heterogeneous. This twofold process poses great challenges, as cities have to manage the multi-faceted integration of their arriving newcomers into society and urban life, as well as ensure continued social cohesion.[1]

Strong integration policies are needed that support urban migrants in finding jobs, living in socially mixed neighborhoods, learning the language, and enabling their children to go to school. In addition to policies related to education, health care, the job market, housing, and finance, the ways that cities are designed and constructed are important elements of integration policy. For example, well-designed urban patterns and functioning public spaces that serve as meeting places for urban dwellers can aid in facilitating interaction, connectivity, and social mixing—all important aspects of cohesive cities.

Various urban planning and design measures can be used to strengthen the relationship between space and social integration, helping to address the challenges that cities face with respect to migration, segregation, and socioeconomic polarization. At the national level, programs and frameworks can

Franziska Schreiber is Project Manager at adelphi, an independent think tank and public policy consultancy based in Berlin, Germany. **Alexander Carius** is Cofounder and Managing Director of adelphi.

enable actions in cities and neighborhoods to improve the social and economic conditions of residents, as examples from Germany, Denmark, India, and South Africa illustrate. At the local level, city-wide and neighborhood planning can develop compact, well-connected and integrated urban patterns that facilitate social interaction and integration, as illustrated by case studies from Berlin, Germany; Guangzhou, China; Medellín, Colombia; and Oslo, Norway. Planning and design approaches that support "inclusive cities" and greater social cohesion include land-use planning, integrated land-use and transport planning, upgrading street networks, and public-space design.

Tackling Growing Urban Challenges

In our increasingly urbanized world, cities function as a melting pot for people with differing cultural backgrounds, religions, interests, and social status. In this context, cities and municipalities face the twin challenges of not only absorbing the influx of people from diverse social and ethnic backgrounds, but also counteracting the trend of rising socioeconomic polarization and the segregation of cities into privileged and disadvantaged neighborhoods.

These two challenges are often intertwined and need to be approached holistically. Although research indicates that "no intrinsic link between deprivation and ethnic heterogeneity" exists, there is ample evidence that poorly managed urban migration results in the marginalization and segregation of people with different backgrounds. Questions related to the impact of immigration and ethnic diversity on the social fabric in cities are being debated in countries across the globe. Such discussions have been particularly prominent in the context of the refugee crisis in the European Union, where hundreds of thousands of refugees from conflict-torn and fragile regions, such as Afghanistan, Eritrea, Iran, Iraq, Syria, and the Western Balkans, are seeking asylum.[2]

Although many cities and municipalities are demonstrating courage, flexibility, and creativity in organizing ad hoc accommodation, care, and food for new migrants, the long-term challenge will be to ensure their full integration into society and to create acceptance among the local population. The latter is related to the rise of xenophobia and to fears about the consequences of uncontrolled, overwhelming migration, such as added competition in the labor market or a decline of social cohesion. Such fears have arisen in many European countries in response to the influx of refugees, and local governments need to take these concerns seriously in order to counteract the prevailing perception of migration as a "problem."

In addition to managing the integration of immigrants, cities and municipalities must provide sufficient infrastructure to accommodate their growing and diversifying populations and to avoid the emergence of new inequalities in urban areas while fostering social cohesion. For example, local governments must meet the increasing demand for housing and provide sufficient infrastructure and basic services, such as electricity, water, sanitation, health care, and education. Cities in developing and emerging countries, in particular, often lack the capacity to meet these needs and are confronted with the sprawl of informal settlements and slums (and thus an intensification of social and spatial segregation). Between 1990 and 2012, the share of the urban population living in slum areas in developing regions increased from 35 percent to 46 percent.[3]

The huge demand for housing is a challenge in developed countries as well, where rental prices are rising rapidly and the amount of social housing is declining, with adverse impacts on the social structure in neighborhoods. According to a government-conducted housing survey, the social housing stock in the United Kingdom has declined from 5.5 million homes in 1980/81 to 3.8 million homes in 2010/11, suggesting that people increasingly face difficulties in accessing adequate, affordable, and secure housing. Although the United Kingdom was once a forerunner in providing public housing, this achievement has been undermined by recent polices, such as the "Right to Buy," under which millions of social-housing units were sold. As waiting lists for social housing lengthen due to the slow construction of new houses, not even half of the demand for this housing is being met, and the degree of spatial segregation between the rich and poor in U.K. cities is increasing.[4]

Policies are needed at the national and local levels to support integration and to counteract segregation through infrastructure measures. However, the work of urban planners and designers also can contribute greatly to social cohesion. Even though the reorganization of space to create more-integrated urban patterns (for example, socially and functionally mixed areas that are well connected and easily accessible) and physical interventions (such as urban design measures in public spaces) cannot solve the roots of social and economic problems, they can aid in creating more-inclusive cities. Karin Peters and her colleagues at Wageningen University in the Netherlands argue that "interactions in daily life between people across ethnic divides are one way of creating social cohesion, because they provide the basis for bonds between individuals." It therefore is important to consider what (and how) planning and design measures at different scales, including the national, city,

ghborhood levels, can foster social interaction and integration in
works.[5]

ties and countries have successfully implemented inclusive national
lans, policies, and measures that provided a "spatial fix" to social
d initiated positive locational dynamics. In Colombia, the city
edellín implemented an innovative public transport system to connect
poor and formerly inaccessible districts with the rest of the city, helping to
enhance quality of life, attract tourists, and reduce the level of crime in these
areas; however, this move did not solve the fundamental roots of poverty of
many residents. The International Organization for Migration notes that, to
achieve the greatest impact, "effective national and international instruments
and institutions also need to be put in place." Planning and design measures
should be embedded into a broader urban-cohesion policy, which involves a
range of policy approaches in the areas of education, health care, employment,
housing, and finance.[6]

From Exclusion to Interaction to Cohesion

There is a common perception that the quality of public and civic life is in
alarming decline worldwide. Since the 1970s, economic inequality has grown,
resulting in socioeconomic polarization and spatial segregation, especially in
urban areas. More than two-thirds of the urban population lives in cities where
the income gap has widened sharply in the past three decades. The level of
income inequality in these cities often surpasses the United Nations alert line of
0.4, based on the so-called Gini coefficient, which ranges from 0 (everyone has
the same income) to 1.0 (maximum inequality of income). (See Figure 18–1.)[7]

In cities of developing and emerging countries, informal and illegal settlements accommodate up to 80 percent of the urban population, and the
urban divide is often reflected in the spatial configuration of the city. But rising income inequality is also a challenge in developed countries. According
to recent studies, the degree of segregation by income has risen in 11 of 13
major European cities, including Madrid and Vienna, as well as in 27 of the
30 largest major metropolitan areas in the United States, such as Houston and
Los Angeles.[8]

Although the reasons for this trend are manifold, several key processes can
be identified. In developed countries, the main factors driving segregation
in cities are globalization, the withdrawal of government support, economic
restructuring, and the lack of investment in social housing. The transition from

Figure 18–1. Most Unequal Cities by Income, Selected Cities in the Developing World, 1992–2008

a manufacturing to a service-based economy has led to a dramatic change in the job market. Fewer employment contracts are unlimited, many people work under precarious conditions and need more than one job to survive, and the service sector is not able to accommodate all the workers that lost their jobs in the context of de-industrialization.[9]

Moreover, cities and municipalities are cutting down on social expenditures and public services, while reducing or stopping investments in social and affordable housing, with the result being a rapidly decreasing low-income housing stock. This comes mostly at the expense of already disadvantaged population groups and exacerbates the separation of low- and high-income groups in urban areas. Those residents who have the resources move to neighborhoods with better schools, while others who cannot afford the rising rents are displaced to the edge of the city. Consequently, global and local restructuring processes are closely intertwined and result in spatial patterns that reflect and accelerate inequality and exclusion in cities. (See Chapter 7.)[10]

The reasons for segregation in cities of developing and emerging countries relate mainly to the rise of the middle class, racial discrimination, provision of secure tenure, and economic liberalization, as well as to world-class city aspirations that often result in massive infrastructure and urban renewal projects with large-scale displacement of low-income or illegal groups. This prevailing trend of socioeconomic exclusion and spatial fragmentation has adverse consequences for the urban realm. As UN-Habitat explains: "[It] is impacting negatively on social cohesion and reduces the economic vibrancy and the overall prosperity of the city, including the quality of life of the citizens. Informal settlements and disconnected peripheries, dysfunctional public space and increasing insecurity are often the apparent results."[11]

Recognition of the negative impacts of exclusion and segregation calls for measures that "foster the development of a harmonious society in which all groups have a sense of belonging, participation, inclusion, recognition and legitimacy," according to researchers Gerard Boucher and Yunas Samad. Urban planners and designers can play an important role in this context, helping to support "inclusive cities" that value all people and their needs equally. The concept of inclusive cities often is approached through the lens of a particular marginalized group, such as the elderly, children, slum dwellers, migrants, the unemployed, or disabled people. Social cohesion is an important component of the inclusive city and is based on the notion of community building, cooperation, and social relations among persons of different socioeconomic and ethnic backgrounds. Urban planning and design measures at different scales can contribute to forming social ties and interaction—a prerequisite for social cohesion—and help to create a feeling of belonging in increasingly diverse and fragmented cities.[12]

National Urban Planning Programs and Frameworks

Socioeconomically deprived neighborhoods and city districts, which often are characterized by a high concentration of migrants and their descendants, cannot be understood in isolation. Their roots lie far beyond the local context. National and regional programs are needed to provide a framework for jumpstarting local initiatives and to allocate financial means for these initiatives to work. As UN-Habitat has observed, "[r]ecent experiences have clearly shown that social integration, inclusion and cohesion can be promoted through interventions at different scales."[13]

National programs are only effective, however, if they are well-designed and are supported by institutional and governance structures. A review of four national planning programs implemented in Germany, Denmark, India, and South Africa demonstrates that numerous factors determine their success on the ground. (see Box 18–1.) These factors include: the selection process for

---

Box 18–1. A Review of Four National Urban Planning Programs

Social City Program

Germany's Social City Program was established in 1999 with the objective of stabilizing and upgrading socially and economically deprived urban areas. It seeks to achieve social cohesion in often ethnically heterogeneous neighborhoods through an integrated approach that combines physical and social interventions in the target areas. As an important element of the federal urban development policy, the program was equipped with €150 million ($160 million) in 2015 (a significant increase from previous years) and had funded 659 actions in 390 cities as of the end of 2014.

National Urban Renewal Program

Based on the observation that poverty is increasingly urbanizing, South Africa's National Urban Renewal Program was established in 2001 as a 10-year initiative to promote socio-political, economic, and spatial integration of selected urban areas. The program focused primarily on exclusion areas (socially, economically, and racially) and supported eight urban districts in six cities, which were characterized by high levels of crime, poor connectivity to surrounding neighborhoods, high unemployment rates and inequality, and shortage of formal housing stock. Measures implemented under the program ranged from enhancing employment opportunities to enhancing access to the areas through better transportation services and improving education, local economies, and social capital.

Kvarterløft Program

Denmark's national urban regeneration program, Kvarterløft, ran from 1997 to 2007 and was later followed by the financially reduced Omradefornyelse. The area-based program was set up with the aim of addressing increasing social problems and the spatial concentration of immigrants and refugees. The program combined measures targeting both people and places and fostered coordinated and integrated approaches among different public sectors and by involving the local community.

Jawaharlal Nehru National Urban Renewal Mission

In contrast to the national programs mentioned above, India's Jawaharlal Nehru National Urban Renewal Mission did not apply an area-based approach but was launched in 2005 (and ran until 2015) with the goal of redeveloping entire cities and towns (65 in total) by making them more equitable, livable, and economically productive. With an investment of $20 billion, the program focused on upgrading infrastructure services and providing basic services to the urban poor. Its implementation faced numerous challenges, however, due to a lack of planners trained to realize integrated approaches, a shortfall in strengthening local governance, and a delay in financial flows from the national to the state governments.

Source: See endnote 14.

deprived areas, the need for an integrated approach that combines physical and social measures, building local capacity, providing adequate financial resources, and conducting monitoring and evaluation.[14]

First, the selection process of targeted areas is critical to the success of interventions carried out at the local level. Yet the decision to declare a neighborhood "deprived" is often in different hands. In the case of South Africa and India, the national programs were centrally driven, which meant that the target areas were chosen top-down by either the national or state governments, without any consultation of local actors. Yet involving the local level and applying a bottom-up selection process—for example, by asking municipalities or communities to submit an expression of interest for participating in the program—are crucial for creating ownership. A proposal submitted by neighborhoods or cities simultaneously indicates awareness and their openness to change, thus increasing the chances of success in the long run. Both the German and Danish programs set up such an application process, which also helped them gain greater visibility and impact.

Part of the Jawaharlal Nehru National Urban Renewal Mission (JNNURM) was funding for thousands of transit buses, including this one in Pune.

Rovan Vaz

Second, the goal of social cohesion and interaction cannot be achieved solely through physical interventions, such as renovating residential buildings, improving lighting in public spaces, and reducing the number of housing units to combat vacancy. Rather, physical measures need to be combined with social measures aimed at improving living conditions in districts, such as creating new employment opportunities, providing better social and cultural facilities, and designing attractive public spaces that invite residents to stay and interact. The German program seeks to achieve exactly this: to upgrade the built environment while enhancing the situation of the local residents. Activities funded through the program range from modernization of buildings and the living environment; to supporting business start-ups, training, and education initiatives; to promoting language learning and fostering ethnic entrepreneurship and self-employment of immigrants and their descendants.[15]

Other programs, such as the Danish and South African ones, also highlight the need for combined social and physical efforts and place particular emphasis on the participation of local residents. On paper, such programs stress the need for combining physical and social interventions; however, their practical implementation remains controversial and widely criticized. Financially, infrastructure measures and physical upgrading have been the predominant focus of these programs, whereas social initiatives and civic participation remain underrepresented. Denmark's Kvarterløft program, for example, claimed to focus on social initiatives and participation, yet more than 90 percent of the financial resources have been spent on physical improvements.[16]

Third, cities and municipalities often lack sufficient financial and personnel capacities and face weak coordination among different planning departments. The latter, in particular, poses a huge obstacle to the goal of simultaneously implementing social and physical measures, which can create fruitful synergies. For example, when designing a new public space, it would be advantageous to also consider how this could be coupled with providing space for local shops and a new community center, to create a vibrant place of interaction. However, coordination and communication is often insufficient, not only within the city administration, but also between different levels of administration. This can lead to different levels of administration having different understandings of the objectives to be achieved through the program, resulting in incoherence during implementation. The Indian and South African cases demonstrate that well-trained personnel, as well as structures and reforms at the local level, are needed to ensure that national objectives can be translated into local action.[17]

Fourth, because financial constraints often are major barriers to the implementation of concrete actions in cities, a national scheme that provides financial support for personnel and capacity building resources in cities can help in realizing concrete projects at the local level. Moreover, such a scheme can help finance the creation of the new institutions and agencies that may be necessary to manage and coordinate national programs on the ground. As part of the Social City program in Berlin, neighborhood management offices were gradually introduced in the target areas after 2005, with the overall goal of empowering local residents and involving them in decision-making processes and the development of their area. Neighborhood councils, consisting of and elected by local actors, decide how and for what projects the funds from the program can be used, and they also maintain the dialogue with the neighborhood management teams and the governmental administration. The neighborhood management offices facilitate networking and communication among existing nongovernmental organizations, businesses, and other social and cultural initiatives in the area to bundle and mobilize local resources.[18]

Fifth, monitoring and evaluation are key to tracking progress on the implementation of a program, identifying gaps, making adjustments where needed, and reviewing progress. However, systematic monitoring and evaluation often are not mandatory, and a lack of reliable data makes such initiatives difficult. Lessons learned from Denmark demonstrate the need to develop realistic indicators for measuring a program's progress and success, especially for aspects such as participation and empowerment, as well as establishing a continuous monitoring system.[19]

In sum, national programs can provide an important framework for local initiatives to work on the ground, yet respective mechanisms and structures must be in place to ensure greater impact. Although the four national programs discussed above were able to achieve positive change in the target areas, these were related mainly to upgrading the built environment, such as by renovating buildings and improving public plazas and other spaces. Social needs were often overlooked, and measures targeting the socioeconomic status of residents (for example, access to jobs, education, mobility, culture) were too limited. Many of the initiatives also had too short a time frame to create long-term change. It is crucial that national programs have a long-term scope and be based on continued political commitment. Interventions at the local level also need a scope that goes beyond the target areas in order to avoid stigmatizing them without ultimately remedying the situation.[20]

City-wide and Neighborhood Planning

National urban planning frameworks can—if designed properly—serve as a catalyst for local action to upgrade socioeconomically deprived urban areas. While national programs deliver pivotal framework conditions, their ultimate success rests on initiatives at the city and neighborhood levels. Key to designing sustainable and inclusive neighborhoods and cities is the ability to read and understand their language. Urban planners and designers need to carefully observe and analyze people's behavior in the urban realm and to design streets, public spaces, and entire neighborhoods accordingly. As author Jan Gehl puts it, to create "cities for people" or "people-friendly" cities, urban planning has to apply the human dimension that is focused on creating city spaces as *meeting places* for urban dwellers.[21]

A variety of planning and design measures allow for compact, well-connected, and integrated urban patterns that promote social cohesion in cities and provide spaces of encounter and social interaction. Among these are: land-use planning, the promotion of mixed-use areas with good access to public transport (via transit-oriented development), the rearrangement of street patterns, and public space design.

Land-use Planning for Balanced Urban Development

Land-use planning provides an important tool to guide and influence the development of cities. The consideration of not only economic aspects, but also environmental and social values, in land-use planning is necessary to allow for balanced and sustainable urban development. For example, community gardens fulfill important sociocultural functions and contribute greatly to social cohesion and food security. (See Box 18–2.) Yet in times of neoliberal city practices and enduring privatization of public land and properties in cities around the globe, such grassroots initiatives usually lack sufficient financial resources to continue.[22]

The recent wave of privatization has profound impacts on the urban realm and fails to acknowledge the increasing sociocultural complexity characterizing contemporary cities. It gradually diminishes the availability of spaces where new forms of social relations potentially could be formed. The selling of public assets, such as former school buildings, is often a shortsighted strategy that could cause unforeseen problems, as shown by the experience of Berlin. To consolidate the city's financial situation, the Berlin government established a property fund in 2001 to generate revenues through sales of city-owned land and properties in an auctioning process, without consideration of other aspects, such as the social value of initiatives. Some 400–500 public assets were sold annually, greatly reducing the number of city-owned properties. Yet when the unprecedented influx of refugees prompted a need for large-scale accommodations, the city government was forced to buy back buildings at much higher prices. Taking a more holistic and balanced approach in handling city-owned land and properties is key to preserving non-commodified spaces in cities and to retaining an adequate capacity to react in times of crisis. (See Chapter 16.)[23]

Box 18–2. Pro Huerta: Urban Agriculture and Food Security in a Changing World

As urban populations swell, many cities are struggling to ensure food security and adequate nutrition in the face of challenges such as climate change, economic and natural disasters, farmland degradation, and the immense barriers that the urban poor face to accessing fresh, nutritious food. Experiences with the Pro Huerta ("Pro Garden") program in Argentina and Haiti suggest that there are effective ways not only to improve nutrition, but also to shore up social resilience among vulnerable populations.

Buenos Aires and Rosario, Argentina

Argentina's National Institute of Agricultural Technology approved the Pro Huerta program in 1990 as a means to address the serious economic and food security challenges affecting the country, including a dramatic jump in food prices in Buenos Aires. Pro Huerta was formally adopted under the National Food Security Plan in 2003, and, in 2011, the government pledged more than $10 million to expand the program.

Pro Huerta helps Buenos Aires's poorest populations diversify their diets, access fresh food, lower their food budgets, and increase their incomes. The program is designed to boost self-sufficiency by providing the tools necessary to build food gardens, including seed kits, chickens and rabbits, and training in pest control, animal husbandry, and organic gardening methods. By late 2015, the program had helped set up more than 56,000 family gardens—supplementing the diets of some 350,000 people, or nearly 11.5 percent of the city's population—as well as more than 900 school gardens and 500 community gardens. A family garden can produce 200 kilograms of vegetables annually, enough for a five-person family.

Pro Huerta launched in Argentina's third largest city, Rosario, in February 2002. At that time, roughly 60 percent of the city's population was living below the poverty line, and food staples had quadrupled in cost, leading to theft and rioting. The Rosario government's Urban Agriculture Program and a local group, CEPAR, partnered to pilot the Pro Huerta model, offering tools and seeds to 20 gardening groups. By 2004, 800 community gardens were growing food for 40,000 residents.

The Pro Huerta program was successful in repurposing vacant land—comprising more than one-third of Rosario's land area—for gardens. The city has since updated its land-use laws to include urban farming and is building a green belt of parks and multi-scale gardens. Pro Huerta also has created venues for direct marketing to the public and has set up cooperatives that prepare and sell produce, soups, jams, and natural cosmetics. By 2004, 10,000 low-income households in Rosario were selling enough produce to lift themselves above the poverty line. An estimated two-thirds of participants were women.

In 2013, as the city's economy improved, participation dropped to some 1,800 residents, almost 14 percent of them full-time producers. The Pro Huerta program has been

Box 18–2. continued

replicated in 88 percent of Argentina's municipalities, with more than 630,000 gardens and 130,000 farms providing food for over 3.5 million people nationally. A network of 20,000 promoters manages the program, participating in agro-ecological fairs and working with thousands of institutions and organizations across Argentina.

Haiti

The Pro Huerta program also has spread to Brazil, Colombia, Guatemala, Venezuela, and Haiti, a country that suffers from widespread poverty, inadequate nutrition, and high dependence on food imports. Haiti launched Pro Huerta in 2005 with support from the Argentine Fund for Horizontal Cooperation, adapting the program to the local context and using local leadership to manage it. Argentinian experts trained a team of Haitian agricultural engineers, who then taught a network of volunteer promoters—mostly women—how to provide trainings within their communities. Between 2005 and 2008, these efforts helped establish 16,086 family gardens, 2,700 school gardens, and 1,900 community gardens.

In addition to producing food, Pro Huerta has resulted in the creation of resilient social networks that have helped communities respond to disruptions. In 2008, after Hurricanes Gustav and Ike destroyed thousands of gardens across Haiti, the program bounced back thanks to robust community cohesion. By the end of 2009, 1,843 promoters, 11,465 gardens, and more than 80,000 participants were active in the program. Following the 2010 Haiti earthquake, Pro Huerta was instrumental in fighting the cholera outbreak, providing more-nutritious diets for susceptible populations, offering expertise on food handling, and building special water storage facilities and sand filters to avoid disease transmission.

According to a Pro Huerta survey, 93 percent of program participants in Haiti improved their food situation, 86 percent of households were able to access a greater variety and quantity of food, and household spending was halved to just 33 percent of monthly income. In 2014, the Union of South American Nations pledged $3 million to extend Pro Huerta to 2016, with the goal of nearly doubling participation to 220,000. Haiti hopes to extend the program to 1 million participants by 2019.

Lessons Learned

In Argentina, the Pro Huerta program was predominantly a response to short-term economic disturbance; however, the model also performs strongly in countries, such as Haiti, that face continuous threats to food security. In a world where food supply and access are increasingly affected by variations in climate, environmental conditions, equity, and natural and economic disasters, urban agriculture programs such as Pro Huerta can be used to empower underserved communities, providing them with the tools to build a healthier life and to help them cope with future turmoil and change.

Kristina Solheim, Program Manager, goNewHavengo
Source: See endnote 22.

Integrating Land-use and Transport Planning to Foster Social Cohesion:
The Example of Transit-oriented Development (TOD)

At the neighborhood level, integrated land-use and transport planning—coupled with the creation of high-density mixed-use areas—facilitates demographic, socioeconomic, and cultural diversity. In particular, transit-oriented development (TOD) has become a popular planning approach to create inclusive, connected communities through spatial planning. Regulatory and incentive mechanisms, such as local planning schemes, educational campaigns, and incentives for developers and communities, are crucial for successful implementation. TOD is based on the principle of designing high-quality mixed-use areas around transit stations to enhance access to public transport and pedestrian- and cyclist-friendly environments while reducing dependence on private cars. Areas that prioritize walking and cycling typically are characterized by higher levels of social interaction and help residents who are unable to afford a car to overcome transport poverty. (See Chapter 11.)[24]

The design of mixed-use areas follows the idea of creating "urban villages" where residents are provided with housing, transportation, community and recreational facilities and services, public spaces, and retail within a short distance. To facilitate community diversity and social cohesion, these services and facilities should cater to the needs of different social groups with varying interests and demands. TOD therefore should be designed and managed in a way that allows for diversity in housing (for example, in design, form, tenure, and affordability), land-use, employment, and retail, and that provides multiple public and open spaces as focal points for the community. Safeguarding community diversity over the long run requires long-term investments in social housing and community infrastructure. Further, developing a TOD precinct requires a continuous participatory planning process that targets a diversity of groups to build ownership and a shared sense of identity.[25]

Well-designed TOD offers numerous environmental benefits. The continued rise in transport volumes not only leads to increasing traffic congestion, but also contributes to environmental and health challenges such as rising air pollution and greenhouse gas emissions. Designing neighborhoods based on the principle of walkability and cycling as well as good access to public transport is urgent. TOD holds tremendous potential in countries like China. (See Chapter 7.) The country's third largest city, Guangzhou, has invested massively in a highly efficient bus rapid transit (BRT) system and is building new promenades and bicycle lanes to encourage walking and cycling.

Guangzhou's BRT system, which is the first worldwide that is fully integrated with a metro system, carries more than 800,000 passengers daily and has significantly reduced traffic jams and vehicle kilometers traveled. Thanks to multiple sub-stops and passing lanes at each station, average bus speeds increased from about 15 kilometers per hour to about 22 kilometers per hour—an attractive and speedy alternative to individual motorized transport. The network of small, walking-oriented streets surrounding the city's Shipaiqiao station is being complemented by new high-density commercial and residential developments, helping to revitalize the entire area. Shipaiqiao station and its surrounding area are now easily

Dedicated BRT lanes on Tianhe Road, Guangzhou.

accessible by public transport and have become a prime location for shopping, working, living, and strolling.[26]

Upgrading Street Networks to Reintegrate Neighborhoods

Streets can have a great impact on the vitality and integration of a given area. They are not only a means of transportation, but also a fundamental shared public space that facilitates numerous social, cultural, and economic activities and allows people to interact. Well-designed street patterns that facilitate connectivity and mobility can counteract socio-spatial segregation and help to re-integrate areas into city structures.

UN-Habitat promotes a street-led approach to the citywide transformation and regeneration of slum areas in many developing and emerging countries. The absence of streets and open spaces segregates and disconnects slums from the rest of the city. Done right, the upgrading of street networks in slums can bring advantages including security of land tenure, future consolidation of settlements, optimization of land use, poverty reduction, and increased social interactions among residents. However, upgrading of street networks also requires political will and needs to be based on a strong participatory planning process. The latter is a necessary precondition not only to create ownership, but also to conduct a reliable inventory of the physical configuration and socio-spatial structure of a settlement. Participatory planning also helps to inform the design of area-based plans and street patterns that capture the "multiple functions of streets based on nuances of everyday practices of street life and people's aspirations."[27]

The Integrated Program for the Improvement of Squatter Areas (PRIMED) in Medellín, Colombia, provides a good example of the benefits of well-designed street networks. Based on strong political will and the desire to counteract spatial exclusion and promote social development in deprived areas, the program facilitated the application of an innovative public transport system based on cable cars connecting the target areas with the rest of the city. (See Chapter 4.) The first cable car line was implemented in the poor and densely populated northeastern district, which was characterized by minimal road infrastructure and thus a lack of accessibility. The program was not limited to the implementation of the transport system, however, but also combined urban upgrading measures, including interventions in public spaces, social housing, and other social infrastructure, which were realized in a participatory manner.[28]

Impact studies reveal that these combined interventions helped to upgrade Medellín's densely populated and low-income neighborhoods and integrate them into the city's fabric. They also boosted the quality of life of the urban poor by enhancing accessibility for local residents and outsiders alike, improving air quality, counteracting stigmatization of these areas, and providing local residents with a sense of social and political inclusion. Levels of violence and crime in the neighborhoods surrounding the cable car lines dropped significantly, which helped to revitalize public life. In addition to social and mobility aspects, the PRIMED program considered environmental outcomes. The Metro Company, which evaluates the environmental performance of the cable car system and monitors the reduction in greenhouse gas emissions, concludes that the hydroelectric aerial cable cars could help to reduce up to 121,029 tons of carbon dioxide between 2010 and 2016, compared to the fossil fuel-operating vehicles that the system replaces.[29]

Although the upgrading of street networks can support development, foster integration, and bring environmental benefits, it often requires demolition and relocation to make space for the construction of new streets or an aerial cable car public transport system. This tradeoff was evident in the case of Medellín and had to be negotiated within the community. Overall, however, improved street networks have great potential to integrate entire neighborhoods into city systems and to improve the quality of life of local residents.

Public Space Design

Public spaces allow people to meet and interact on ostensibly neutral ground. They provide a democratic space for different social groups to participate in civic activities. Especially in developing and emerging countries, where urban inhabitants often live in densely populated housing areas with few economic resources, public spaces form a fundamental part of community life. Urban parks, in particular, serve as a vital public space where everyday experiences are shared and negotiated among different social and ethnic groups, and where numerous opportunities for intercultural interaction exist.

Extensive city improvements, such as upgrading street networks, are costly and time-consuming. However, small interventions in public spaces—such as improvements to bench seating, providing movable chairs, closing streets to car use, and laying new pavement to encourage pedestrian traffic—can make a huge difference and help reinforce daily life in a fast and cost-efficient manner. For example, the location of street furniture has a compelling effect on how public space is used and accepted and how long people tend to stay and interact with strangers.[30]

The post-industrial waterfront promenade at Aker Brygge in Oslo is a good example of how urban design can influence social interaction. As part of a broader

A sunny July day on the waterfront promenade at Aker Brygge in Oslo.

neighborhood renovation project in the 1990s, old benches on the promenade were replaced with Parisian-style double park benches, and the overall seating capacity was increased. Consequently, the number of people sitting in the area more than doubled, and social interactions among strangers multiplied.

Some two decades later, the same architects were tasked with adapting and renewing the area, again with an emphasis on encouraging social interaction and diversity. They developed a "site-specific concept for street furniture and 'staying,'" which aided in creating numerous opportunities to sit, lie, eat, read, or chat with acquaintances or strangers. The pedestrian and bicycle path was reorganized to create wider, more generous public spaces, and sun loungers and comfortable benches were installed, inviting people to sunbathe and lie down. Although the provision of sufficient seating opportunities within cities and neighborhoods is crucial, other factors—such as views and orientation toward street activities, as well as movability of seating options—determine the vitality of a place.[31]

Conclusion

Socioeconomic polarization and spatial segregation have become prevailing trends in cities worldwide, with adverse impacts on quality of life and social cohesion. As cities become increasingly diverse, these trends often have an ethnic component as well. Many socioeconomically deprived areas are characterized by a high concentration of migrants, making their multi-faceted integration into city life more challenging. Consequently, finding solutions to counteract disparities and inequalities while strengthening relations and interactions among socially and ethnically diverse groups has become an urgent matter.

Although urban planners and designers cannot solve the roots of exclusion and inequality per se, they can aid in increasing the accessibility and integration of deprived areas and provide spaces that increase the chances of interaction and the forming of social relations among people from differing ethnic backgrounds. National urban planning programs offer a useful framework for local initiatives to kick off and work on the ground. Applying an integrated approach that effectively combines social and physical measures, coupled with a bottom-up selection process, capacity building, the establishment of governance structures, the provision of financial resources, and monitoring and evaluation is key for the success of national programs.

At the city and neighborhood levels, numerous approaches and measures have been tested globally to overcome socio-spatial segregation and exclusion. In particular, the creation of mixed-use and socially mixed areas—coupled with good access to public transport, housing diversity, and sufficient provision of vibrant public spaces that facilitate inter-ethnic encounters—are

promising ways to enhance social cohesion. Approaches and planning principles, such as socioeconomically balanced land-use planning, transit-oriented
development, and upgrading street patterns, have been successful in building
well-connected, compact, and integrated urban patterns that allow for sustainable urban development. Well-designed public spaces also can serve as
a key locus where new forms of sociability can emerge. Urban planners and
designers have the tools and instruments at hand to contribute greatly to social
cohesion in cities, yet political will and the participation of a broad array of
stakeholders, including local residents, is a fundamental precondition to the
success of any measure.[32]

CITY VIEW

Durban, South Africa

Debra Roberts and Sean O'Donoghue

Durban Basics

Municipal population: 3.4 million
Municipal area: 2,297 square kilometers
Population density: 1,498 inhabitants per square kilometer
Source: See endnote 1.

Netting sardines from the Durban beach during the annual sardine migration along the coast.

Community Ecosystem-Based Adaptation to Climate Change

The city of Durban, also known as eThekwini Municipality, is located in South Africa's KwaZulu–Natal province and situated within the Maputo-Pondoland-Albany "biodiversity hotspot," one of just 35 such hotspots worldwide. Durban's population has grown by 1.1 percent, or 660,000 residents, since 2001. More than 70 percent of the population is African, and large numbers of them are afflicted by poverty.[2]

Many cities in the developing world lack the capacity to adapt in the face of emerging climate variability. These cities often do not have, or fail to properly maintain, "gray" infrastructure such as drains, sewers, and roads. Meanwhile, rapid urban growth and poor planning have led to the degradation or destruction of "green" infrastructure such as wetlands, forests, grasslands, and productive soils. Many cities are caught in a perfect storm of population growth, escalating adaptation needs, and substantial development deficits created by a shortage of human and financial resources, increasing levels of informality, poor governance, environmental degradation, biodiversity loss, poverty, and growing inequality. In this context, "ecosystem-based adaptation" is an appealing concept because it embraces the notion of using ecosystems to aid people and save the resources on which they depend.[3]

In South Africa, these challenges have been exacerbated by a legacy of formalized racial division that has created widespread social, economic, and environmental injustice, inequity, and exclusion. In the case of Durban, this has manifested itself in an urban form that perpetuates a system where the poorest and most vulnerable residents live far from jobs and services, often in compromised environmental conditions. This nexus of human need and environmental risk continues to pose a significant challenge to city planners charged with achieving equity and sustainability for all in post-apartheid South Africa.

Climate Threats and Obstacles to Adaptation

Recent efforts to develop a systematic conservation plan for Durban have revealed serious threats to the city's ecosystems. As the planet warms, temperatures in Durban are likely to increase by 1.5–2.5 degrees Celsius (°C) by 2065 and by 3–5°C by 2100. Projections suggest an increase in aggregated rainfall by 2065, and up to 500 millimeters more rainfall annually by 2100. This is likely to manifest itself in more-frequent extreme rainfall events and higher stream flow intensity, with prolonged dry spells between rainfall events. The projected impacts include an increase in extreme weather overall, the erosion and loss of topsoil, a rise in vector-borne diseases, species extinctions, and potential reductions in agricultural yields.[4]

Durban's capacity to adapt to these climate-related changes will need to be enhanced substantially despite the following impediments:

- short-term political and development needs overriding long-term concerns such as sustainability and adaptability;
- the lack of finances and skilled human resources to undertake adaptation planning and implementation;
- ineffective links between government and community structures that prevent an adequate assessment and response to community-level risk;
- an economic development model that is locked into a standard manufacturing paradigm rather than transitioning to a more-adaptable green economy;
- lack of political will and understanding of the critical relationship between climate change and biodiversity issues;
- the impact of the global recession embedding the business-as-usual model of urban planning and management rather than favoring flexibility and innovation;
- decreasing opportunities for ecosystem-based adaptation due to extensive transformation of natural habitat; and
- lack of a legal mandate for climate protection planning and action at the local level.

Municipal and Community Adaptation

Durban initiated a citywide Municipal Climate Protection Program (MCPP) in 2004, which included a strong and early focus on adaptation. The adaptation work stream has three main components: municipal adaptation (activities linked to the key line functions of local government), community-based adaptation (activities focused on improving the adaptive capacity of local communities), and a series of urban management interventions that address specific climate change challenges (such as the urban heat-island effect, increased stormwater runoff, water conservation, and sea-level rise).[5]

Within each of these three components, a number of projects focus on ecosystem-based adaptation, following a "learning-by-doing" model of development and implementation. As a result, local-level adaptation is proving to be an incremental, iterative, and non-linear process that relies on experimentation, flexibility, and innovation as the means of achieving progress.

Within the realm of municipal adaptation, it became clear that a sectoral approach to adaptation planning was far more successful than a broad strategy, as it facilitated the development of focused champions who could carry the adaptation message back to colleagues within their respective sectors. The planning process identified 47 possible interventions within the health, water, and disaster management sectors (which were selected as pilot areas). A cost-benefit study, initiated in 2011, sought to prioritize these interventions in terms of human rather than financial benefit. This model was deemed more appropriate for a city of the Global South.[6]

The community-based adaptation component highlighted some important research issues and identified the lack of social cohesion as a key limiting factor in achieving meaningful adaptation. So far, little actual community transformation has resulted. More work is required, but the potential is limited by the lack of appropriate human and financial resources available to the city's Environmental Planning and Climate Protection Department (EPCPD).

In 2015, Durban developed and approved its first combined mitigation and adaptation climate change strategy. Implementation plans include the mainstreaming of sustainability initiatives that will align with a greening program associated with Durban's hosting of the 2022 Commonwealth Games.

Acquiring and Protecting Conservation Areas

The Durban Metropolitan Open Space System (D'MOSS) covers some 75,000 hectares of open space, including estuarine areas, forests, wetlands, grasslands, and woodlands. Formally integrated into the municipality's planning schemes in 2010, D'MOSS is designed to protect Durban's biological diversity and to ensure sustainable ecosystem services, which are seen as a critical tool for climate change adaptation. The natural infrastructure complements human-built infrastructure—such as wetlands reducing the need for expensive stormwater infrastructure—helping to provide poor and vulnerable populations with a safety net against natural disasters and potential economic shocks related to climate change.[7]

Between 2002 and 2015, EPCPD acquired 591 hectares of land with high biodiversity value, much of it adjacent to nature reserves, thereby maximizing the ecological integrity of the landscape and enhancing connectivity. Another key conservation measure is the use of "special rating areas," which apply an additional levy on property taxes in a given area to improve land management. For example, the 354 hectare Giba Gorge Environmental Precinct was established in 2009 as a pilot project for invasive alien plant control, fire management, and pollution monitoring. Durban also has worked to ensure that biodiversity conservation considerations were integrated into town planning schemes that historically have been at odds with environmental objectives.

Reforestation Projects and "Treepreneurs"

In parts of the city where biodiversity and ecosystem assets have been lost, it is necessary to expand and enhance conservation lands. In 2008, during the lead-up to the 2010 FIFA Football World Cup in Durban, the Buffelsdraai community reforestation initiative was launched, involving 521 hectares of land that previously had been cleared for sugarcane cultivation. The Wildlands Conservation Trust trained residents from some of Durban's poorest and most vulnerable communities to become "treepreneurs," or individuals who collect native seeds from local forests. Since its inception, the project has created some 43 permanent jobs, 16 part-time jobs, and 389 temporary jobs for members of the Buffelsdraai and Osindisweni communities.[8]

In addition, some 583 community treepreneurs are engaged in producing and trading trees. The treepreneurs grow locally sourced native seedlings for the project, earning credit that can be exchanged at quarterly "tree stores" for items such as food, building materials, and other pre-ordered goods, or to cover school fees. Early indications suggest that the direct socioeconomic impact on the communities is significant, bringing improved educational opportunities and food security.[9]

A second reforestation project was established in 2009 on 250 hectares of communal land at Inanda Mountain, an area of severe forest degradation resulting from high levels of wood harvesting for firewood and building materials and from uncontrolled fires. Activities have centered on clearing and controlling invasive alien plants and managing their natural replacement by native species. Tree seedlings, produced by 76 treepreneurs, also are planted.

This model has been rolled out to a third site and has prompted the development of the Community Ecosystem-Based Adaptation (CEBA) concept, which highlights the mutually beneficial and positively reinforcing relationship between ecosystems and human communities. The CEBA concept extends the treepreneur model into a catchment-wide process, whereby ecosystem restoration and maintenance provide a range of "ecopreneur" opportunities for poorer residents—including tree propagation, alien plant removal, riparian bank restoration, and recyclable materials collection, as well as training opportunities. The CEBA model is attractive because it is acceptable to both private sponsors and political leadership, and it combines both mitigation and adaptation in a no-regrets, easily replicable approach.[10]

Green Roof Pilot Project

Durban's Green Roof Pilot Project, initiated in 2008 on an existing municipal building, is designed to explore the diverse benefits of green roofs to the city, such as reducing stormwater runoff, bringing native plants and animals back into the city center, and lowering roof and indoor temperatures. Crop-planting trials suggest that green roofs also can contribute to improved urban food security. The success of this project has encouraged the municipality's Architecture Department to become more involved in championing the concept of green roofs on municipal-owned buildings.[11]

Expanded Public Works Projects

Two expanded public works projects focus on ecosystem management and the control of invasive alien plants. Rising temperatures and higher carbon dioxide concentrations are likely to increase opportunities for invasive species, threatening both biodiversity and valuable ecosystem services. The Working for Ecosystems program, initiated at the national level in 2006 and now funded by the EPCPD, employs 185 people and provides training to community members as well as staff from other municipal departments involved in the control of invasive species. The Working on Fire program, set up in 2009, has 43 staff and aims to alleviate poverty and develop skills by employing people to manage fires and undertake invasive plant control, mostly in priority areas of high biodiversity. Due to the program's success, a three-year contract extension between the municipality and the implementing agent was signed in late 2015.

Institutional Change and International Networking

Durban has used a multi-pronged approach to mainstream the need for climate protection within municipal operations. This has included institutional restructuring (such as the creation of the EPCPD's Climate Protection Branch), the inclusion of the Municipal Climate Protection Program as a deliverable in the city's key strategic planning document (the Integrated Development Plan), aligning the development of the Municipal Adaptation Plans with existing work streams, and developing a combined strategy for climate change adaptation and mitigation.

In 2011, Durban hosted the annual Conference of the Parties of the United Nations Framework Convention on Climate Change, affording a strategic opportunity to advance the adaptation agenda as an urgent priority for African cities and to profile Durban's adaptation work more widely. Working with partners—including the South African Local Government Association, the South African Cities Network, the National Department of Environmental Affairs, and ICLEI–Local Governments for Sustainability—Durban's EPCPD organized an adaptation-focused international local government convention at the event. This led to the Durban Adaptation Charter, which has been signed by 341 mayors and local government leaders, representing 1,069 cities from 45 countries, half of them African. Among other things, the Charter commits local governments to ensure that adaptation strategies are aligned with mitigation strategies, to recognize the needs of vulnerable communities, to prioritize the role of functioning ecosystems as core municipal green infrastructure, and to seek innovative funding mechanisms.[12]

Debra Roberts heads the Environmental Planning and Climate Protection Department of eThekwini Municipality in Durban. **Sean O'Donoghue** manages the Climate Protection Branch within the Department. Both are Honorary Research Associates at the University of KwaZulu-Natal.

This City View is based largely on two articles: Debra Roberts et al., "Exploring Ecosystem-based Adaptation in Durban, South Africa: "Learning-by-doing" at the Local Government Coal Face," *Environment & Urbanization* 24, no. 1 (2012): 167–95; and Debra Roberts and Sean O'Donoghue, "Urban Environmental Challenges and Climate Change Action in Durban, South Africa," *Environment & Urbanization* 25, no. 2 (2013): 299–319.

CHAPTER 19

Urbanization, Inclusion, and Social Justice

James Jarvie and Richard Friend

The developing world—Asia and Africa, in particular—is going through an unprecedented wave of urbanization. This is causing a fundamental transformation of ecological landscapes, social values, and economic relations. Small cities, with populations of around 500,000 or less, are seeing the most rapid change. Much of it is risky. Cities, too often, are expanding into hazardous locations along coasts, deltas, floodplains, and river basins. At the same time, climate change is exacerbating the risks posed by many of the hazards that these fast-growing cities are creating.[1]

Urbanization and climate change are core governance challenges. Increasingly, global responses to addressing these challenges are being framed in terms of building urban resilience, with an emphasis on strengthening cities' ability to deal with shocks and crises. However, too much of the effort on the ground is focusing on the technical aspects of urban planning and infrastructure investment, without fully realizing that the foundation to lasting solutions is political—and transformative in nature.

Urbanization is unfolding in many cities and countries without effective representation, transparency, or accountability, driven by a diverse range of powerful, yet sometimes murky, political and economic interests. Those interests focus on short-term political and financial gains, often intensifying inequality and with little consideration of alternative, climate-compatible futures. The longer-term instabilities being set up are rarely on any political agenda.

James Jarvie is with Mercy Corps' Asian Cities Climate Change Network. **Richard Friend** is an independent scholar and Co-Investigator of the Urban Climate Resilience in Southeast Asia (UCRSEA) Partnership, funded by the International Development Research Centre and the Social Sciences and Humanities Research Council of Canada.

Urbanization and Climate Change

The combination of urbanization and climate change presents a complex and dangerous problem, particularly in Asia. During cyclone seasons, agencies are preparing for more-frequent coastal disasters, recalling devastating events such as Cyclone Nargis, which killed more than 138,000 people in Myanmar in 2008, and Typhoon Haiyan, which killed thousands in the Philippines in 2013. In 2015 alone, five "super typhoons" were recorded, including Typhoon Chan-hom, which required the evacuation of millions of people in China and caused nearly $1 billion in economic losses. National and municipal governments need to focus increasingly on disaster mitigation and to devote funding and other resources to take action. If the urbanization–climate nexus in vulnerable areas is not addressed, the impacts of weather-related disasters will increase, killing many more people and resulting in larger economic losses. It is literally a matter of life and death.[2]

Addressing this nexus will require vastly improved decision-making processes, as places that were once quiet coastal backwaters become vulnerable urban centers. Enlightened decision makers will need to rein in diverse and powerful, yet competing, sets of interests and values, while marshaling the know-how to understand and adapt to greater degrees of climate-related uncertainty and risk. Yet making this work on the ground is hugely challenging, because those responsible for solving these problems frequently also are responsible for creating them in the first place.

Urban Governance

The greatest opportunity for addressing these challenges should lie at the city level, where local governments and citizens best understand their own needs and vulnerabilities. Yet city-level governance is, in many ways, least able to cope with the rate of change. The level of administrative and financial authority at the city level is not always clear, and capacity is often low. Local requirements typically determine the tasks of local government, yet vulnerabilities and climate shocks often create impacts that cascade beyond administrative boundaries.

Gaps in urban governance are most obvious in failures in land-use planning and in the provision of core infrastructure and services to urban residents. Land-use planning frequently is merely a process of retrospectively mapping changes on the ground that already have occurred. This often is called "weak governance," but, instead, it should be seen as being designed to serve the political and economic interests of powerful elites. Being able to control the designation of land-use zoning opens up opportunities for speculative land investment.

Asia, which has seen the greatest level of recent global urbanization, offers clear examples. In Thailand, the impacts of the Bangkok floods of 2011 were shaped largely by a history of urbanization and industrial development in critical flood-prone areas, in defiance of earlier land-use plans. As is common across Asia, the ruling imperative was one of "buy low and sell high." The greatest financial returns from land-use speculation were to be won in the conversion of low-value land that was largely flood-prone and/or agricultural land. The largest wetland in Vientiane, Laos, the That Luang Marsh, is both a symbol of national identity and an important source of drainage for the city. Yet, despite protests and concerns about potential flood risks, it currently is being converted into an industrial park.

International airports also are located in flood-prone areas. Thailand's King Cobra Swamp is now known internationally as Suvannabhumi Airport, and Indonesia's major airport, Soekarno Hatta, also sits in a swamp where flooding frequently hinders access. These examples of placing expensive, critically important infrastructure in highly vulnerable locations are not accidents of policy. They result from intentional decision making by exclusive alliances of economic and political interests that focus on short-term gains and that have little consideration for the long-term risks threatening citizens, their assets, and their futures.

The inability of cities to provide needed urban services is all too common. Across sub-Saharan Africa, cities are growing at unprecedented rates, and essential services have not been made available to the bulk of the people. Some city governments, unable to keep up with growth, have relied on international institutions for support. A World Bank program in Tanzania provided more than $60 million for grants, administrative support, and community infrastructure improvement to support informal areas of Dar es Salaam, representing 70 percent of the population. Further projects followed, but only about 14 percent of the overall population was reached. The Tanzanian government is unlikely to be able to afford to scale up the program further, and no other donor can match the World Bank's reach. International donors, even when they do provide assistance to governments that are struggling financially, still lack the resources to contribute in a way that has large-scale, sustainable impact.[3]

Dependence on international aid is not a sustainable approach to improved urbanization. Improved and transparent national budgeting for city development is needed, including reductions in budget misallocations and corruption. Similar situations are found across more-urbanized Asia. In many Indian cities, barely 60 percent of the population has access to safe water and sanitation. Capital cities such as Kathmandu, Nepal, are well known for their failure to provide basic services such as water, energy, and waste management. Such failures are exposed painfully in the aftermath of disasters, such as Nepal's April 2015 earthquake. Kathmandu also illustrates the truism that service provision often depends on wealth and influence, as residents who are able to afford their own small-scale water and energy systems can bypass erratic or non-existent state services.[4]

Problematic electricity infrastructure in Kathmandu.

Lyle Rosbotham

The failure of cities to provide critical systems and services to residents opens opportunities for private sector interests to invest in urban public facilities. In Jakarta, Indonesia, the water supply system is spatially and socially differentiated, serving predominantly middle and upper class areas, the business district, and industrial zones. It largely excludes lower-income households, which must access water via shallow and deep wells, private household wells or rainwater collection systems, water vendors, bottled water, standpipes, and/or private localized networks that are connected to deep wells. The privately owned options are more expensive than public services, placing a further burden on the poor. Slum residents earning around $2 a day often pay half of this income to vendors carrying potable water for drinking and cooking into areas off the water grid. Worldwide, many essential public services are becoming privatized, and, in many places, the private sector is even driving the process of making public urban policy.[5]

Local governments rarely have self-sufficient power and knowledge to create or adequately manage city planning frameworks that safeguard citizens and assets for the long term. Even where core services such as water, energy, and transport are strained or broken, new infrastructure such as malls and factories are built and plugged into these services, straining them even further, without contribution to their improvement. Power brownouts and blackouts, water shortages, hyper-congested roads, and bad air quality too often are the outcomes. In 2012, more than 700 million people in India suffered from two consecutive blackouts as electricity grids were unable to meet demand, sparking fear of riots.[6]

Globally, an estimated 150 million people live in cities that have perennial water shortages. An estimated 886 million live in cities with seasonal shortages of at least one month per year, and more than 1 billion urban citizens are expected to face serious water shortages by 2050. New Delhi, India, has the world's most dangerous urban air quality because of high concentrations of particulate matter, contributing to India's ranking as the country with the highest death rate from chronic respiratory disease. Most cities in Asia and Africa do not have the monitoring systems in place to be able to assess air quality, let alone manage emissions to ensure a safe environment. Jakarta and its satellite cities, which only now are starting to build a mass transit system for a combined population of 28 million, have been voted the world's most traffic-congested urban area—more than five times as congested as Rotterdam in the Netherlands, the world's least-congested city.[7]

Investment Without Inclusion

When gaps in service provision and basic infrastructure are recognized, additional investment is called for. Yet what this investment buys, who benefits, and who pays for it rarely are clear. Accessing project investment, project planning, and implementation are governance issues in which different stakeholders will have different perspectives and priorities. Collusion between political and business interests, shielded by murky policy-making mechanisms, leads to finance agendas being guided by narrow interests while opening opportunities for corruption. Close scrutiny of such dealings is crucial, but there are obvious difficulties, as those involved in criminal practices are unwilling to publicly discuss their operations. Motivations, however, are easier to discern.[8]

In Indonesia, the basis for corruption has been studied closely because of recent, frequent prosecutions of politicians at multiple levels of government. One problem that has been revealed is that reductions in state funding for political parties, in the face of continuing high costs for achieving political office, has led to party income coming increasingly from corporate contributions. These contributions are, effectively, investments by companies for which the payoffs are contracts for infrastructure projects, trade monopolies, and state bank loans. Under these conditions, political parties are renting their political functions to corporate donors.[9]

The power of corporate interests in urbanization is no accident. Trade and capital flows drive urbanization, and cities both contribute to and channel huge shares of national gross domestic product. Further, city governments have split—and often conflicting—responsibilities. They function as managers with regulatory responsibilities, but also as entrepreneurs with a responsibility for, and interest in, attracting finance and investment. In the latter role, they access revenues from business transactions while diminishing their regulatory duties.[10]

As prices increase for rural land on the outskirts of cities—which is needed for city expansion—so do opportunities for speculation and corruption. Sometimes, this process of investment and repurposing land is cloaked in loose arguments about city development, but, virtually everywhere, it coincides with evictions and relocations, as the increased value of urban land and property drives this land out of reach for most urban citizens. Much of the urban arena thereby becomes privatized. The whole process of urban policy and planning can shift, discretely or obviously, from the domain of the government to a barely regulated private sector ruled by private interests. New funding streams that focus on the infrastructure promoting economic growth, and that fail to consider the needs and interests of people not benefiting from it, further risk accelerating dangerous trends of social inequality.[11]

Phnom Penh, Cambodia, is the capital city of one of the poorest countries in Asia. Given its small size and recent emergence from severe civil conflict, the city has received a disproportionately large amount of foreign investment and international donor support in relation to the national budget. International donor funding helped with the development of a master plan to coordinate public and private investment in the city to 2020, but it is not succeeding. Within the opaque, centralized structure of the Cambodian government, competing groups operate among a mixture of international donors and private investors from China, South Korea, and Malaysia, as well as Cambodia itself. State actors continue to build closed public-private partnerships and to implement projects without reference to the master plan. The social cost has been high: more than 40 relocation sites outside the city host residents who have been forcibly evicted from the city center to make room for new infrastructure investments, including a casino, shopping malls, and condominiums—contributing to city gentrification.[12]

Across much of the Global South, the benefits to ordinary citizens of foreign investments and the large projects that they support, as well as the politics surrounding their approval, will remain opaque. This is exacerbated by the fact that most foreign direct investment, donor funds, and loans for cities are channeled through national agencies. Investment practices remain top-down and are implemented without discussion with the citizens who will be taking on the resulting debt.[13]

This compounds a key emerging issue: across the world, city government authorities work within limited administrative boundaries and cannot compete with the transnational power of trade and investment. In countries such as Indonesia, where the government has undergone much decentralization since the late 1990s, patterns of decentralization do not enable cities to attract and control the financing needed to be able to plan inclusive, self-determined futures. In studies of 10 cities across Africa with widely varying populations, cities were challenged by a lack of financial resources to maintain and develop services and infrastructure for their general populations, yet they still were able to attract private investment for the engines of economic growth. Inter-governmental cooperation is required urgently to address this, yet it remains largely absent. Decentralization often means transferring responsibilities to local-level institutions while only rarely facilitating the creation of new institutions or the devolution of the fiscal autonomy required for responding to new and additional responsibilities.[14]

A New Urban Agenda

As urbanization increasingly leaves the poor behind, the international community is starting to pay attention. A United Nations (UN) Conference on Housing and Sustainable Urban Development will be held in October 2016, and a New Urban Agenda, to be adopted and implemented through the 2030 Agenda for Sustainable Development, is taking form. It includes a commitment to governance that incorporates the language of inclusion, participation, and resilience. At the same time, the newly adopted UN Sustainable Development Goals (SDGs) include SDG 11: "Make cities and human settlements inclusive, safe, resilient, and sustainable." It is a call for action by 2030 to "ensure universal access to adequate, safe, and affordable housing and basic services; to enhance inclusive and sustainable urbanization and capacity for participatory, integrated, and sustainable human settlement planning and management; and to increase the number of communities adopting and implementing policies that embrace inclusion, resource efficiency, mitigation and adaptation strategies that enable resilience to climate change."[15]

Even though recognition is growing of the significance of governance in shaping urban futures, there are concerns about how this urban agenda is being framed and how it keeps to historical commitments to rights such as access to information, services, and land tenure. Civil society organizations are calling increasingly for these to be recognized to address the need for enforcement and accountability of national and municipal governments.[16]

Commitments to specific rights need to be reaffirmed, deepened, and strengthened, as well as placed firmly at the heart of the new urban agenda. The Right to the City idea recognizes cities as collective endeavors and affirms residents' rights as citizens to quality of life, safe environments, and public spaces, as well as to housing and social and cultural services. Of particular importance, it also recognizes residents' rights to shape and design the cities in which they live. The Right to the City was a central feature of earlier global inter-governmental commitments, as expounded in the 1996 Habitat II conference and refined further in the 2004 World Charter of the Right to the City, which is being applied in the Brazil City Statute and the Montreal Charter.

The Right to the City is in line with other international commitments on environmental rights, among them the Access Rights defined in Principle 10 of the 1992 Rio Declaration, which include access to information, access to participation in decision making, and access to redress and remedy. Although governments of the world have made binding commitments to uphold these access rights, performance has been mixed. Yet where commitments have been ratified in national law, such as in Europe's 1988 Aarhus Convention (on "Access to Information, Public Participation in Decision-making and Access to Justice in Environmental Matters"), the impacts have been significant.

Rights issues are made more complicated by the retrenchment of democracy in many parts of the world. There are few functioning democracies in rapidly urbanizing countries, where civil society, the press, and the courts are under pressure. In Asia, recent backward steps have been drastic and marked. In 2015 alone, Thailand was taken over by a military regime that has clamped down on civil society voices, and, in Cambodia and India, laws were passed that restrict the ability of nongovernmental organizations (NGOs) and other civil society groups to operate.

Governing conditions increasingly favor small, powerful groups with limited interest in transparency, accountability, and checks and balances. The public, meanwhile, is denied access to information about environmental monitoring or the zoning and investment plans that guide private sector and state actions. Space for public participation is constrained. The consequences are no surprise: across the developing world, city development planning often unfolds at a cost to the unrepresented poor. In Jakarta, improvement of city drainage has resulted in conflict with, and displacement of, slum communities that have been established for generations. In Brazil, investment in Olympic facilities is being used as a pretext for social cleansing, according to residents of the country's *favelas*. In Qatar, preparations for the World Cup reportedly have caused some 1,200 fatalities among migrant workers. The expansion of public transport infrastructure in Bangkok has led to evictions, pushing poorer residents away from gentrifying areas, while the new transport systems are priced beyond their means.[17]

Skytrain station in central Bangkok.

Urban poverty is increasingly recognized as having been underestimated in both its scale and its depth. It reveals itself not only in income deficits, but in other dimensions as well. The underestimation is partly a matter of gaps in the methods and indicators for measuring urban poverty, but also of the limited statistics regarding who are urban residents. Assessments of poverty based on indicators of income and expenditure often fail to account for the diverse economic and social needs of urban residents. Assessments that use the Multidimensional Poverty Index (MPI), developed by the Oxford Poverty & Human Development Initiative and the United Nations Development Programme (UNDP), take a more holistic approach to assessing poverty, including rights to and participation in planning and decision-making processes, access to and control over public goods and services, and environmental safety—as well as income and expenditure.[18]

Application of such approaches can provide a useful balance to more commonly used assessments. For example, in Ho Chi Minh City, Vietnam, UNDP has found that 16 percent of residents qualify as poor through an MPI lens, whereas, by the official definition, almost no one is poor. Income was not a good poverty indicator because of the high cost of access to core services such as water, toilets, and health care compared to rural settings. Migrants often are overlooked in surveys, despite being among the poorest residents. When urban people apply their own indicators of poverty, the ability to influence one's life and decision-making processes often is improved. Understanding urban poverty more clearly is a critical first step in building equity and inclusion, but it often is ignored. Yet, ultimately, the challenge is not merely reducing poverty, but improving well-being, rights, and prosperity.[19]

This has implications for several funded programs that currently are being rolled out, most prominently India's Smart Cities program. Some $15 billion will be spent on building 100 "smart" cities in the country and rejuvenating another 500 over five years. The intent is to make these cities more livable and inclusive than current Indian municipal centers and to drive economic growth. Yet their proposed functions as special economic zones, exempt from taxes and labor laws under privatized governance, make it clear that these cities will serve elites. Some will be extended gated communities, non-inclusive almost by definition. Such initiatives highlight the type of gaps that civil society voices are advocating to close, in order to increase the role of social inclusion and justice in urban development throughout the world.[20]

Until those gaps are closed, inequity will rise both within and among cities. Investments will be biased toward centers that can best guarantee risk-free returns. Urban agendas, including India's Smart Cities program, are likely to divert capital from poorer, less-competitive urban centers. If investments continue to be made as isolated and project-based decisions, national patterns of urbanization will risk creating winners and losers among their urban centers. Iconic, resilient cities will exist alongside poorer, unresilient neighbors.[21]

Transforming the Urban Agenda

With cities recognized as being at the forefront of addressing global climate change, it is clear that urbanization of the future will need to be very different from urbanization of the past, and from current trajectories. There is an urgent need for a transformative urban future that is socially just, inclusive, and ecologically viable. The biggest challenge to a transformative urban agenda is improving governance to achieve sustainability goals in places where it is currently dysfunctional, corrupt, inefficient, and/or incompetent, even though all required policies and regulations are nominally present. Transformations will need to be founded on open and informed debate as well as on engagement between people and their networks and with government planning processes.

Moving away from weak governance systems that are influenced heavily by powerful cliques will be hard and will not be possible everywhere, yet there are places where citizen representatives have formed networks, committees, and other advocacy groups alongside government "champions" seeking to improve government accountability. Where these have become established, efforts are being made to replicate and scale up successes at a peer level across cities, and through national-level engagement. Through the Asian Cities Climate Change Resilience Network, a nine-year initiative launched in 2008, the Rockefeller Foundation has invested $59 million to help shape national and municipal urbanization policy in 10 cities in India, Indonesia, Thailand and Vietnam, and its practices are now being copied by a further 50 cities across the region.[22]

The language of engagement among citizens, governments, and the private sector also needs reframing. Terms such as "resilience" need to be understood as part of building an inclusive urban agenda. Urban-focused and civil society-based networks in the developing world are concerned that the term "resilience" has been co-opted to mean "building resilience of cities," without sufficient definition of what is meant by "city." Legitimate questions include: What exactly is resilience being built for? Who will benefit? Who is authorizing the proposed measures?[23]

A greater focus on rights-based approaches needs to facilitate processes through which desperately needed city investments can be made in inclusive, transparent, and accountable terms. This includes strengthening regulatory frameworks while promoting investment. Cities are driven by trade and finance and, in turn, drive the economies of almost every country in the world. The need for them to continue to serve society in this way is fundamental to civilization. Yet, at the same time, cities serve their citizenry—not the other way around. Public systems and services should be adequately funded, accountable, and preferably under public ownership. (See Chapter 16.)

The challenge is immense. International development donors and NGOs calling for change in the trajectories of urbanization have minuscule amounts of capital to invest in programs and projects compared to those controlling the process. This is a challenge that citizens also must take on themselves. Addressing inclusion and social justice in the developing world's new and expanding cities will need to be done via persuasion, advocacy, and finding common ground with finance and investment mechanisms that too rarely are engaged.

Notes

World's Cities at a Glance

1. United Nations (UN), "Urban Population at Mid-Year by Major Area, Region and Country, 1950-2050," *World Urbanization Prospects: 2014 Revision* (New York: 2014).

2. UN, "Percentage of Population at Mid-Year Residing in Urban Areas by Major Area, Region and Country, 1950–2050," *World Urbanization Prospects: 2014 Revision*.

3. UN, "Average Annual Rate of Change of the Urban Population, by Major Area, Region, and Country, 1950–2050," *World Urbanization Prospects: 2014 Revision*.

4. UN, "Number of Cities Classified by Size Class of Urban Settlement, Major Area, Region and Country, 1950–2030," *World Urbanization Prospects, 2014 Revision*.

5. Stefan Bringezu, *Assessing Global Land Use: Balancing Consumption with Sustainable Supply*, A Report of the Working Group on Land and Soils of the International Resource Panel (Paris: UN Environment Programme (UNEP), Division of Technology, Industry and Economics (DTIE), 2014).

6. Karen C. Seto et al., "A Meta-Analysis of Global Urban Land Expansion," *PLoS ONE* 6, no. 8 (2011); Bringezu, *Assessing Global Land Use*; Joan Clos, "Foreword," in Mark Swilling et al., *City-Level Decoupling: Urban Resource Flows and the Governance of Infrastructure Transitions*, A Report of the Working Group on Cities of the International Resource Panel (Paris: UNEP, 2013).

7. Richard Dobbs et al., *Urban World: Mapping the Economic Power of Cities* (New York: McKinsey Global Institute, March 2011).

8. Richard Dobbs et al., *Urban World: Cities and the Rise of the Consuming Class* (New York: McKinsey Global Institute, 2012).

9. Swilling et al., *City-Level Decoupling*; UN-Habitat, *State of the World's Cities Report 2010/2011* (Nairobi: 2010). GNP is an economic statistic that is equal to GDP plus any income earned by residents from overseas investments minus income earned within the domestic economy by overseas residents.

10. Achim Steiner, "Foreword," in Swilling et al., *City-Level Decoupling*.

11. Swilling et al., *City-Level Decoupling*.

12. Dobbs et al., *Urban World: Cities and the Rise of the Consuming Class*.

13. Christopher A. Kennedy et al., "Energy and Material Flows of Megacities," *Proceedings of the National Academy of Sciences* 112, no. 19 (2015): 5,985–90; UN, "Toward Sustainable Cities," Chapter 3 in *World Economic and Social Survey 2013* (New York: 2013).

14. Ibid.; UN-Habitat, *Streets as Public Spaces and Drivers of Urban Prosperity* (Nairobi: 2013).

15. UN, "Toward Sustainable Cities"; World Health Organization (WHO), *Why Urban Health Matters* (Geneva: 2010).

16. WHO, "7 Million Premature Deaths Annually Linked to Air Pollution," press release (Geneva: March 25, 2014); WHO, "Household Air Pollution and Health," fact sheet no. 292 (Geneva: March 2014).

Chapter 1. Imagining a Sustainable City

2. Deep Decarbonization Pathways Project, *Pathways to Deep Decarbonization: 2015 Report* (Paris: Sustainable Development Solutions Network and Institute for Sustainable Development and International Relations, December 2015); Jörgen Larsson and Lisa Bolin, *Low-carbon Gothenburg 2.0: Technological Potentials and Lifestyle Changes* (Gothenburg, Sweden: Mistra Urban Futures, 2014); Jennie Moore and William E. Rees, "Getting to One-Planet Living," in Worldwatch Institute, *State of the World 2013: Is Sustainability Still Possible?* (Washington, DC: Island Press, 2013).

Chapter 2. Cities in the Arc of Human History: A Materials Perspective

1. This rough estimate reflects vagueness regarding the transition from villages to cities and the gradual pace of the spread of cities. Jericho, for example, is estimated to be 11,000 years old, while cities in Mesopotamia are estimated to have arisen around 7,500 years ago. Thus, 10,000 years is a rough, rounded approximation for the original rise of cities. The figure of 12,000 is from Helmut Haberl et al., "A Socio-Metabolic Transition Toward Sustainability? Challenges for Another Great Transformation," *Sustainable Development* 19, no. 1 (2011): 1–14.

2. Mark Swilling et al., *City-Level Decoupling. Urban Resource Flows and the Governance of Infrastructure Transitions* A Report of the Working Group on Cities of the International Resource Panel (Paris: United Nations Environment Programme, 2013); Haberl et al., "A Socio-Metabolic Transition Toward Sustainability?"

3. Leslie White, *The Science of Culture: A Study of Man and Civilization* (Clinton Corners, NY: Percheron Press, 2005).

4. Rolf Peter Sieferle, "Sustainability in a World History Perspective," in Brigitta Benzing and Bernd Herrmann, eds., *Exploitation and Overexploitation in Societies Past and Present. IUAES-Intercongress 2001 Goettingen* (Münster, Germany: LIT Publishing House, 2000), 123–42. Box 2–1 from Marina Fischer-Kowalski, Fridolin Krausmann, and Irene Pailua, "A Sociometabolic Reading of the Anthropocene: Modes of Subsistence, Population Size and Human Impact on Earth," *The Anthropocene Review* 1, no. 1 (2014): 8–13. Figure 2–1 from Sieferle, "Sustainability in a World History Perspective."

5. Two to four times from Fischer-Kowalski, Krausmann, and Pailua, "A Sociometabolic Reading of the Anthropocene"; 0.01 percent from Marina Fischer-Kowalski and Helmut Haberl, *Socioecological Transitions and Global Change: Trajectories of Social Metabolism and Land Use* (Cheltenham, U.K.: Edward Elgar, 2007); passive solar existence from Rolf Peter Sieferle, *The Subterranean Forest: Energy Systems and the Industrial Revolution*, translated from the German by Michael P. Osman (Cambridge, U.K.: White Horse Press, 2001); Haberl et al., "A Socio-Metabolic Transition Toward Sustainability?"

6. Richard Wrangham, *Catching Fire: How Cooking Made Us Human* (New York: Basic Books, 2009); Lewis Mumford, *The City in History: Its Origins, Its Transformations, and Its Prospects* (New York: Harcourt, Brace, and World, 1961), 10.

7. Box 2–2 from the following sources: Tim de Chant, "Hunter-Gatherers Show Human Populations Are Hard-Wired for Density," *Scientific American* blog, August 16, 2011; Marcus J. Hamilton et al., "Nonlinear Scaling of Space Use in Human Hunter-gatherers," *Proceedings of the National Academy of Sciences* 104, vol. 11 (2007): 4,765–69; Michael Batty and Peter Ferguson, "Defining City Size," *Environment and Planning B: Planning and Design* 38, no. 5 (2011): 753–56.

8. Sieferle, *The Subterranean Forest*; Mumford, *The City in History*.

9. Haberl et al., "A Socio-Metabolic Transition Toward Sustainability?"

10. Fischer-Kowalski, Krausmann, and Pailua, "A Sociometabolic Reading of the Anthropocene"; Table 2–1 from Haberl et al., "A Socio-Metabolic Transition Toward Sustainability?"

11. Vaclav Smil, *Energy in Nature and Society. General Energetics of Complex Systems* (Cambridge, MA: MIT Press, 2008); Fridolin Krausmann et al., "The Global Sociometabolic Transition: Past and Present Metabolic Profiles and Their Future Trajectories," *Journal of Industrial Ecology* 12, no. 5–6 (2008): 637–56.

12. Figure 2–2 from Kees Klein Goldewijk, Arthur Beusen, and Peter Janssen, "Long-term Dynamic Modeling of Global Population and Built-up Area in a Spatially Explicit Way: HYDE 3.1," *The Holocene* 20, no. 4 (2010): 565–73; data available at ftp://ftp.pbl.nl/hyde/supplementary/population/table_4.xls; 2050 projection from United Nations, *World Population Prospects* (New York: 2015).

13. Mumford, *The City in History*, 37.

14. Table 2–2 from Ian Morris, *Social Development* (Palo Alto, CA: Stanford University, October 2010).

15. Mumford, *The City in History*, 33.

16. Ibid., 34.

17. Thorkild Jacobsen and Robert M. Adams, "Salt and Silt in Ancient Mesopotamian Agriculture," *Science* 128, no. 3334 (1958): 1,251–58; Haberl et al., "A Socio-Metabolic Transition Toward Sustainability?"

18. Krausmann et al., "The Global Sociometabolic Transition."

19. Marina Fischer-Kowalski, Fridolin Krausmann, and Barbara Smetschka, "Modelling Transport as a Key Constraint to Urbanisation in Pre-industrial Societies," in Simron Singh et al., *Long Term Socio-Ecological Research: Studies in Society-Nature Interactions Across Spatial and Temporal Scales* (New York: Springer, 2014); Krausmann et al., "The Global Sociometabolic Transition."

20. Table 2–3 from Morris, *Social Development*; Krausmann et al., "The Global Sociometabolic Transition."

21. Krausmann et al., "The Global Sociometabolic Transition."

22. Mumford, *The City in History*, 359.

23. Ibid., 408.

24. "Subway," *Encyclopedia Britannica*, www.britannica.com/technology/subway, updated March 13, 2015; "Skyscraper," *Encyclopedia Britannica*, www.britannica.com/technology/skyscraper, updated April 22, 2015.

25. Krausmann et al., "The Global Sociometabolic Transition."

26. Fridolin Krausmann et al., "Growth in Global Materials Use, GDP and Population During the 20th Century," *Ecological Economics* 68, no. 10 (2009): 2,696–2,705. Table 2–4 from Haberl et al., "A Socio-Metabolic Transition Toward Sustainability?"

27. Krausmann et al., "The Global Sociometabolic Transition."

28. Figure 2–3 from Tertius Chandler, *Four Thousand Years of Urban Growth: An Historical Census* (Lewiston, NY: Saint David's University Press, 1987).

29. Krausmann et al., "The Global Sociometabolic Transition."

30. Fischer-Kowalski, Krausmann, and Pailua, "A Sociometabolic Reading of the Anthropocene"; Peter Victor, "Questioning Economic Growth," *Nature* 468 (November 18, 2010): 370–71.

31. Table 2–5 from Fischer-Kowalski, Krausmann, and Pailua, "A Sociometabolic Reading of the Anthropocene."

32. Ernst Ulrich von Weizsäcker et al., *Factor Five: Transforming the Global Economy Through 80% Improvements in Resource Productivity* (London: Earthscan, 2009).

33. Krausmann et al., "The Global Sociometabolic Transition."

Chapter 3. The City: A System of Systems

1. International Institute for Applied Systems Analysis (IIASA), "Urban Energy Systems," Chapter 18 in *Global Energy Assessment: Toward a Sustainable Future* (Cambridge, U.K.: Cambridge University Press, 2012); population share from United Nations (UN) Department of Economic and Social Affairs, "World Urbanization Prospects," electronic database, http://esa.un.org/unpd/wup/CD-ROM/. Direct final energy is the energy supplied to the consumer for heating, cooling, and lighting, not including energy embedded in the imports of manufactured goods. Urban shares of energy sources from International Energy Agency (IEA), *World Energy Outlook 2008* (Paris: 2008). The discussion here is based on two sources with the best city-level global coverage: the *Global Energy Assessment* and the *World Energy Outlook 2008*, both cited in this note.

2. Table 3–1 from the following sources: gross domestic product (GDP) per capita is a Worldwatch calculation based on regional groupings from IIASA, "Urban Energy Systems" and on national GDP data from World Bank, "World Development Indicators," electronic database (Washington, DC: December 2015); urban share of population is a Worldwatch compilation based on regional groupings from IIASA, "Urban Energy Systems" and on national urban shares from UN, "Urban Population at Mid-Year by Major Area, Region and Country, 1950–2050," *World Urbanization Prospects: 2014 Revision* (New York: 2014); urban energy per person is a Worldwatch calculation based on data from IIASA, "Urban Energy Systems" and from UN, "Urban Population at Mid-Year by Major Area, Region and Country, 1950–2050"; urban share of total final energy use from IIASA, "Urban Energy Systems."

3. Percentages are Worldwatch calculations based on regional groupings from IIASA, "Urban Energy Systems" and on national GDP data from World Bank, "World Development Indicators"; urban energy per person is a Worldwatch calculation based on data from IIASA, "Urban Energy Systems" and from UN, "Urban Population at Mid-Year by Major Area, Region and Country, 1950–2050."

4. UN-Habitat, "Energy Consumption in Cities," in *State of the World's Cities, 2008-09* (Nairobi: 2008). Figure 3–1 from U.S. Department of Energy, *International Energy Outlook 2013* (Washington, DC: 2013).

5. Alexander Ochs and Shakuntala Makhijani, *Sustainable Energy Roadmaps: Guiding the Global Shift to Domestic Renewables*, Worldwatch Report 187 (Washington, DC: Worldwatch Institute, 2012); IIASA, "Urban Energy Systems."

6. IEA, *Transition to Sustainable Buildings: Strategies and Opportunities to 2050* (Paris: 2013); Ochs and Makhijani, *Sustainable Energy Roadmaps*.

7. IIASA, "Urban Energy Systems."

8. Ibid.; Center for Energy and Climate Solutions, "Cogeneration/ Combined Heat and Power (CHP)," fact sheet (Arlington, VA: March 2011).

9. IIASA, "Urban Energy Systems."

10. Felix Creutzig et al., "Global Typology of Urban Energy Use and Potentials for an Urbanization Mitigation Wedge," *Proceedings of the National Academy of Sciences* 112, no. 20 (2015): 6,283–88; IIASA, "Urban Energy Systems."

11. Share of 75 percent from Mark Swilling et al., *City-Level Decoupling: Urban Resource Flows and the Governance of Infrastructure Transitions*, A Report of the Working Group on Cities of the International Resource Panel (Paris: UN Environment Programme (UNEP), 2013). Table 3–2 from Stefan Giljum et al., "Global Patterns of Material

Flows and Their Socio-Economic and Environmental Implications: A MFA Study on All Countries World-Wide from 1980 to 2009," *Resources* 3, no. 1 (2014): 319–39. Data here refer to the most common measure of materials consumption, Domestic Material Consumption (DMC), which is calculated as Domestic Extraction Used (DEU) plus net direct imports.

12. Giljum et al., "Global Patterns of Material Flows and Their Socio-Economic and Environmental Implications"; Ulrich Kral et al., "The Copper Balance of Cities: Exploratory Insights into a European and an Asian City," *Journal of Industrial Ecology* 18, no. 3 (2014): 432–44.

13. Luis M. A. Bettencourt et al., "Growth, Innovation, Scaling, and the Pace of Life in Cities," *Proceedings of the National Academy of Sciences* 104, no. 17 (2007): 7,301–06.

14. Ibid.

15. Ibid.

16. Luis Bettencourt and Geoffrey West, "A Unified Theory of Urban Living," *Nature* 467 (October 21, 2010): 912–13.

17. Bettencourt et al., "Growth, Innovation, Scaling, and the Pace of Life in Cities."

18. Table 3–3 from Daniel Hoornweg and Perinaz Bhada-Tata, *What a Waste: A Global Review of Solid Waste Management* (Washington, DC: World Bank, 2012).

19. Ibid.; Daniel Hoornweg, Perinaz Bhada-Tata, and Christopher Kennedy, "Peak Waste: When Is It Likely to Occur?" *Journal of Industrial Ecology* 19, no. 1 (2015): 117–28.

20. Christopher A. Kennedy et al., "Energy and Material Flows of Megacities," *Proceedings of the National Academy of Sciences* 112, no. 19 (2015): 5,985–90.

21. Thomas Graedel et al., *Recycling Rates of Metals: A Status Report* (Paris: UNEP and International Resource Panel, 2011); UNEP and International Environmental Technology Centre, "Policy Brief on E-waste: What, Why and How" (Osaka, Japan: May 13, 2013).

22. Steve Jennings, *Food in an Urbanized World: The Role of City Region Food Systems in Resilience and Sustainable Development* (London: International Sustainability Unit, April 2015); grapes share is a Worldwatch calculation based on production and export data from IndexMundi, "Agricultural Production, Supply, and Distribution," www.indexmundi.com/agriculture.

23. Sustainable Cities Institute, *Bringing Nutritious, Affordable Food to Underserved Communities: A Snapshot of Healthy Corner Store Initiatives in the United States* (Washington, DC: National League of Cities, February 2014).

24. Thomas Reardon et al., *Urbanization, Diet Change, and Transformation of Food Supply Chains in Asia* (East Lansing, MI: Michigan State University, Global Center for Food Systems Innovation, May 2014).

25. Urban expansion from Eugenie Birch with Alexander Keating, *Feeding Cities: Food Security in a Rapidly Urbanizing World*, conference report, Penn Institute for Urban Research, University of Pennsylvania, March 13–15, 2013; World Health Organization (WHO), "Obesity," fact sheet no. 311 (Geneva: August 2014); food environment from T. H. Chan School of Public Health, Harvard University, "Globalization," www.hsph.harvard.edu/obesity-prevention-source/obesity-causes/globalization-and-obesity/.

26. Table 3–4 from Jennings, *Food in an Urbanized World*.

27. Navin Ramamkutty et al., "Farming the Planet: 1. Geographic Distribution of Global Agricultural Lands in the Year 2000," *Global Biogeochemical Cycles* 22, no. 1 (2008); 19–29 percent from Consultative Group for International Agricultural Research, "Food Emissions: Supply Chain Emissions," Big Facts series, https://ccafs.cgiar.org/bigfacts/#theme=food-emissions&subtheme=supply-chain.

28. WHO, *Diet, Nutrition, and the Prevention of Chronic Disease: Report of a Joint WHO/FAO Expert Consultation* (Geneva: 2002); Mario Herrero et al., "Biomass Use, Production, Feed Efficiencies, and Greenhouse Gas Emissions from Global Livestock Systems," *Proceedings of the National Academy of Sciences* 110, no. 52 (2013): 20,888–93; Low Carbon Oxford, *Foodprinting Oxford: How to Feed a City*, study commissioned by the Oxford City Council (Oxford, U.K.: 2013).

29. UN Food and Agriculture Organization (FAO), "Boosting Food Security in Cities Through Better Markets, Reduced Food Waste," press release (Rome: May 28, 2015); Aarayman Arjun Singhal and Adam Lipinski, "How Food Waste Costs Our Cities Millions," World Resources Institute blog, April 16, 2015.

30. FAO, *Food Wastage Footprint: Impacts on Natural Resources, Summary Report* (Rome: 2013).

31. Ibid.

32. Rich Pirog et al., *Food, Fuel, and Freeways: An Iowa Perspective on How Far Food Travels, Fuel Usage, and Greenhouse Gas Emissions* (Ames, IA: Leopold Center for Sustainable Agriculture, 2001).

33. Christopher L. Weber and H. Scott Matthews, "Food-Miles and the Relative Climate Impacts of Food Choices in the United States," *Environmental Science and Technology* 42, no. 10 (2008): 3,508–13; Zambia from Jennings, *Food in an Urbanized World*.

34. Julie C. Padowski and Steven M. Gorelick, "Global Analysis of Urban Surface Water Supply Vulnerability," *Environmental Research Letters* 9, no. 10 (2014); UN World Water Assessment Programme, *Water for a Sustainable World, The United Nations World Water Development Report 2015* (Paris: UNESCO, 2015).

35. UN World Water Assessment Programme, *Water for a Sustainable World*.

36. Janet G. Hering et al., "A Changing Framework for Urban Water Systems," *Environmental Science and Technology* 47, no. 19 (2013): 10,721–26.

37. Ibid.; Glen T. Daigger, "Sustainable Urban Water and Resource Management," *The Bridge: Linking Engineering and Society* (National Academy of Engineering), Spring 2011.

38. NRW from Alexander Danilenko et al., *The IBNET Water Sanitation and Supply Blue Book, 2014* (Washington, DC: World Bank, 2014); Asit K. Biswas and Cecilia Tortajada, "Water Supply of Phnom Penh: An Example of Good Governance," *International Journal of Water Resources Development* 26, no. 2 (2010): 157–72.

39. David Sedlak, *Water 4.0: The Past, Present, and Future of the World's Most Vital Resource* (New Haven: Yale University Press, 2014).

40. Share of 75 percent and Table 3–5 from Toshio Sato et al., "Global, Regional, and Country Level Need for Data on Wastewater Generation, Treatment, and Use," *Agricultural Water Management* 130 (December 2013): 1–13; National Research Council, *Water Reuse: Potential for Expanding the Nation's Water Supply Through Reuse of Municipal Wastewater* (Washington, DC: National Academies Press, 2012).

41. Sedlak, *Water 4.0*; household level from National Academies of Science, Engineering, and Medicine, *Using Graywater and Stormwater to Enhance Local Water Supplies: An Assessment of Risks, Costs, and Benefits*, prepublication version (Washington, DC: National Academies Press, December 16, 2015).

42. Florida from Sedlak, *Water 4.0*; Namibia from J. Lahnsteiner and G. Lempert, "Water Management in Windhoek, Namibia," *Water Science & Technology* 55, no. 1–2 (2007): 441–48; Orange County from Sarah Yang, "Time Is Now for a New Revolution in Urban Water Systems," *Berkeley News*, February 14, 2014.

43. Sedlak, *Water 4.0*.

44. Ibid.

45. Daigger, "Sustainable Urban Water and Resource Management."

Chapter 4. Toward a Vision of Sustainable Cities

1. "Charter of European Cities & Towns Towards Sustainability (Aalborg Charter), May 27, 1994, www.sustaina blecities.eu/fileadmin/content/JOIN/Aalborg_Charter_english_1_.pdf; "Leipzig Charter on Sustainable European Cities," draft of May 2, 2007, http://ec.europa.eu/regional_policy/archive/themes/urban/leipzig_charter.pdf; State of Victoria, Australia, *Melbourne 2030: Planning for Sustainable Growth* (Melbourne: October 2002); Singapore Ministry of the Environment and Water Resources and Ministry of National Development, *Our Home, Our Environment, Our Future: Sustainable Singapore Blueprint 2015* (Singapore: 2015.)

2. Ernst Ulrich von Weizsäcker et al., *Factor Five: Transforming the Global Economy Through 80% Improvements in Resource Productivity* (London: Earthscan, 2009).

3. Mark Swilling et al., *City-Level Decoupling: Urban Resource Flows and the Governance of Infrastructure Transitions*, A Report of the Working Group on Cities of the International Resource Panel (Paris: United Nations Environment Programme, 2013), 49.

4. European Commission, Directorate-General for the Environment, *Scoping Study to Identify Potential Circular Economy Actions, Priority Sectors, Material Flows and Value Chains* (Brussels: August 2014).

5. Berkeley Public Library, "Tool Lending Library," https://www.berkeleypubliclibrary.org/locations/tool-lend ing-library.

6. Arnold Tukker et al., *Environmental Impact of Products (EIPRO): Analysis of the Life Cycle Environmental Impacts Related to the Final Consumption of the EU-25* (Brussels: European Commission Joint Research Centre, Institute for Prospective Technological Studies, May 2006).

7. Tammy Zboral, *Sustainable Cities: 10 Steps Forward. Municipal Action Guide* (Washington, DC: June 2010); European Union, *Making Our Cities Attractive and Sustainable: How the EU Contributes to Improving the Urban Environment* (Brussels: 2010).

8. Office of the Mayor, City of Los Angeles, "Mayor Villaraigosa Announces Completion of Largest LED Street Light Replacement Program," press release (Los Angeles: June 18, 2013).

9. Peter Droege, *100% Renewable Energy—And Beyond—for Cities* (Hamburg: HafenCity University and World Future Council Foundation, March 2010).

10. Ibid.

11. Danish Ministry of the Environment, "10 Principles for Sustainable City Governance" (Copenhagen: September 25, 2007).

12. Zachary Christin, Tracy Stanton, and Lola Flores, *Nature's Value from Cities to Forests: A Framework to Measure Ecosystem Services Along the Urban-Rural Gradient* (Tacoma, WA: Earth Economics, 2014); Timothy Beatley, *Biophilic Cities: Integrating Nature into Urban Design and Planning* (Washington, DC: Island Press, 2011), 8.

13. Beatley, *Biophilic Cities*, 8.

14. Ibid., 6.

15. Philip Johnstone et al., *Liveability and the Water-Sensitive City: Science-Policy Partnership for Water Sensitive Cities* (Clayton, Victoria, Australia: Cooperative Research Centre for Water Sensitive Cities, Monash University, August 2012).

16. Robert McDonald and Daniel Shemie, *Urban Water Blueprint: Mapping Conservation Solutions to the Global Water Challenge* (Washington, DC: The Nature Conservancy, 2014).

17. Ibid.

18. Table 4–3 from Beatley, *Biophilic Cities*, 8.

19. ResilientCity, "Urban Design Principles," www.resilientcity.org/index.cfm?id=11928, viewed October 22, 2015.

20. Luis M. A. Bettencourt, "The Kind of Problem a City Is: New Perspectives on the Nature of Cities from Complex Systems Theory," in Dietmar Offenhuber and Carlo Ratti, *Decoding the City: How Big Data Can Change Urbanism* (Berlin: De Gruyter, 2014), 168–79.

21. Patrick M. Condon, *Seven Rules for Sustainable Communities* (Washington, DC: Island Press, 2010).

22. Ibid.; Zboral, *Sustainable Cities*.

23. ResilientCity, "Urban Design Principles."

24. Sustainable Cities Institute, "Sustainable Strategies," www.sustainablecitiesinstitute.org/topics/land-use-and-planning/land-use-and-planning-sustainability-strategies.

25. Matthew Carmona et al., *Public Places, Urban Spaces* (Burlington, MA: Elsevier, 2010).

26. Zboral, *Sustainable Cities*.

27. Michael Mandel, *Connections as a Tool for Growth: Evidence from the LinkedIn Economic Graph* (Mountain View, CA: LinkedIn Corporation, November 2014).

28. Megan Heckert and Jeremy Mennis, "The Economic Impact of Greening Urban Vacant Land: A Spatial Difference-in-differences Analysis," *Environment and Planning A* 44 (2012): 3,010–27; Bianca Berragan, "Los Angeles' Tens of Thousands of Vacant Lots: Mapped," *Curbed Los Angeles*, May 15, 2015; Peleg Kremer and Zoé Hamstead, "Transformation of Urban Vacant Lots for the Common Good: An Introduction to the Special Issue," *Cities and the Environment* 8, no. 2 (2015).

29. City of Pickering, Ontario, Canada, "Sustainable Placemaking," https://www.pickering.ca/en/living/sustainable placemaking.asp, viewed November 13, 2015.

30. Julio D. D'Avila, ed., *Urban Mobility and Poverty: Lessons from Medellín and Soacha, Colombia* (Medellín: Universidad Nacional de Colombia, 2013).

31. Project for Public Spaces, *Placemaking and the Future of Cities* (New York: 2012); Coby Joseph, "Medellin Metrocable Improves Mobility for Residents of Informal Settlements," TheCityFix.com, August 26, 2014; D'Avila, ed., *Urban Mobility and Poverty*.

32. Joseph, "Medellin Metrocable Improves Mobility for Residents of Informal Settlements"; Julio D. D'Avila, "Going Up in Medellín: What Can We Learn from the City's Aerial Cable-car Lines?" *DPU News*, March 2013; Magdalena Cerdá et al., "Reducing Violence by Transforming Neighborhoods: A Natural Experiment in Medellín, Colombia," *American Journal of Epidemiology* 175, no. 10 (2012): 1,045–53.

33. Project for Public Spaces, "Bryant Park, NY: Publicly Owned, Privately Managed, and Financially Self-Supporting," www.pps.org/reference/mgmtbryantpark, viewed November 13, 2015.

34. Susan Silberberg et al., *Places in the Making: How Placemaking Builds Places and Communities* (Cambridge, MA: MIT Department of Urban Studies, 2013).

35. World Health Organization, *Health Indicators of Sustainable Cities in the Context of the Rio+20 UN Conference on Sustainable Development. Initial Findings from a WHO Expert Consultation, 17–18 May 2012* (Geneva: 2012).

36. Scott Cloutier, Lincoln Larson, and Jenna Jambeck, "Are Sustainable Cities 'Happy' Cities? Associations Between Sustainable Development and Human Well-being in Urban Areas of the United States," *Environment, Development and Sustainability* 16, no. 3 (2014): 633–47.

37. Danish Architecture Centre & Cities, "Beijing: Exercise Opportunities for All," January 21, 2014, www.dac.dk/en/dac-cities/sustainable-cities/all-cases/health/beijing-exercise-opportunities-for-all/.

38. Fábio Veras Soares, Rafael Perez Ribas, and Rafael Guerreiro Osório, "Evaluating the Impact of Brazil's Bolsa Familia: Cash Transfer Programs in Comparative Perspective," *Latin American Research Review* 45, no. 2 (2010).

39. United Nations Economic and Social Council, Commission on the Status of Women, *Economic Empowerment of Women. Report of the Secretary General* (New York: November 28, 2011).

40. Joel Rogers and Satya Rhodes-Conway, *Cities at Work: Progressive Local Policies to Rebuild the Middle Class. Report Summary* (Washington, DC: Center for American Progress Action Fund, February 2014).

41. Veras Soares, Perez Ribas, and Guerreiro Osório, "Evaluating the Impact of Brazil's Bolsa Familia."

42. Abraham H. Maslow, "A Theory of Human Motivation," *Psychological Review* 50, no. 4 (1943): 370–96; Clayton P. Alderfer, "An Empirical Test of a New Theory of Human Need," *Organizational Behavior and Human Performance* 4, no. 2 (1969): 142–75.

43. Table 4–8 from Johnstone et al., *Liveability and the Water-Sensitive City.*

44. Author's calculations based on data in Living Victoria Ministerial Advisory Council, *Living Melbourne, Living Victoria Roadmap* (Melbourne: Government of Victoria, Department of Sustainability and Environment, 2011).

45. Johnstone et al., *Liveability and the Water-Sensitive City.*

46. UN-Habitat, *Planning Sustainable Cities: Global Report on Human Settlements 2009* (London: Earthscan, 2009); Alexa Kasdan, Erin Markman, and Pat Convey, *A People's Budget: A Research and Evaluation Report on Participatory Budgeting in New York City* (New York: Community Development Project at the Urban Justice Center with the PBNYC Research Team, 2014).

47. URBACT, *Social Innovation in Cities: URBACT II Capitalisation* (St. Denis, France: April 2015).

48. David Satterthwaite et al., *Tools and Methods for Participatory Governance in Cities*, paper presented at the 6th Global Forum on Reinventing Government: Towards Participatory and Transparent Governance, Seoul, South Korea, May 24–27, 2005.

Chapter 5. The Energy Wildcard: Possible Energy Constraints to Further Urbanization

1. William Meyer, "Urban Legends," November 2014, http://news.colgate.edu/scene/2014/11/urban-legends.html.

2. International Energy Agency (IEA), *Key World Energy Statistics 2015* (Paris: 2015), 7.

3. Richard Heinberg, "Goldilocks Zone for Oil Prices Is Gone for Good," *Reuters*, March 24, 2015.

4. Robert Rapier, "Boom to Bust – 5 Stages of the Oil Industry," *Energy Trends Insider*, November 4, 2015; for Hughes' reports, see http://shalebubble.org; J. David Hughes, *Tight Oil Reality Check: Revisiting the U.S. Department of Energy Play-by-Play Forecasts Through 2040 from Annual Energy Outlook 2015* (Santa Rosa, CA: Post Carbon Institute, 2015); Art Berman, "Only 1% of the Bakken Play Breaks Even at Current Oil Prices," *Forbes*, November 3, 2015.

5. J. David Hughes, *Shale Gas Reality Check: Revisiting the U.S. Department of Energy Play-by-Play Forecasts Through 2040 from Annual Energy Outlook 2015* (Santa Rosa, CA: Post Carbon Institute, 2015).

6. Hughes, *Shale Gas Reality Check.*

7. IEA, *Key World Energy Statistics 2015*, 37.

8. Gordon Conway, *The Doubly Green Revolution: Food for All in the Twenty-First Century* (Ithaca, NY: Comstock

Publishing Associates, 1998). Figure 6–1 from Patrick Canning et al., *Energy Use in the U.S. Food System* (Washington, DC: U.S. Department of Agriculture, Economic Research Service, March 2010).

9. Ugo Bardi et al., "Turning Electricity into Food: The Role of Renewable Energy in the Future of Agriculture," *Journal of Cleaner Production* 53 (August 15, 2013): 224–31; Richard Heinberg and Michael Bomfod, *The Food and Farming Transition: Toward a Post-Carbon Food System* (Santa Rosa, CA: Post Carbon Institute, 2009).

10. David Biello, "Will Organic Food Fail to Feed the World?" *Scientific American*, April 25, 2012.

11. Figure 6–2 from Oak Ridge National Laboratory, *Transportation Energy Data Book: Edition 34* (Oak Ridge, TN: 2015), Table 2.7.

12. David Biello, "Bio-Jet Fuel Struggles to Balance Profit with Sustainability," *Scientific American*, December 5, 2011; Steve Conner, "Why the World Is Running Out of Helium," *The Independent* (U.K.), October 22, 2011.

13. Population Reference Bureau, "Human Population: Urbanization," Lesson Plans, July 2009, www.prb.org/Publications/Lesson-Plans/HumanPopulation/Urbanization.aspx.

14. Joseph Tainter, *The Collapse of Complex Societies* (New York: Cambridge University Press, 1988).

15. Ibid.

16. See, for example, The Greenhorns website, www.thegreenhorns.net, and the Transition Network website, https://www.transitionnetwork.org.

Chapter 6. Cities and Greenhouse Gas Emissions: The Scope of the Challenge

1. World Bank, "Urban Development: Overview," www.worldbank.org/en/topic/urbandevelopment/overview, viewed December 17, 2015.

2. Table 6–1 from Daniel Hoornweg, Lorraine Sugar, and Claudia Lorena Trejos Gómez, "Cities and Greenhouse Gas Emissions: Moving Forward," *Environment and Urbanization* 23, no. 1 (2011): 207–27; Christopher Kennedy et al., "Greenhouse Gas Emissions from Global Cities," *Environmental Science & Technology* 43, no. 19 (2009): 7,297–7,302.

3. Felix Creutzig et al., "Global Typology of Urban Energy Use and Potentials for an Urbanization Mitigation Wedge," *Proceedings of the National Academy of Sciences* 112, no. 20 (2015): 6,283–88.

4. C. A. Kennedy, N. Ibrahim, and D. Hoornweg, "Low-Carbon Infrastructure Strategies for Cities," *Nature Climate Change* 4 (2014): 343–46.

5. For varying economic and emissions profiles of different cities, see Chris Sall and Jigar V. Shah, *The Role of Industry in Forging Green Cities* (Washington, DC: Institute for Industrial Productivity, March 2015); for discussions of other distinguishing city characteristics, see also Christopher A. Kennedy et al., "Energy and Material Flows of Megacities," *Proceedings of the National Academy of Sciences* 112, no. 19 (2015): 5,985–90, and Kennedy, Ibrahim, and Hoornweg, "Low-carbon Infrastructure Strategies for Cities."

6. C40 Cities and Arup, *Climate Action in Megacities. C40 Cities Baseline and Opportunities. Volume 2.0* (London: February 2014).

7. Ibid.

8. Ibid.

9. Christopher Kennedy et al., "Methodology for Inventorying Greenhouse Gas Emissions from Global Cities," *Energy Policy* 38, no. 9 (2009): 4,828–37; STAR Communities, "The Rating System," www.starcommunities.org/rating-system; World Resources Institute (WRI), C40 Cities, and ICLEI–Local Governments for Sustainability,

Global Protocol for Community-Scale Greenhouse Gas Emissions Inventories: Executive Summary (Washington, DC: WRI, 2014).

10. World Bank, "Planning and Financing Low-Carbon, Livable Cities," September 26, 2013, www.worldbank.org /en/news/feature/2013/09/25/planning-financing-low-carbon-cities; C40 Cities and Arup, *Climate Action in Megacities*.

11. Creutzig et al., "Global Typology of Urban Energy Use and Potentials for an Urbanization Mitigation Wedge."

12. Global Commission on the Economy and Climate, "Chapter 2: Cities," in *Better Growth, Better Climate. The New Climate Economy Report* (Washington, DC: WRI, 2014); Creutzig et al., "Global Typology of Urban Energy Use and Potentials for an Urbanization Mitigation Wedge."

13. Lorraine Sugar and Christopher Kennedy, "A Low-Carbon Infrastructure Plan for Toronto, Canada," *Canadian Journal of Civil Engineering* 40, no. 2 (2013): 86–96.

14. Creutzig et al., "Global Typology of Urban Energy Use and Potentials for an Urbanization Mitigation Wedge."

15. Scott Nyquist, "Peering into Energy's Crystal Ball," *McKinsey Quarterly*, July 2015; Per-Anders Enkvist et al., "A Cost Curve for Greenhouse Gas Reduction," *McKinsey Quarterly*, February 2007. Currency exchange reflects rate of €1 = \$1.0836.

16. Creutzig et al., "Global Typology of Urban Energy Use and Potentials for an Urbanization Mitigation Wedge"; David Banister, "Cities, Mobility, and Climate Change," *Journal of Transport Geography* 19, no. 6 (2011): 1,538–46.

17. Ibid.

18. Xuemei Bai, "Emerging Patterns of Urban Sustainability in Asia," *The Bridge on Urban Sustainability* 41, no. 1 (Spring 2011): 35–42; Xuemei Bai et al., "Enabling Sustainability Transitions in Asia: The Importance of Vertical and Horizontal Linkages," *Technological Forecasting and Social Change* 76, no. 2 (2009): 255–66.

19. Gordon McGranahan, Deborah Balk, and Bridget Anderson, "The Rising Tide: Assessing the Risks of Climate Change and Human Settlements in Low-Elevation Coastal Zones," *Environment & Urbanization* 19, no. 1 (2007): 17–37.

20. Vaclav Smil, *Energy Transitions: History, Requirements, Prospects* (Santa Barbara: Praeger, 2010).

Chapter 7. Urbanism and Global Sprawl

1. Figure of 80 percent from Global Commission on the Economy and Climate, *Better Growth, Better Climate; The New Climate Economy Report* (Washington, DC: World Resources Institute, 2014); 70 percent from World Bank, *Systems of Cities: Harnessing Urbanization for Growth and Poverty Alleviation* (Washington, DC: 2009) and from Intergovernmental Panel on Climate Change, "Summary for Policymakers," in *Climate Change 2014, Mitigation of Climate Change. Contribution of Working Group III to the Fifth Assessment Report of the Intergovernmental Panel on Climate Change* (New York and Cambridge, U.K.: Cambridge University Press, 2014); population shares from UN-Habitat, *Realizing the Future We Want for All: Report to the Secretary General* (Nairobi: 5 July 2012).

2. Chris Busch and CC Huang, *Quantitative Insights into Urban Form and Transportation Solutions* (San Francisco: Energy Innovation: Policy and Technology, LLC, 16 October 2014).

3. Figure of 90 percent from World Bank, *Systems of Cities*; 86 percent from Chris Busch and CC Huang, *Cities for People: Insights from the Data* (San Francisco: Energy Innovation: Policy and Technology, LLC, April 2015); Global Commission on the Economy and Climate, *Better Growth, Better Climate*.

4. One twentieth to one hundredth from World Bank, *Systems of Cities*; per capita emissions from World Bank, Table 3.8 in *World Development Indicators 2015*, http://wdi.worldbank.org/tables; 2050 target from Deep

Decarbonization Pathways Project, *Pathways to Deep Decarbonization* (New York and Paris: Sustainable Development Solutions Network and Institute for Sustainable Development and International Relations, September 2014).

5. Global Commission on the Economy and Climate, *Better Growth, Better Climate*; Busch and Huang, *Quantitative Insights into Urban Form and Transportation Solutions*.

6. Busch and Huang, *Quantitative Insights into Urban Form and Transportation Solutions*; Shlomo Angel, *Planet of Cities* (Cambridge, MA: Lincoln Institute of Land Policy, 2012).

7. Busch and Huang, *Quantitative Insights into Urban Form and Transportation Solutions*; Peter Calthorpe, "Weapons of Mass Urban Destruction," ForeignPolicy.com, 13 August 2012.

8. Calthorpe, "Weapons of Mass Urban Destruction."

9. Ibid.

10. Ibid.

11. Ibid.

12. Ibid.

13. Ibid.; 1.2 million from Busch and Huang, *Cities for People: Insights from the Data*. Figure 7–1 from Global Commission on the Economy and Climate, *Better Growth, Better Climate*, 35.

14. Calthorpe, "Weapons of Mass Urban Destruction"; World Health Organization, *Global Status Report on Road Safety 2015* (Geneva: 2015), 110.

15. Ibid.

16. Ibid.

17. Ibid.

18. TOD results from author's personal communication with partners in China, 2015.

19. Busch and Huang, *Quantitative Insights into Urban Form and Transportation Solutions*.

20. Jinan from Busch and Huang, *Cities for People: Insights from the Data*; population growth from Calthorpe, "Weapons of Mass Urban Destruction."

21. Calthorpe, "Weapons of Mass Urban Destruction."

22. Centro Mario Molina and Calthorpe Analytics, *Zona Metropolitana del Valle de Mexico. Regional Scenarios and Modeling: Project Description and Work Plan* (Mexico City: June 2013); World Bank, *Systems of Cities*.

23. World Bank, *Systems of Cities*.

24. Ibid.; Barney Cohen, "Urbanization in Developing Countries: Current Trends, Future Projections, and Key Challenges for Sustainability," *Technology in Society* 28, nos. 1–2 (2006): 63–80.

25. World Bank, *Systems of Cities*.

26. Centro Mario Molina and Calthorpe Analytics, *Zona Metropolitana del Valle de Mexico*.

27. Mexico gross national income from World Bank, Table 1.1 in World Development Indicators, http://wdi .worldbank.org/table/1.1; Centro Mario Molina and Calthorpe Analytics, *Zona Metropolitana del Valle de Mexico*.

28. Figure of 20 million people from Centro Mario Molina and Calthorpe Analytics, *Zona Metropolitana del Valle de Mexico*; middle class and wealthier from David E. Dowall and David Wilk, *Population Growth, Land Development, and Housing in Mexico City* (Berkeley, CA: University of California at Berkeley, Institute of Urban

and Regional Development, 1989); segregation from Peter M. Ward, "Mexico City," *New York* 159 (1998): 1–86.

29. Figure 7–3 from Centro Mario Molina and Calthorpe Analytics, *Zona Metropolitana del Valle de Mexico*, 20.

30. Ibid.

31. Ibid.

32. Ibid.

33. Centro Mario Molina and Calthorpe Analytics, *Urban Planning Modeling Scenarios: Mexico City Metropolitan Area* (Mexico City: 18 August 2015).

34. Ibid.

35. Figure 7–4 from Ibid.

36. Ibid.

37. Ibid.

38. Ibid.

39. Ibid.

40. Ibid.

41. Figure 7–5 from Ibid., 31.

42. Ibid.; Calthorpe, "Weapons of Mass Urban Destruction."

43. Busch and Huang, *Cities for People: Insights from the Data*; Global Commission on the Economy and Climate, *Better Growth, Better Climate.*

44. Busch and Huang, *Cities for People: Insights from the Data*; Global Commission on the Economy and Climate, *Better Growth, Better Climate*; Centro Mario Molina and Calthorpe Analytics, *Urban Planning Modeling Scenarios: Mexico City Metropolitan Area.*

45. Busch and Huang, *Quantitative Insights into Urban Form and Transportation Solutions*; Calthorpe, "Weapons of Mass Urban Destruction."

City View: Shanghai, China

1. Population and area data from Shanghai Municipal Government website: www.shanghai.gov.cn (in Mandarin).

2. National Development Reform Commission (NDRC), *The 11th Five-Year Plan*, www.gov.cn/english/2006-03/23/content_234832.htm; NDRC, *The 12th Five-Year Plan*, www.cbichina.org.cn/cbichina/upload/fckeditor/Full%20Translation%20of%20the%2012th%20Five-Year%20Plan.pdf.

3. Shanghai Municipal Government, *The 12th Five-Year Plan*, www.shdrc.gov.cn/main?main_colid=498&top_id=398&main_artid=19079 (in Mandarin).

4. Shanghai Municipal Government, *The 11th Five-Year Plan on Energy Development*, September 29, 2006, www.pkulaw.cn/fulltext_form.aspx?Db=lar&Gid=16889806&EncodingName=big5 (in Mandarin).

5. Shanghai Municipal Government, *The 11th Five-Year Plan on Energy Conservation*, January 26, 2007, www.pkulaw.cn/fulltext_form.aspx?Gid=16920548&Db=lar&EncodingName=big5 (in Mandarin); Shanghai Municipal Government, *The 12th Five-Year Plan on Industrial Energy Conservation and Comprehensive Utilization*, www.sheitc.gov.cn (in Mandarin).

6. Shanghai Municipal Government, *The 11th Five-Year Plan on Environmental Protection and Ecological Construction*, November 12, 2007, www.shanghai.gov.cn/nw2/nw2314/nw2319/nw10800/nw11407/nw16795/u26aw12669.html (in Mandarin).

7. Shanghai Municipal Government, *Managing Rules for Carbon Emission Trading*, November 18, 2013, www.shanghai.gov.cn/nw2/nw2314/nw2319/nw11494/nw12654/nw31364/u26aw37414.html (in Mandarin).

8. Shanghai Municipal Government, *New Energy Vehicles Promotion Plan*, June 5, 2014, http://news.china.com.cn/2015lianghui/2015-02/28/content_34912935.htm (in Mandarin).

9. Shanghai Statistical Bureau, *Shanghai Statistical Year Book 2006–2013*, www.stats-sh.gov.cn/data/release.xhtml (in Mandarin).

10. Ibid.; offshore wind data from Wikipedia, https://zh.wikipedia.org/wiki/%E4%B8%8A%E6%B5%B7%E6%9D%B1%E6%B5%B7%E5%A4%A7%E6%A9%8B%E6%B5%B7%E4%B8%8A%E9%A2%A8%E9%9B%BB%E5%A0%B4 (in Mandarin).

11. Shanghai Municipal Government, *The 12th Five-Year Plan on Industrial Development*, December 28, 2011, www.shanghai.gov.cn/nw2/nw2314/nw2319/nw10800/nw11407/nw25262/u26aw30908.html (in Mandarin); 2013 sectoral data from Shanghai Statistical Bureau, *Shanghai Statistical Year Book 2006–2013*.

12. Shanghai Municipal Government, *The 12th Five-Year Plan on Energy Conservation and Climate Change*, March 28, 2012, www.shanghai.gov.cn/nw2/nw2314/nw2319/nw22396/nw22403/u21aw597380.html (in Mandarin).

13. "Shanghai Achieved Its 11th Five-Year Plan's Energy Intensity Target," http://forum.home.news.cn/detail/99771912/1.html (in Mandarin).

14. "Shanghai Carbon Market Overview," www.ce.cn/cysc/ny/gdxw/201512/01/t20151201_7218182.shtml (in Mandarin); Shanghai Statistical Bureau, *Shanghai Statistical Year Book 2006–2013*.

15. "New Energy Vehicles Sales Booming in Shanghai," www.dlev.com/39063.html (in Mandarin).

16. "Shanghai Metro," Wikipedia, https://en.wikipedia.org/wiki/Shanghai_Metro.

17. Shanghai Municipal Government, *The 12th Five-Year Plan on Environmental Protection and Ecological Construction*, April 11, 2012, www.shanghai.gov.cn/shanghai/node2314/node25307/node25455/node25459/u21ai602117.html (in Mandarin).

18. Shanghai Municipal Government, "Regarding Phase III of the Dongtan Eco-city Land Reserve," November 23, 2015, www.shgtj.gov.cn/2011/gcjsxx/xmxx/ghxzyj/201511/t20151130_671266.html (in Mandarin).

19. "Shanghai High-rise Fire Death Toll Rises to 58," *Xinhua News Agency*, November 19, 2010.

20. Shanghai Statistical Bureau, *Shanghai Statistical Year Book 2006–2013*.

21. "Thousands Protest in Shanghai Suburb Over Chemical Plant Fears," *Reuters*, June 27, 2015.

Chapter 8. Reducing the Environmental Footprint of Buildings

1. International Renewable Energy Agency (IRENA), "IRENA Opens Doors on New Permanent Headquarters in Masdar City," press release (Abu Dhabi: June 3, 2015).

2. Navigant Research, "Global Building Stock Database. Commercial and Residential Building Floor Space by Country and Building Type: 2013-2023," https://www.navigantresearch.com/research/global-building-stock-database.

3. European Union from Alessandro Cesale et al., *The Role of Public, Cooperative and Social Housing Providers in the Fair Energy Transition* (Brussels: Housing Europe, May 2015); buildings over 50 years old from European Commission, "Buildings," http://ec.europa.eu/energy/en/topics/energy-efficiency/buildings; United States from

Na Zhao, "The Aging Housing Stock," National Association of Home Builders, August 11, 2015, http://eyeonhous ing.org/2015/08/the-aging-housing-stock-2/; New York City Mayor's Office of Sustainability, "Energy Efficiency," www.nyc.gov/html/planyc/html/sustainability/energy-efficiency.shtml.

4. Economist Intelligence Unit and Siemens, *European Green City Index. Assessing the Environmental Impact of Europe's Major Cities* (Munich: 2009).

5. Mark Roseland, *Toward Sustainable Communities. Solutions for Citizens and Their Governments* (Gabriola Island, Canada: New Society Publishers, 2012), 196.

6. International Energy Agency (IEA), *Energy Efficiency Market Report 2015. Market Trends and Medium-Term Prospects* (Paris: 2015).

7. Ibid.

8. Stephanie Vierra, "Green Building Standards and Certification Systems," Whole Building Design Guide, October 27, 2014, https://www.wbdg.org/resources/gbs.php.

9. Lily Mitchell, "Green Star and NABERS: Learning from the Australian Experience with Green Building Rating Tools," presentation at the Fifth Urban Research Symposium, Marseille, France, June 28–30, 2009.

10. BREEAM, "What Is BREEAM?" www.breeam.com/about.jsp?id=66, viewed October 20, 2015.

11. U.S. Green Building Council (USGBC), "USGBC Announces International Rankings of Top 10 Countries for LEED Green Building," press release (Washington, DC: July 22, 2015).

12. Karim Elgendy, "Estidama vs BREEAM vs LEED," Carboun Middle East Sustainable Cities, April 17, 2010, www.carboun.com/sustainable-urbanism/comparing-estidama's-pearls-rating-method-to-leed-and-breeam/. Table 8–1 based on the following sources: Vierra, "Green Building Standards and Certification Systems"; BREEAM, "What Is BREEAM?"; USGBC, "USGBC Announces International Rankings of Top 10 Countries for LEED Green Building"; Green Globes, "About Green Globes," www.greenglobes.com/about.asp; Green Rating for Integrated Habitat Assessment, "About GRIHA," www.grihaindia.org/index.php?option=com_content&view=ar ticle&id=73; Institute for Building Efficiency, "Green Building Rating Systems: China," fact sheet (Milwaukee, WI: September 2013); Business Sweden, "Certifications," www.business-sweden.se/en/Trade/international-markets /americas/Brazil/Environmental-Technology-Initiative/Green-Building/Green-building-market-overview/Certi fications/; Vietnam Green Building Council, "LOTUS Green Building Rating & Classification System," http://vgbc .org.vn/index.php/pages/lotus-rating-tool; see also websites of the organizations administering each rating system.

13. Jonathan Hiskes, "The Case for Super-ambitious Living Buildings. A Talk with Jason McLennan," *Grist*, September 30, 2010; International Living Future Institute, "Bullitt Center," http://living-future.org/bullitt-center-0; SITES website, www.sustainablesites.org.

14. Roseland, *Toward Sustainable Communities*, 207–08.

15. Box 8–1 based on the following sources: Mostra Convegno Expocomfort (MCE) Asia, "Asia's Green Building Industry & Growth," www.mcexpocomfort-asia.com/About-MCE-Asia/Regional-Industry-Outlook/, viewed August 21, 2015; USGBC, *Green Building Economic Impact Study* (Washington, DC: September 2015); U.S. Census Bureau, "Annual Value of Construction Put in Place, 2008-2014," https://www.census.gov/construction/c30/histor ical_data.html. Table 8–2 from IEA, *Energy Efficiency Market Report 2015*; China construction market calculated from 2013 data in AECOM, *Asia Construction Outlook 2014* (Singapore: 2014). AECOM puts total Chinese construction at $1.78 trillion; excluding infrastructure construction (37 percent of total), building construction comes to $1.12 trillion.

16. U.S. Environmental Protection Agency, *Sustainable Design and Green Building Toolkit for Local Governments* (Washington, DC: June 2013).

17. Andy Gouldson et al., *Accelerating Low-Carbon Development in the World's Cities*, New Climate Economy Working Paper (Washington, DC: Global Commission on the Economy and Climate, 2015).

18. Justin Gerdes, "Copenhagen's Ambitious Push to Be Carbon Neutral by 2025," *Yale Environment 360*, April 11, 2013.

19. Gabriela Weber de Morais, "Citizens Contributing to Urban Sustainability in Vauban, Germany," in Mark Swilling et al., *City-Level Decoupling: Urban Resource Flows and the Governance of Infrastructure Transitions. Case Studies from Selected Cities*, A Report of the Working Group on Cities of the International Resource Panel (Paris: United Nations Environment Programme, 2013), 19–21.

20. San Francisco Department of the Environment, "Green Building," www.sfenvironment.org/buildings-environments/green-building, viewed September 21, 2015.

21. C40 Cities, "Case Study: Seoul's Building Retrofit Program," December 2014, www.c40.org/case_studies/seoul-s -building-retrofit-program.

22. Tokyo Metropolitan Government and C40 Cities Climate Leadership Group, "Chapter 4: Experiences from Frontrunner Cities," in *Urban Efficiency: A Global Survey of Building Energy Efficiency Policies in Cities* (Tokyo and New York: 2014).

23. Renewable Energy Network for the 21st Century (REN21), *Renewables 2015 Global Status Report* (Paris: 2015), 97; Bärbel Epp, "China: No Sales Permit Without Solar," SolarThermalWorld.org, August 21, 2014; Rizhao from REN21, Institute for Sustainable Energy Policies, and ICLEI–Local Governments for Sustainability, *Global Status Report on Local Renewable Energy Policies* (Paris: 2011), 46.

24. ICLEI, *Barcelona, Spain. Using Solar Energy – Supporting Community Self Sufficiency*, ICLEI Case Studies 173 (Bonn: December 2014).

25. Ibid.

26. ICLEI and IRENA, "São Paulo, Brazil. Local Government Regulation: Ordinances and Laws to Promote Renewable Energy" (Abu Dhabi: 2013); Associação Brasileira de Refrigeração, Air Condicionado, Ventilação e Aquecimento (ABRAVA), "Relatório de Pesquisa. Produção de Coletores Solares para Aquecimento de Água e Reservatórios Térmicos no Brasil. Ano de 2014" (Brasilia: Departamento Nacional de Aquecimento Solar, May 2015).

27. Ibid.

28. Roseland, *Toward Sustainable Communities*, 205–06; USGBC, World Green Building Council (WGBC), and C40 Cities, *Green Building City Market Briefs* (Washington, DC: February 2015).

29. International Labour Organization (ILO), *Sustainable Development, Decent Work and Green Jobs* (Geneva: 2013); Vanessa Kriele, "Brazil Offers New Green Building Credit Terms," SolarThermalWorld.org, September 29, 2015.

30. USGBC, WGBC, and C40 Cities, *Green Building City Market Briefs*.

31. Johannesburg from ILO, *Working Towards Sustainable Development* (Geneva: 2012); Natalie Mayer, "Energy-Efficient Housing Upgrades for the Poor in Cape Town, South Africa," in Swilling et al., *City-Level Decoupling*, 74–76; Holle Linnea Wlokas and Charlotte Ellis, *Local Employment Through the Low-pressure Solar Water Heater Roll-out in South Africa* (Cape Town: Energy Research Centre, University of Cape Town, 2013).

32. Fuel poverty from Cesale et al., *The Role of Public, Cooperative and Social Housing Providers in the Fair Energy Transition*; European Federation of Public, Cooperative and Social Housing from Housing Europe, "About Us," www.housingeurope.eu/section-37/about-us.

33. "E3SoHo Final European Workshop: ICT-enabled Energy Efficiency in European Social Housing," www.euro cities.eu/eurocities/events/E3SoHo-Final-European-Workshop-ICT-enabled-Energy-Efficiency-in-European -Social-Housing-; ENCERTICUS, "ESESH: Saving Energy in Social Housing with ICT," http://med-encerticus.eu /it/link/esesh-saving-energy-in-social-housing-with-ict.asp.

34. U.S. Department of Energy, Office of Energy Efficiency and Renewable Energy, "Cool Roofs Are Ready to Save Energy, Cool Urban Heat Islands, and Help Slow Global Warming," fact sheet (Washington, DC: undated).

35. Laura Wisland, "How Many Homes Have Rooftop Solar? The Number Is Growing…," *The Equation* (Union of Concerned Scientists blog), September 4, 2014.

36. Table 8–3 from International Green Roof Association (IGRA), "Green Roof Types," www.igra-world.com /types_of_green_roofs/.

37. "Why don't all public buildings have green roofs? Or all large private buildings (e.g. businesses)? Would this be a good idea? What would it take to make it happen and to make it worthwhile?" TheNatureofCities.com, August 12, 2015.

38. Wolfgang Ansel and Roland Appl, *Green Roof Policies – An International Review of Current Practices and Future Trends* (Nürtingen, Germany: IGRA, 2015).

39. Dorthe Rømø, "Green Roofs Worldwide," PowerPoint presentation (Copenhagen: 2012), www.scp-knowledge .eu/sites/default/files/Rømø%202012%20Green%20roofs%20worldwide_0.pdf; more than 80 cities from Green-roofs.com, "Industry Support," www.greenroofs.com/Greenroofs101/industry_support.htm, viewed September 21, 2015; Stuttgart from IGRA, *Green Roof News*, no. 2 (2015): 11, and from Rømø, "Green Roofs Worldwide."

40. "France Decrees New Rooftops Must Be Covered in Plants or Solar Panels," *Agence France-Presse,* March 19, 2015; Hidalgo from IGRA, *Green Roof News*, 3.

41. C40 Cities, "Case Study: Nature Conservation Ordinance Is Greening Tokyo's Buildings," March 18, 2015, www.c40.org/case_studies/nature-conservation-ordinance-is-greening-tokyo-s-buildings.

42. City Planning Division, Toronto, "Green Roofs," http://www1.toronto.ca/wps/portal/contentonly?vgnextoid =3a7a036318061410VgnVCM10000071d60f89RCRD; Green Roofs for Healthy Cities, *2014 Annual Green Roof Industry Survey* (Toronto: May 2015).

43. USGBC, WGBC, and C40 Cities, *Green Building City Market Briefs.*

44. Passive House Institute, "Passive House Requirements," http://passiv.de/en/02_informations/02_passive-house -requirements/02_passive-house-requirements.htm.

45. International Passive House Association, "Passive House Legislation," www.passivehouse-international.org /index.php?page_id=176; Munich from REN21, *Renewables 2015 Global Status Report.*

46. Box 8–2 from Marie-Pierre Establie D'Argencé, Sylvaine Herold, and Henri Le Marois, "Lessons from the Project 'Employment Centres and Sustainable Development' in France," in Organization for Economic Co-oper-ation and Development (OECD) and European Centre for the Development of Vocational Training (CEDEFOP), *Greener Skills and Jobs*, OECD Green Growth Studies (Paris: 2014).

47. USGBC, WGBC, and C40 Cities, *Green Building City Market Briefs.*

48. Rachel Young, "Global Approaches: A Comparison of Building Energy Codes in 15 Countries," in American Council for an Energy-Efficient Economy, *Proceedings 2014 ACEEE Summer Study on Energy Efficiency in Buildings* (Washington, DC: 2014).

49. Online Code Environment & Advocacy Network, "Code Status: International Non-Residential," http://energy codesocean.org/code-status-international-non-residential.

50. European Commission, "Buildings"; Concerted Action Energy Performance of Buildings website, www.epbd -ca.eu.

51. Coalition for Energy Savings, *Implementing the EU Energy Efficiency Directive: Analysis of Member States Plans to Implement Article 5* (Brussels: May 2015).

City View: Freiburg, Germany

1. Data from City of Freiburg, Amt für Bürgerservice und Informationsverarbeitung, *Statistisches Jahrbuch 2015* (Freiburg: November 2015).

2. City of Freiburg, "Green City Brochure," https://www.freiburg.de/pb/site/Freiburg /get/640888/Green-City -Brochure_English.pdf.

3. European Sustainable Cities Platform, "The Aalborg Charter," www.sustainablecities.eu/aalborg-process/char ter, viewed December 2015.

4. Council for Sustainable Development, Oberbürgermeisterdialog "Nachhaltige Stadt," www.nachhaltigkeitsrat .de/en/projects/projects-of-the-council/nachhaltige-stadt.

5. City of Freiburg, "1. Freiburger Nachhaltigkeitsbericht 2014," www.freiburg.de/ pb/site/Freiburg/get/761949 /Freiburger_Nachhaltigkeitsbericht_2014.pdf.

6. City of Freiburg, "G-13/147: Nachhaltigkeitsmanagement, hier: a) Steuerung des Kommunalen Nach-haltigkeitsprozesses und Weiterentwicklung des Handlungskonzeptes b) Verknüpfung Neues Kommunales Haushaltsrecht und Nachhaltigkeitszielsystem: Schlüsselprodukte."

7. City of Freiburg, "G-12/089: Einrichtung eines Fonds 'Bildung für nachhaltige Entwicklung.'"

8. City of Freiburg, "Stadt der UN-Weltdekade 'Bildung für nachhaltige Entwicklung,'" www.freiburg.de/pb /,Lde/206227.html?QUERYSTRING=UNESCO+Bildung.

9. City of Freiburg, "Handlungsprogramm Wohnen, Bevölkerungsprognose und Wohnungsmarktanalyse," www .freiburg.de/pb/,Lde/767365.html?QUERYSTRING=Altersstruktur.

10. City of Freiburg, "G-15/126: Aktionsplan für ein inklusives Freiburg 2015/2016."

11. City of Freiburg, "G-14/047: Klimaschutzbilanz für die Jahre 2010 und 2011 und Fortschreibung des Klima-schutzkonzeptes (Klimaschutzziele, Konzessionsabgabe und Maßnahmenplan der Verwaltung)."

12. Freiburger Stadtbau, "Leuchtturmprojekt des nachhaltigen Bauens," press release (Freiburg: April 21, 2007).

13. City of Freiburg, "G-15/009: Klimaschutz in Gewerbe und Industrie, hier: Vorstellung des Klimaschutzteilkon-zept für den Green Industry Park Freiburg."

14. City of Freiburg, "G-08/031: Verkehrsentwicklungsplan (VEP) 2020 – Verabschiedung des Endberichtes mit Zielen, Schwerpunkten und Maßnahmenprogramm."

15. City of Freiburg, "Vauban," www.freiburg.de/pb/,Lde/208732.html.

16. Ibid.; City of Freiburg, *Tiefbauamt, VEP 2020 Ausstellungstafel Verkehrsberuhigung* (Freiburg: 2008).

17. City of Freiburg, "G-15/137: Perspektivplan hier: Drei Denkrichtungen für die zukünftige Freiraum- und Sied-lungsentwicklung," www.perspektivplan-freiburg.de.

Chapter 9. Energy Efficiency in Buildings: A Crisis of Opportunity

1. Oswaldo Lucon et al., "Buildings," in Ottmar Edenhofer et al., eds., *Climate Change 2014: Mitigation of Climate Change. Contribution of Working Group III to the Fifth Assessment Report of the Intergovernmental Panel on Climate Change* (Cambridge, U.K. and New York: Cambridge University Press, 2014), 675; The Holy See, "Encyclical Letter Laudato Si' of the Holy Father Francis on Care for Our Common Home" (Vatican City: May 24, 2015).

2. Greg Kats and Andrew Seal, "Buildings as Batteries: The Rise of 'Virtual Storage,'" *The Electricity Journal* 25, no. 10 (2012): 59–70.

3. Capital E, "CO2toEE," https://cap-e.com/industry-transformation/co2toee.

4. Figure 9–1 from Jason Channell et al., *Energy Darwinism II: Why a Low Carbon Future Doesn't Have to Cost the Earth* (London: Citi Global Perspectives & Solutions, 2015).

5. International Energy Agency (IEA), *Energy Efficiency Market Report 2015* (Paris: 2015); Institute for European Environmental Policy, *Review of Costs and Benefits of Energy Savings: Task 1 Report 'Energy Savings 2030'* (London: 2013); Birol quotes from Jocelyn Timperley, "Ten Billion Tonnes CO_2 Emissions Saved Thanks to Energy Efficiency Over Past 25 Years, Says IEA," BusinessGreen.com, October 9, 2015.

6. World Energy Council, *World Energy Perspective: Energy Efficiency Policies: What Works and What Does Not* (London: 2013); ODYSSEE-MURE, *Synthesis: Energy Efficiency Trends and Policies in the EU: An Analysis Based on the ODYSSEE and MURE Databases*, September 2015.

7. Ibid.

8. Efficiency Valuation Organization, "History," www.evo-world.org/index.php?option=com_content&view=article&id=40&Itemid=163&lang=en. See also linked archival documents at Wikipedia, "International Performance Measurement and Verification Protocol," https://en.wikipedia.org/wiki/International_performance_measurement_and_verification_protocol. Disclosure: The author served as the Founding Chairman of IPMVP.

9. Building Performance Institute Europe, *Europe's Buildings Under the Microscope: A Country-by-Country Review of the Energy Performance of Buildings* (Brussels: October 2011).

10. Graham S. Wright and Katrin Klingenberg, *Climate-Specific Passive Building Standards*, prepared for the National Renewable Energy Laboratory (NREL) on behalf of the U.S. Department of Energy (DOE) Building America Program, Office of Energy Efficiency and Renewable Energy (EERE) (Golden, CO: NREL, July 2015).

11. Marika Rošā, Claudio Rochas, and Nicholas Stancioff, F3Horizon 2020 Programme, "SUNShINE," presentation, Brussels, April 28–29, 2015, www.managenergy.net/lib/documents/1369/original_SUNSHINE_Marika_Rosa.pdf?1431000214; Steven Fawkes, personal communication with author, October 2015.

12. U.S. General Services Administration, "Green Building Advisory Committee," www.gsa.gov/portal/category/102591; U.S. Environmental Protection Agency, "The Social Cost of Carbon," http://www3.epa.gov/climatechange/EPAactivities/economics/scc.html. Disclosure: The author serves as Chair of this Federal Advisory Committee.

13. John Shonder, *Energy Savings from GSA's National Deep Energy Retrofit Program* (Washington, DC: September 2014); DOE, EERE, "Buildings," http://energy.gov/eere/efficiency/buildings; Paul Torcellini et al., *Main Street Net-Zero Energy Buildings: The Zero Energy Method in Concept and Practice* (Golden, CO: NREL, July 2010).

14. Future Communities, "Hammarby Sjostad, Stockholm, Sweden, 1995 to 2015," www.futurecommunities.net/case-studies/hammarby-sjostad-stockholm-sweden-1995-2015.

15. Renovate America website, https://renovateamerica.com; PACE Funding website, www.pacefunding.com.

16. Amy Westervelt, "Why the Military Hates Fossil Fuels," *Forbes*, February 2, 2012.

17. For many examples, see C40 Cities, "C40 Cities," www.c40.org/cities; Gregory Kats, President, Capital E, presentation at Institute of Medicine Workshop on Bringing Public Health into Urban Revitalization, Washington, DC, November 10, 2014, http://iom.nationalacademies.org/Activities/Environment/Environmental-HealthRT/2014-NOV-10/Videos/Welcome%20and%20Session%201/3-Kats-Video.aspx.

18. Greg Kats and Keith Glassbrook, *Affordable Housing Smart Roof Report* (Washington, DC: 2015). Figure 9–2 from Gregory Kats, *Greening Our Built World: Costs, Benefits, and Strategies* (Washington, DC: Island Press, 2009). The figure shows the present value of 20 years of estimated impacts based on the study data set and synthesis of relevant research. Note that there is significantly greater uncertainty, and less consensus, around methodologies for estimating health and societal benefits.

19. Percentages based on author experience.

20. Sean Cahill, D.C. Building Industry Association, personal communication with author, June 2015.

21. Box 9–1 based on the following sources: EnerNex Corporation, *Eastern Wind Integration and Transmission Study*, prepared for NREL (Golden, CO: 2011); Jim Eyer and Garth Corey, *Energy Storage for the Electricity Grid: Benefits and Market Potential Assessment Guide*, a study for the DOE Energy Storage Systems Program (Albuquerque, NM: Sandia National Laboratories, February 2010); AtSite website, http://atsiteinc.com; Kats and Seal, "Buildings as Batteries: The Rise of 'Virtual Storage'"; National Research Council, *Rising to the Challenge: U.S. Innovation Policy for the Global Economy* (Washington, DC: National Academies Press, 2012); quotes from speech by Dorothy Robyn, Deputy Under Secretary of Defense for Installations and Environment, ICF International, Washington, DC, April 19, 2012; U.S. Department of Defense, *Annual Energy Management Report, Fiscal Year 2010* (Washington, DC: July 2011). See also testimony of Gregory Kats before the U.S. House of Representatives Committee on Oversight and Government Reform, July 2012, at https://cap-e.com/media/; Admiral Mike Mullen, Chairman of the Joint Chiefs of Staff, speech at Energy Security Forum, Washington, DC, October 13, 2010, www.jcs.mil/speech.aspx?id=1472; U.S. Department of Defense, *Quadrennial Defense Review Report* (Washington, DC: February 2010).

22. Greg Kats et al., *Energy Efficiency Financing Models and Strategies: Pathways to Scaling Energy Efficiency Financing from $20 Billion to $150 Billion Annually*, prepared by Capital E for The Energy Foundation (Washington, DC: March 2012).

City View: Melbourne

1. City of Melbourne, "Melbourne in Numbers," www.melbourne.vic.gov.au/AboutMelbourne/Statistics/Pages/MelbourneSnapshot.aspx; population data are preliminary for 2014.

2. Ibid.

3. City of Melbourne, "Adapting to Climate Change: Heatwaves and Days of Extreme Heat," https://www.melbourne.vic.gov.au/Sustainability/AdaptingClimateChange/Pages/Heatwaves.aspx.

4. City of Melbourne, "Adapting to Climate Change," https://www.melbourne.vic.gov.au/Sustainability/AdaptingClimateChange/Pages/AdaptingClimateChange.aspx.

5. City of Melbourne, "Melbourne's Urban Forest Strategy—Making a Great City Greener," www.melbourne.vic.gov.au/Sustainability/UrbanForest/Pages/About.aspx.

6. Ibid.; City of Melbourne, "Melbourne's Urban Forest," infographic, www.melbourne.vic.gov.au/Sustainability/UrbanForest/Documents/Urban_Forest_infographic.pdf.

7. City of Melbourne, *Total Watermark – City as a Catchment: Update 2014* (Melbourne: 2014).

8. City of Melbourne, "Stormwater Harvesting," https://www.melbourne.vic.gov.au/ParksandActivities/Parks/Pages/StormwaterHarvesting.aspx.

9. City of Melbourne, *Zero Net Emissions by 2020: Update 2014* (Melbourne: 2014).

10. City of Melbourne, "Zero Net Emissions," https://www.melbourne.vic.gov.au/Sustainability/CouncilActions/Pages/ZeroNetEmissions.aspx.

11. City of Melbourne, "Carbon Neutral Council Operations 2013-14," https://www.melbourne.vic.gov.au/Sustainability/CouncilActions/Pages/CarbonNeutral.aspx.

12. City of Melbourne, "About 1200 Buildings," https://www.melbourne.vic.gov.au/1200buildings/Pages/About1200Buildings.aspx; City of Melbourne, "Carbon Neutral Council Operations 2013-14"; City of Melbourne, "Melbourne Wins World Climate Leadership Award," press release (Melbourne: September 23, 2014).

13. Enterprise Melbourne, "CitySwitch Green Office Program," https://www.melbourne.vic.gov.au/enterprisemelbourne/environment/Pages/CitySwitch.aspx; City of Melbourne, "Carbon Neutral Council Operations 2013-14."

14. City of Melbourne, "Council House 2," www.melbourne.vic.gov.au/Sustainability/CH2/aboutch2/Pages/Factsandfigures.aspx; City of Melbourne, "Library at the Dock's Sustainable Design on Display," press release (Melbourne: July 25, 2014); Green Building Council of Australia, "Green Star," www.gbca.org.au/green-star/.

15. City of Melbourne, *Homes for People: Housing Strategy 2014-2018* (Melbourne: January 2015).

16. City of Melbourne, *Understanding the Quality of Housing Design* (Melbourne: February 2013); Paul Myors, Rachel O'Leary, and Rob Helstroom, *Multi Unit Residential Buildings Energy & Peak Demand Study* (Melbourne: Energy Australia, October 2005).

17. City of Melbourne, "Sustainable Solutions for Apartments," https://www.melbourne.vic.gov.au/Sustainability/WhatCanIDo/Pages/SustainableLivingintheCity.aspx; C40 Cities, "The 2015 Finalists," www.c40.org/custom_pages/awards_finalists.

18. City of Melbourne, "High-rise Apartment Recycling," https://www.melbourne.vic.gov.au/ForResidents/WasteRecyclingandNoise/householdgarbage/Pages/Highriserecycling.aspx; City of Melbourne, *Waste and Resource Recovery Plan 2015–18* (Melbourne: 2015).

19. City of Melbourne, "Zero Net Emissions."

20. Ibid.

21. City of Melbourne, "City of Melbourne Transport Strategy," https://www.melbourne.vic.gov.au/AboutCouncil/PlansandPublications/strategies/Pages/transportstrategy.aspx; City of Melbourne, *Transport Strategy 2012 – Planning for Future Growth* (Melbourne: 2012).

22. City of Melbourne, "Melbourne Metro Rail Project," https://www.melbourne.vic.gov.au/BuildingandPlanning/FutureGrowth/ExternalProjects/Pages/MetroProjectRailLink.aspx; City of Melbourne, *Transport Strategy 2012*.

23. City of Melbourne, *Draft Bicycle Plan 2016–2020* (Melbourne: 2015); The Good Wheel Project, www.thesqueakywheel.com.au/good-wheel.

24. City of Melbourne, "Zero Net Emissions"; City of Melbourne, *Waste and Resource Recovery Plan 2015–2018*.

25. Ibid.

26. City of Melbourne, *Future Melbourne – A Bold, Inspirational and Sustainable City, Executive Summary* (Melbourne: July 2008); City of Melbourne, "Future Melbourne 2026," https://www.melbourne.vic.gov.au/AboutCouncil/PlansandPublications/Pages/FutureMelbourne.aspx.

27. City of Melbourne, "100 Resilient Cities," www.melbourne.vic.gov.au/AboutMelbourne/Resilient_Cities/Pages/ResilientCities.aspx.

Chapter 10. Is 100 Percent Renewable Energy in Cities Possible?

1. City of Vancouver, *Renewable City Strategy* (Vancouver: November 2015); Renewable Energy Policy Network for the 21st Century (REN21), *Renewables 2015 Global Status Report* (Paris: 2015).

2. REN21, *Renewables 2015 Global Status Report.*

3. International Renewable Energy Agency (IRENA), *Statute of the International Renewable Energy Agency (IRENA)* (Abu Dhabi: 2009); Nuclear Energy Institute, "Protecting the Environment – Nuclear Energy Institute," www.nei.org/Issues-Policy/Protecting-the-Environment, viewed November 19, 2015.

4. Michael Cooper and Dalia Sussman, "Nuclear Power Loses Public Support in New Poll," *New York Times*, March 22, 2011; Damian Carrington, "Dip in Nuclear Power Support After Fukushima Proves Shortlived, *The Guardian* (U.K.), January 18, 2012; "Japan Turns on Nuclear Power Four Years After Fukushima," *Al Jazeera*, August 11, 2015; Landeszentrale für politische Bildung Baden-Württemberg, "Die Energiewende 2011," www .lpb-bw.de/energiewende.html, viewed November 19, 2015.

5. REN21, *Renewables 2015 Global Status Report*; Rowena Mason, "Most of Britain's Major Cities Pledge to Run on Green Energy by 2050," *The Guardian* (U.K.), November 23, 2015. Note that targets and achievements are continually evolving. Table 10–1 from the following sources: City of Vancouver, *Renewable City Strategy*; REN21, *Renewables 2015 Global Status Report*; Malmö stadsbyggnadskontor, *Energistrategi för Malmö* (Malmö: 2009); Go 100% Renewable Energy website, go100percent.org; City of Austin, *Community Climate Plan* (Austin: 2015); City of Amsterdam, 2040 *Energy Strategy* (Amsterdam: February 2010).

6. Zachary Shahan, "California Now Has 1 Gigawatt of Solar Power Installed," CleanTechnica.com, November 11, 2011; John Farrell and Matt Grimley, *Public Rooftop Revolution* (Washington, DC: Institute for Local Self-Reliance, June 2015).

7. China and Chandigarh from REN21, *Renewables 2015 Global Status Report*; Rhonda Winter, "Israel's Special Relationship with the Solar Water Heater," *Reuters*, March 18, 2011; Bärbel Epp, "Austria: Solar Thermal to Breathe New Life into Vienna's Urban Development," SolarThermalWorld.org, November 4, 2014.

8. Justin Gerdes, "Copenhagen's Ambitious Push to be Carbon Neutral by 2025," *Yale Environment 360*, April 11, 2013; Lily Riahi et al., *District Energy in Cities: Unlocking the Potential of Renewable Energy and Energy Efficiency* (Paris: United Nations Environment Programme, 2015).

9. ICLEI–Local Governments for Sustainability and IRENA, *Dezhou, China: Green Economic Development with Renewable Energy Industries* (Abu Dhabi: 2012).

10. City of Melbourne, "Melbourne Unites to Support Renewable Energy," press release (Melbourne: November 30, 2015).

11. Indian Ministry of New and Renewable Energy (MNRE), "Solar/Green Cities," http://mnre.gov.in/schemes /decentralized-systems/solar-cities/; MNRE, "State-wise Status of Solar Cities as on 19.08.2015," http://mnre.gov .in/file-manager/UserFiles/State-wise-status-of-Solar-Cities.pdf.

12. REN21, *Renewables 2015 Global Status Report*; Community Power Project, "What Is Community Power?" (Brussels: December 2013); American Public Power Association, "Public Power: Shining a Light on Public Service" (Washington, DC: May 2013).

13. Renewable Cities, *Final Report Global Learning Forum, May 13–15, Vancouver, B.C.* (Vancouver: 2015).

14. Sam Orr, Ayman Fahmy, and Dinos Hadjiloizou, "Leading Canada by Example: University of British Columbia Invests in Greener Infrastructure," *District Energy* (International District Energy Association), Second Quarter 2014; Advanced Manufacturing Office, Office of Energy Efficiency and Renewable Energy, U.S. Department of Energy, "Minimize Boiler Short Cycling Losses," Steam Tip Sheet #16 (Washington, DC: January 2012).

15. Renewable Cities, "In Conversation – Leshan Moodliar (Durban) & Danielle Murray (Austin)," YouTube video, July 7, 2015, https://youtu.be/CO87ZZHX6SA.

16. Danielle Murray, "Fair Rate Setting for a Renewable Future," presentation at Renewable Cities Global Learning Forum, May 13–15, Vancouver, Canada, http://forum.renewablecities.ca/presentations/sessions/Renewable-Cities-Danielle-Murray.pptx.

17. U.S. Environmental Protection Agency, "Solar Power Purchase Agreements," http://www3.epa.gov/greenpower/buygp/solarpower.htm.

18. John Duda, "Energy, Democracy, Community," *Medium*, August 3, 2015.

19. Ibid.

20. City of Copenhagen Technical and Environmental Administration, *CPH 2025 Climate Plan* (Copenhagen: September 2012); Laura K. Khan et al., "Recommended Community Strategies and Measurements to Prevent Obesity in the United States," *Morbidity and Mortality Weekly Report*, July 24, 2009.

21. Will Sloan, "Revving Up the Electric Car," *Ryerson University News*, July 28, 2014, www.ryerson.ca/news/news/General_Public/20140728-revving-up-the-electric-car.html; California Independent System Operator, "What the Duck Curve Tells Us About Managing a Green Grid," 2013, https://www.caiso.com/Documents/FlexibleResourcesHelpRenewables_FastFacts.pdf.

22. Renewable Cities, *Final Report Global Learning Forum*.

City View: Vancouver, Canada

1. Data and write-up from internal documents from the City of Vancouver.

Chapter 11. Supporting Sustainable Transportation

1. Rachael Nealer, David Reichmuth, and Don Anair, *Cleaner Cars from Cradle to Grave* (Cambridge, MA: Union of Concerned Scientists, November 2015).

2. Figure 11–1 from Todd Litman, *Analysis of Public Policies That Unintentionally Encourage and Subsidize Urban Sprawl*, paper commissioned by LSE Cities at the London School of Economics and Political Science, on behalf of the Global Commission on the Economy and Climate for the New Climate Economy Cities Program (Victoria, BC: Victoria Transport Policy Institute, 2015).

3. Stephen M. Wheeler, "Built Landscapes of Metropolitan Regions: An International Typology," *Journal of the American Planning Association* 81, no. 3 (2015): 167–90.

4. Ibid.

5. Ibid.

6. Philipp Rode et al., *Accessibility in Cities: Transport and Urban Form*, NCE Cities Paper 03 (London: LSE Cities at the London School of Economics and Political Science, 2014); density data and Figure 11–2 from Peter Newman and Jeffrey Kenworthy, *The End of Automobile Dependence: How Cities Are Moving Away from Car-Based Planning* (Washington, DC: Island Press, 2015).

7. Luis Zamorano and Erika Kulpa, "People-Oriented Cities: Mixed-Use Development Creates Social and Economic Benefits," World Resources Institute blog, July 23, 2014; Luis Zamorano, "The Perfect Storm: One Country's History of Urban Sprawl," TheCityFix.com, March 5, 2014.

8. Joan Clos, "A New Paradigm for Urban Planning," Climate Leader Papers, May 24, 2012, www.climateactionprogramme.org/climate-leader-papers/a_new_paradigm_for_urban_planning; Deutsche Gesellschaft für Tech-

nischeZusammenarbeit (GTZ), "Informal Public Transit. Recommended Reading and Links," June 2010, www.sutp .org/component/phocadownload/category/85-rl-ipt?download=145:rl-ipt-en; Robert Cervero, "Informal Transit: Learning from the Developing World," *Access Magazine* (Spring 2001).

9. Michael Kimmelman, "Express Bus Service Shows Promise in New York," *New York Times*, July 19, 2015; Seth Freed Wessler, *Filling the Gaps: COMMUTE and the Fight for Transit Equity in New York City* (Oakland, CA: Applied Research Center, 2010); Robert Hickey et al., *Losing Ground. The Struggle of Moderate-Income Households to Afford the Rising Costs of Housing and Transportation* (Washington, DC and Chicago: Center for Housing Policy and Center for Neighborhood Technology, October 2012).

10. Juan Miguel Velásquez, "How to Orient Cities for People, Not Cars," GreenBiz.com, March 25, 2015; Heshuang Zeng, "On the Move: Limiting Car Usage in Industrialized Economies," TheCityFix.com, November 6, 2013; Heshuang Zeng, "On the Move: Reducing Car Usage and Ownership in China, Latin America, and Other Developing Economies," TheCityFix.com, November 7, 2013; ICLEI–Local Governments for Sustainability, *Mexico City's Green Plan: EcoMobility in Motion*, ICLEI Case Studies 120 (Bonn: November 2010); Lulu Xue, "4 Lessons from Beijing and Shanghai Show How China's Cities Can Curb Car Congestion," World Resources Institute blog, April 19, 2015; "Urban Access Regulation in Europe," http://urbanaccessregulations.eu; 226 cities from Dario Hidalgo, "Sustainable Mobility Trends Around the World," Embarq, March 10, 2014, www.slideshare.net/EMBARQNetwork /embarq-trends-2014-dario-hidalgo; London from "Streetwise," *The Economist*, September 5, 2015. Box 11–1 from International Association of Public Transport (UITP), "3 Really Simple Steps: How to Reduce Congestion and Pollution, Generate Revenue and Overhaul Your City," May 19, 2014, www.uitp.org/news/3-steps-milan, and from ICLEI, *Milan, Italy. The Ecopass Pollution Charge and Area C Congestion Charge – Comparing Experiences with Cordon Pricing over Time*, ICLEI Case Studies 157 (Bonn: July 2013).

11. Kanika Jindal, "In Photos: Bhopal Becomes India's Fifth City to Join the Car-Free Raahgiri Movement," TheCityFix.com, October 8, 2014; Institute for Transportation and Development Policy (ITDP), "ITDP Welcomes Clayton Lane as New CEO," October 5, 2015, https://www.itdp.org/itdp-welcomes-clayton-lane-as-new-ceo/; Gwladys Fouche and Terje Solsvik, "Oslo Aims to Make City Center Car-free Within Four Years," *Reuters*, October 19, 2015.

12. Susan A. Shaheen and Adam P. Cohen, "Growth in Worldwide Carsharing. An International Comparison," in *Transportation Research Record: Journal of the Transportation Research Board, No. 1992* (Washington, DC: Transportation Research Board of the National Academies, 2007), 81–89.

13. Ibid.; Navigant Research, "Carsharing Programs," https://www.navigantresearch.com/research/carsharing-programs; Statista, "Number of Vehicles in the Global Car Sharing Market from 2006 to 2014 (in 1,000)," www.statista .com/statistics/415322/car-sharing-number-of-vehicles-worlwide/; Statista, "Number of Car Sharing Users Worldwide from 2006 to 2014 (millions)," www.statista.com/statistics/415636/car-sharing-number-of-users-worldwide/; Navigant Research, "Carsharing Services Will Surpass 12 Million Members Worldwide by 2020," press release (Boulder, CO: August 22, 2013).

14. Heshuang Zeng, "On the Move: Car-Sharing Scales Up," TheCityFix.com, December 18, 2013; ITDP, "ITDP Welcomes Clayton Lane as New CEO."

15. Shared-Use Mobility Center (SUMC), "SUMC to Help Lead $1.6 Million Low-Income Carsharing Pilot in LA, July 24, 2015, http://sharedusemobilitycenter.org/news/sumc-to-help-lead-1-6-million-low-income-carsharing-pilot-in-la/.

16. Zeng, "On the Move: Car-Sharing Scales Up"; Andrew Nusca, "Enterprise Acquires PhillyCarShare," ZDNet .com, August 9, 2011.

17. Zeng, "On the Move: Car-Sharing Scales Up."

18. Simone Pathe, "Uber the Unfair? Are Ride-sharing Firms Exploiting Deregulation?," *PBS Newshour*, October

2, 2014; Avi Asher-Schapiro, "Against Sharing," *Jacobin*, September 19, 2014; Liz Alderman, "Uber's French Resistance," *New York Times*, June 3, 2015; Mark Scott and Melissa Eddy, "German Court Bans Uber Service Nationwide," *New York Times*, September 2, 2014; Steven Hill, *Raw Deal. How the "Uber Economy" and Runaway Capitalism Are Screwing American Workers* (New York: St. Martin's Press, 2015); Mark Scott, "BlaBlaCar, a Ride-Sharing Start-Up in Europe, Looks to Expand Its Map," *New York Times*, July 2, 2014.

19. ICLEI, *Bremen, Germany. A Role Model for Car-Sharing Is Targeting 20,000 Users by 2020*, ICLEI Case Studies 159 (Bonn: August 2013). Euro to U.S. dollar conversions reflect average exchange rate for 2015 (January to mid-November).

20. Figure of 718 excludes airport shuttles, commuter/mainline rail lines, entertainment parks, and funiculars. Another 92 "heritage tram" and "other" systems are not included in the total. Information appears to be current as of early 2013. Light Rail Transit Association, "A World of Trams and Urban Transit," www.lrta.org/world/worldind .html.

21. UITP, *Statistics Brief. World Metro Figures* (Brussels: October 2014); Mircea Steriu, Statistics Manager, UITP, Brussels, personal communication with author, August 31, 2015; 2014 and 2015 from Wikipedia, "List of Metro Systems," https://en.wikipedia.org/wiki/List_of_metro_systems, viewed October 6, 2015. Figure 11–3 based on these two sources.

22. UITP, *Statistics Brief. World Metro Figures*; Wikipedia, "List of Metro Systems."

23. "New Subway (Metro) Systems Cost Nearly 9 Times as Much as Light Rail," *Light Rail Now*, February 13, 2014, https://lightrailnow.wordpress.com/2014/02/13/new-subway-metro-systems-cost-nearly-9-times-as-much -as-light-rail/. Table 11–1 adapted from Cledan Mandri-Perrott with Iain Menzies, *Private Sector Participation in Light-Rail-Light Metro Transit Initiatives* (Washington, DC: World Bank, 2010).

24. ITDP, "What is BRT?" https://www.itdp.org/library/standards-and-guides/the-bus-rapid-transit-standard/what -is-brt/.

25. Aileen Carrigan et al., *Social, Environmental and Economic Impacts of BRT Systems. Bus Rapid Transit Case Studies from Around the World* (Washington, DC: World Resources Institute and EMBARQ, December 2013).

26. Global BRT Data website, brtdata.org. Figure 11–4 from Ibid.

27. Table 11–2 from Ibid; C40 Cities and Arup, *Climate Action in Megacities. C40 Cities Baseline and Opportunities. Volume 2.0* (New York: February 2014).

28. Bogotá from Carrigan et al., *Social, Environmental and Economic Impacts of BRT Systems*; Buenos Aires from ITDP, "Five City Transport Transformations That May Surprise You," May 18, 2015, https://www.itdp.org/five-city -transport-transformations-that-may-surprise-you/; Francesca Perry, "Everyone Praises Green Copenhagen. But What If Your City Has 20m People?," *The Guardian* (U.K.), April 2, 2015; Mexico City from Stephanie Valgañón and Geovana Royacelli, "Movilidad: la enfermedad y el remedio," CiudadanosENRED.com, May 5, 2014.

29. Global BRT Data website.

30. ITDP, "Five City Transport Transformations That May Surprise You"; UN-Habitat, *International Guidelines on Urban and Territorial Planning. Towards a Compendium of Inspiring Practices* (Nairobi: April 2015), 23.

31. Johannesburg from Andy Gouldson et al., *Accelerating Low-Carbon Development in the World's Cities*, New Climate Economy Working Paper (Washington, DC: Global Commission on the Economy and Climate, 2015); Ibidun Adelekan, "A Simple Approach to BRT in Lagos, Nigeria," in Mark Swilling et al., *City-Level Decoupling: Urban Resource Flows and the Governance of Infrastructure Transitions. Case Studies from Selected Cities* (Paris: United Nations Environment Programme, 2013), 45–48.

32. Box 11–2 from ITDP, "The BRT Standard," https://www.itdp.org/library/standards-and-guides/the-bus-rapid

-transit-standard/, and from ITDP, *Best Practice in National Support for Urban Transportation. Part 1: Evaluating Country Performance in Meeting the Transit Needs of Urban Populations* (New York: 2014).

33. ITDP, *Best Practice in National Support for Urban Transportation.*

34. Stefanie Swanepoel, "The Climate Action Plan of Portland, Oregon," in Swilling et al., *City-Level Decoupling,* 58–61; road closures from "Streetwise," *The Economist.*

35. ICLEI, *Freiburg, Germany. Cycling 2020 – A Concept Fit for the Future,* ICLEI Case Studies 156 (Bonn: July 2013); Sven Eberlein, "Universal Principles for Creating a Sustainable City," Planetizen.com, August 11, 2011; Gabriela Weber de Morais, "Citizens Contributing to Urban Sustainability in Vauban, Germany," in Swilling et al., *City-Level Decoupling,* 19–21.

36. ICLEI, *Freiburg, Germany;* Weber de Morais, "Citizens Contributing to Urban Sustainability in Vauban, Germany."

37. Christian Tang Jensen, "Making Politicians Invest in Bicycle Infrastructure," Cycling Embassy of Denmark, June 30, 2015, www.cycling-embassy.dk/2015/06/30/making-politicians-invest-in-bicycle-infrastructure/; "In Almost Every European Country, Bikes Are Outselling New Cars," *National Public Radio,* October 24, 2013.

38. Justin Gerdes, "Copenhagen's Ambitious Push to be Carbon Neutral by 2025," *Yale Environment 360,* April 11, 2013; ICLEI, *Münster, Germany. Cycling and Public Transport: The Way Forward,* ICLEI Case Studies 158 (Bonn: August 2013); "Ten Cycling Cities to Discover in Europe," *Huffington Post UK,* July 20, 2015.

39. The Copenhagenize Index, http://copenhagenize.eu/index/index.html; Priscila Pacheco, Luísa Zottis, and Sergio Trentini, "How Two Community Groups Are Successfully Fostering Bike Culture in Brazil," TheCityFix.com, August 26, 2015; Luísa Zottis, "Using Bikes to Improve Mobility in Rio de Janeiro's Favelas," TheCityFix.com, August 19, 2015.

40. ICLEI, *Bogotá, Colombia. Building a Plan to Transform Non-Motorized Transport in Bogotá,* ICLEI Case Studies 165 (Bonn: August 2013); Buenos Aires from Perry, "Everyone Praises Green Copenhagen. But What If Your City Has 20m People?" and from ITDP, "Five City Transport Transformations That May Surprise You"; ICLEI, *Mexico City's Green Plan: EcoMobility in Motion.*

41. Peter Midgley, "On the Move: The Swift, Global Expansion of Bicycle-sharing Schemes," TheCityFix.com, December 4, 2013; 2014 data from Susan A. Shaheen et al., *Public Bikesharing in North America During a Period of Rapid Expansion: Understanding Business Models, Industry Trends and User Impacts* (San Jose, CA: Mineta Transportation Institute, October 2015); U.S. cities from SUMC, "5 Bike Sharing Trends to Watch This Summer," Eco Watch.com, July 7, 2015.

42. C40 Cities and Arup, *Climate Action in Megacities.*

43. Best-performance cities from Colin Hughes, "Building Towards Better Bike-sharing Systems," TheCityFix.com, February 26, 2014; smartphone apps from "Streetwise," *The Economist;* innovations from Midgley, "On the Move: The Swift, Global Expansion of Bicycle-sharing Schemes" and from SUMC, "5 Bike Sharing Trends to Watch This Summer"; Josh Cohen, "Birmingham's New Bike-Share Will Have Electric-Assist Bicycles," NextCity.org, May 4, 2015.

44. SUMC, "5 Bike Sharing Trends to Watch This Summer."

45. Sustainable Transport Awards, "Winners," http://staward.org/winners.

Chapter 12. Urban Transport and Climate Change

1. Ralph Sims et al., "Transport," in *Climate Change 2014: Mitigation of Climate Change. Contribution of Working Group III to the Fifth Assessment Report of the Intergovernmental Panel on Climate Change* (Cambridge, U.K. and

New York: Cambridge University Press, 2014); International Energy Agency, *Policy Pathways: A Tale of Renewed Cities* (Paris: 2013); Michael A. Replogle and Lewis M. Fulton, *A Global High Shift Scenario: Impacts and Potential for More Public Transport, Walking, and Cycling With Lower Car Use* (New York: Institute for Transportation and Development Policy and University of California at Davis, 2014).

2. Andy Gouldson et al., *Accelerating Low-Carbon Development in the World's Cities*, New Climate Economy Working Paper (Washington, DC: Global Commission on the Economy and Climate, 2015).

3. Figure 12–1 based on data in Replogle and Fulton, *A Global High Shift Scenario*.

4. Gouldson et al., *Accelerating Low-Carbon Development in the World's Cities*.

5. Figure 12–2 from International Transport Forum, "Chapter 4 Preview: Urban Passenger Transport Scenarios for Latin America, China and India," in *ITF Transport Outlook 2015* (Paris: 2015).

6. Ibid.

7. United Nations Framework Convention on Climate Change (UNFCCC), "Background on the UNFCCC: The International Response to Climate Change," http://unfccc.int/essential_background/items/6031.php.

8. UNFCCC, "Clean Development Mechanism," http://unfccc.int/kyoto_protocol/mechanisms/clean_develop ment_mechanism/items/2718.ph, viewed December 29, 2015.

9. Paris Process on Mobility and Climate (PPMC), *Operational Plan – PPMC* (Shanghai: September 9, 2015).

10. Figure 12–3 from SLoCaT Partnership analysis of UNFCCC "INDCs as communicated by Parties," available at http://www4.unfccc.int/submissions/indc/Submission%20Pages/submissions.aspx.

11. Figure 12–4 from Ibid.; Table 12–1 from the following sources: People's Republic of China, *Enhanced Actions on Climate Change* (Beijing: June 30, 2015); Cote d'Ivoire, *Contributions prevues determinees au niveau national de la Cote d'ivoire* (Abidjan: September 30, 2015); Federal Democratic Republic of Ethiopia, *Intended Nationally Determined Contribution (INDC) of the Federal Democratic Republic of Ethiopia* (Addis Ababa: October 6, 2015); Republic of Gabon, *Contribution prévue déterminée au niveau national, Conférence des Parties 21* (Libreville: March 31, 2015); Japan, *Submission of Japan's Intended Nationally Determined Contribution (INDC)* (Tokyo: July 17, 2015); Hashemite Kingdom of Jordan, *Intended Nationally Determined Contribution (INDC)* (Amman, September 30, 2015); Republic of Macedonia, *Submission by the Republic of Macedonia* (Skopje: August 4, 2015); Republic of Korea, *Submission by the Republic of Korea, Intended Nationally Determined Contribution* (Seoul: June 30, 2015).

12. UNFCCC, "Lima-Paris Action Agenda," http://newsroom.unfccc.int/lpaa/; United Nations Environment Programme, *The Emissions Gap Report: Are the Copenhagen Accord Pledges Sufficient to Limit Global Warming to 2° C or 1.5° C? A Preliminary Assessment* (Nairobi: November 2010).

13. UNFCCC, "Transport," http://newsroom.unfccc.int/lpaa/transport/; SLoCaT Partnership in support of the PPMC, "Transport Initiatives Proposed in the Context of an Action Agenda on Transport and Climate Change" (Shanghai: 2015). This count includes all LPAA commitments except those on aviation, maritime transport, and global rail.

14. UNFCCC, "MobliseYourCity: Local Governments in Developing Countries Take High Road to Low Carbon" (Bonn: 2015); UNFCCC, "Transport."

15. C40 Cities, "C40 Latin American Mayors Forum Showcases Region's Bold Climate Leadership," C40 blog, March 27, 2015; C40 Cities, "C40 Clean Bus Declaration Urges Cities and Manufacturers to Adopt Innovative Clean Bus Technologies," C40 blog, May 20, 2015.

16. SLoCaT Partnership, "Secretary General's Climate Summit 2014," http://slocat.net/climatesummit.

17. Ibid.

18. SLoCaT Partnership and World Cycling Alliance, "More Cycling on the Agenda: Our Permanent Mission to the UN," September 29, 2014, www.ecf.com/news/more-cycling-on-the-agenda-our-permanent-mission-to-the -un; European Cyclists' Federation (ECF) and World Cycling Alliance, "Cycling Delivers on the Global Goals" (Brussels: 2015); SLoCaT Partnership, "Secretary General's Climate Summit 2014"; ECF, "Cycling and Public Transport Lobby Join Forces for More Sustainable and Active Mobility," press release (Brussels: September 14, 2015).

19. Global Fuel Economy Initiative website, www.globalfueleconomy.org; Sustainable Energy for All, *Global Energy Efficiency Accelerator Platform: Action Statement and Action Plan* (New York: September 23, 2014).

20. Global Green Freight Action Plan, *The Global Green Freight Action Plan* (Paris: May 2015).

21. California Environmental Protection Agency, "New Initiative Accelerates Global Transition to Zero-Emis-sion Vehicles," press release (Sacramento: August 20, 2015), www.calepa.ca.gov/pressroom/Releases/2015/EV GlobalTran.htm.

22. ATEC ITS website, www.atec-itsfrance.net/home.cfm.

23. PPMC and SLoCaT Partnership, "City Initiatives on Sustainable, Low Carbon Transport" (Shanghai: May 21, 2015).

24. Civitas website, www.civitas.eu.

25. The five networks are the Climate Alliance, Council of European Municipalities and Regions, Fedarene, Euro-cities, and Energy Cities; Jeppe Mikel Jensen, Member of the Covenant of Mayors Office, personal communication with authors, July 10, 2015.

26. European Mobility Week, "European Mobility Week and Do the Right Mix Join Forces in Promoting Sustain-able Urban Mobility," press release (Brussels: April 4, 2015); European Mobility Week, "Participants 2015," www .mobilityweek.eu/cities/.

27. Gouldson et al., *Accelerating Low-Carbon Development in the World's Cities.*

28. UN General Assembly, "Resolution Adopted by the General Assembly on 25 September 2015. Transforming Our World: The 2030 Agenda for Sustainable Development" (New York: September 25, 2015).

City View: Singapore

1. Data from Department of Statistics Singapore, www.singstat.gov.sg, viewed November 15, 2015.

2. World Economic Forum, *The Global Competitiveness Report 2010–2011* (Davos, Switzerland: 2015), 15–16; United Nations Development Programme, *World Development Report 2014: Work for Human Development* (New York: 2015), 160; Mercer LLC, "2014 Quality Of Living Worldwide City Rankings – Mercer Survey," press release (New York: February 19, 2014); Siemens AG, *Asian Green City Index* (Munich: 2011), 10.

Chapter 13. Source Reduction and Recycling of Waste

1. Daniel Hoornweg and Perinaz Bhada-Tata, *What a Waste: A Global Review of Solid Waste Management* (Wash-ington, DC: World Bank, 2012); United Nations, *Municipal Solid Waste Action Statement and Plan*, prepared for Climate Summit 2014: Catalyzing Action, New York, September 23, 2014.

2. Mark Roseland, *Toward Sustainable Communities: Solutions for Citizens and Their Governments* (Gabriola Island, Canada: New Society Publishers, 2012), 100.

3. Table 13–1 from International Labour Organization (ILO), *Working Towards Sustainable Development* (Geneva: 2012), 112.

4. Roseland, *Toward Sustainable Communities*, 102.

5. ILO, *Working Towards Sustainable Development*.

6. C40 Cities, "Solid Waste Management Initiative. Sustainable Solid Waste Systems: Network Overview," www .c40.org/networks/sustainable_solid_waste_systems, viewed October 9, 2015.

7. United Nations, *Municipal Solid Waste Action Statement and Plan*.

8. Office of the City Auditor, Portland, Oregon, "5.33.080 Environmentally Preferable Procurement," February 20, 2013, www.portlandonline.com/auditor/?c=37766&a=441003; City of Portland, Oregon, Planning and Sustainability, "2015 Sustainable City Principles, 2030 Objectives," www.portlandoregon.gov/bps/article/527220; Mary Mazzoni, "Two Years In, Lessons from Portland's Composting Program," Earth911.com, September 17, 2013.

9. Virali Gokaldas, "San Francisco, USA. Creating a Culture of Zero Waste," in Cecilia Allen et al., *On the Road to Zero Waste. Successes and Lessons from Around the World* (Quezon City, The Philippines: Global Alliance for Incinerator Alternatives (GAIA), June 2012).

10. New York City Mayor's Office of Sustainability, "Waste and Recycling," www.nyc.gov/html/planyc/html/sustain ability/waste-recycling.shtml; Buenos Aires from Siemens and C40 Cities Climate Leadership Group, "C40 and Siemens Honor Cities for Leadership in Tackling Climate Change," press release (New York: September 23, 2014); Vancouver from Roseland, *Toward Sustainable Communities*, 107.

11. Aimee Van Vliet, "The Story of Capannori – A Zero Waste Champion," Zero Waste Europe, written for GAIA (Brussels: September 2013); Silvia Giannelli, "Pioneering Italian Town Leads Europe in Waste Recycling," *Inter Press Service*, May 17, 2013.

12. Joan Marc Simon, *The Story of Gipuzkoa* (Brussels: Zero Waste Europe, June 2015).

13. Erika Oblak, *The Story of Ljubljana* (Brussels: Zero Waste Europe, June 2015).

14. Cecilia Allen, *Flanders, Belgium: Europe's Best Recycling and Prevention Program* (Quezon City, The Philippines: GAIA, June 2012).

15. C40 Cities, "Waste Management System Case Study: Oslo, Norway," December 4, 2012, www.c40.org/case _studies/waste-management-system; John Tagliabue, "A City That Turns Garbage into Energy," *New York Times*, April 29, 2013; Sustainable Cities, "Waste and Energy Management, Oslo," www.sustainablecities.eu/local-stories /oslo-waste-management; Stefan Holmerz, "Oslo's Colourful Solution to Waste Management," *Waste Management World* 16, no. 3 (2015).

16. Waste and population data from C40 Cities, "Waste Management System Case Study: Oslo, Norway." Box 13–1 based on the following sources: Nate Seltenrich, "Incineration Versus Recycling: In Europe, A Debate Over Trash," *Yale Environment 360*, August 28, 2013; Helen Russell, "Trash to Cash: Norway Leads the Way in Turning Waste into Energy," *The Guardian* (U.K.), June 14, 2013; Confederation of European Waste-to-Energy Plants, "Waste-to-Energy in Europe in 2013," www.cewep.eu/information/data/studies/m_1459; Eurostat, "Municipal Waste Statistics," July 2015, http://ec.europa.eu/eurostat/statistics-explained/index.php/Municipal_waste_statistics; U.S. Environmental Protection Agency (EPA), *Advancing Sustainable Materials Management: Facts and Figures 2013* (Washington, DC: June 2015), Table ES-1.

17. ICLEI–Local Governments for Sustainability and International Renewable Energy Agency, *Belo Horizonte, Brazil: Waste to Energy for More Effective Landfill Site Management* (Abu Dhabi: 2012); EPA, "Energy Projects and Candidate Landfills," March 4, 2015, http://www3.epa.gov/lmop/projects-candidates/index.html; Ann Ballinger and Dominic Hogg, *The Potential Contribution of Waste Management to a Low Carbon Economy* (Brussels: Zero Waste Europe, Zero Waste France, and ACR+, 2015); C40 News Team, "Waste Management a Priority for C40's African Cities," *National Geographic*, March 3, 2014.

18. Natalie Mayer, "100% Biogas-Fuelled Public Transport in Linköping, Sweden," in Mark Swilling et al., *City-Level Decoupling: Urban Resource Flows and the Governance of Infrastructure Transitions. Annex: Case Studies from Selected Cities*, A Report of the Working Group on Cities of the International Resource Panel (Paris: United Nations Environment Programme, 2013), 84–86.

19. Deutsche Gesellschaft für Internationale Zusammenarbeit (GIZ) and ICLEI, "Lille Metropole, France. Waste to Fuel: Biogas Powered Buses in Lille Metropole," Urban NEXUS Case Study 07 (Bonn: August 2014); European Biofuels Technology Platform, "Biogas/Biomethane for Use as a Transport Fuel," www.biofuelstp.eu/biogas.html.

20. Bellamy Pailthorp, "Renewable Natural Gas from Landfill Fueling Local Buses," *KPIU 88.5 Seattle*, August 16, 2013, www.kplu.org/post/renewable-natural-gas-landfill-fueling-local-buses; Big Blue Bus, "Big Blue Bus, Fueling a Renewable Future One Bus at a Time," press release (Santa Monica: July 15, 2015).

21. ILO, *Working Towards Sustainable Development*.

22. Allen et al., *On the Road to Zero Waste*.

23. Ibid.; Collaborative Working Group on Solid Waste Management in Low- and Middle-income Countries and GIZ, *The Economics of the Informal Sector in Solid Waste Management* (St. Gallen and Eschborn: April 2011).

24. Neil Tangri, "Pune, India. Waste Pickers Lead the Way to Zero Waste," in Allen et al., *On the Road to Zero Waste*.

25. Apiwat Ratanawaraha, "Socialisation of Solid Waste Management in Ho Chi Minh City, Vietnam," in Swilling et al., *City-Level Decoupling*, 81–83.

26. Melanie Samson, ed., *Refusing to Be Cast Aside: Waste Pickers Organising Around the World* (Cambridge, MA: Women in Informal Employment: Globalizing and Organizing (WIEGO), 2009); ILO, *Working Towards Sustainable Development*. Box 13–2 from ILO, idem, and from Sonia M. Dias, *Overview of the Legal Framework for Inclusion of Informal Recyclers in Solid Waste Management in Brazil*, Urban Policies Briefing Note No. 8 (Cambridge, MA: WIEGO, 2011).

27. Oscar Ricardo Schmeiske, "Incentivised Recycling in Curitiba, Brazil," in Swilling et al., *City-Level Decoupling*, 79–81.

28. Ibid.

29. Cecilia Allen, "Buenos Aires City, Argentina. Including Grassroots Recyclers," in Allen et al., *On the Road to Zero Waste*.

30. Ibid.

31. Kristie Robinson, "Buenos Aires Embraces 'Cartoneros' in Push for Zero Waste," Citiscope.org, October 16, 2014; C40 Cities, "Buenos Aires – Solid Urban Waste Reduction Project," City Climate Leadership Awards 2014, www.c40.org/awards/1/profiles/13.

City View: Ahmedabad and Pune

1. Data from Registrar General & Census Commissioner, Ministry of Home Affairs, Government of India, *Census of India, 2011*, http://censusindia.gov.in. Ahmedabad population figures from 2010.

2. Ibid.

3. Akshima T. Ghate and S. Sundar, *Policy Brief – Proliferation of Cars in Indian Cities: Let Us Not Ape the West* (New Delhi: The Energy and Resources Institute, June 2014); Registrar General & Census Commissioner, *Census of India, 2011*.

4. Ministry of Urban Development, Government of India, "Atal Mission for Rejuvenation and Urban

Transformation (AMRUT)," http://amrut.gov.in; Ministry of Urban Development, Government of India, "Smart Cities Mission," http://smartcities.gov.in; Ministry of Housing and Urban Poverty Alleviation, Government of India, "Pradhan Mantri Awas Yojana – Housing for All (Urban)," http://mhupa.gov.in/User_Panel/UserView .aspx?TypeID=1493.

5. Centre for Environment Education (CEE) and IBI Group, *Learnings from Pune Pilot BRTS Project* (Pune: May 2015); Rainbow BRT website, rainbowbrt.in; Institute for Transportation and Development Policy, *Impact of Rainbow BRT* (2015).

6. Janmarg Ahmedabad Bus Rapid Transit System website, www.ahmedabadbrts.org.

7. Sanskriti Menon, "'Streets for People' College Courses," *SUM Net Newsletter* 2, nos. 3–4 (2014); National Association of Street Vendors of India, "Launch of Street Sathi Book and Mobile Apps," http://nasvinet.org/newsite /launch-of-street-sathi-book-and-mobile-apps/.

8. Pune Municipal Corporation (PMC), *Budget 2015-16*, www.punecorporation.org; Sanskriti Menon and Amarnath Avinash Madhale, *Participatory Budgeting in Pune: A Critical Review* (Pune: CEE, 2013).

9. www.OurPuneOurBudget.org; Kunal Kumar, Municipal Commissioner, PMC, "Budget Speech 2016–17."

10. AmdaVadmA website, www.amdavadma.org; Ahmedabad Municipal Corporation, "Chapter 11: Vision Ahmedabad," in *City Development Plan for Ahmedabad*, www.egovamc.com/Citizens/CDP/chapter11.pdf.

11. Tina Parekh and Himanshu Kaushik, "Tankas Help Gujarat Tide Over Water Woes," *Times of India*, May 22, 2005.

12. Gandhinagar Solar Rooftop Programme website, www.egujarat.net/gg/gandhinagar_solar_rooftop.html.

13. "About Indradhanushya," https://indradhanushyapune.wordpress.com/about-indradhanushya/; Mangesh Dighe, Environment Officer, PMC, personal communication with authors, January 2016.

Chapter 14. Solid Waste and Climate Change

1. Senthil Velsivasakthivel and Natarajan Nandini, "Airborne Multiple Drug Resistant Bacteria Isolated from Concentrated Municipal Solid Waste Dumping Site of Bangalore, Karnataka, India," *International Research Journal of Environment Sciences* 3, no. 10 (2014): 43–46.

2. Jenna R. Jambeck et al., "Plastic Waste Inputs from Land into the Ocean," *Science* 347, no. 6223 (2015): 768–71; World Economic Forum, *The New Plastics Economy: Rethinking the Future of Plastics* (Geneva: 2016).

3. S. Solomon et al., eds., *Contribution of Working Group I to the Fourth Assessment Report of the Intergovernmental Panel on Climate Change 2007* (Cambridge and New York: Cambridge University Press, 2007).

4. International Solid Waste Association, *Waste and Climate Change* (Vienna: 2009).

5. Daniel Hoornweg, Perinaz Bhada-Tata, and Christopher Kennedy, "Peak Waste: When Is It Likely to Occur?" *Journal of Industrial Ecology* 19, no. 1 (2015): 117–28.

6. Ibid. Table 14–1 from Daniel Hoornweg and Perinaz Bhada-Tata, *What a Waste: A Global Review of Solid Waste Management* (Washington, DC: World Bank, 2012), 5.

7. Table 14–1 from Hoornweg and Bhada-Tata, *What a Waste*.

8. European Environment Agency (EEA), *Waste Opportunities: Past and Future Climate Benefits from Better Municipal Waste Management in Europe* (Copenhagen: 2011).

9. European Topic Centre on Resource and Waste Management, *Municipal Waste Management and Greenhouse Gases* (Copenhagen: January 2008).

10. Global Methane Initiative, *Global Methane Emissions and Mitigation Opportunities* (Washington, DC: September 2011).

11. EEA, *Waste Opportunities.*

12. Seoul Solution, "Key Policies: Recycling (Smart Waste Management in Seoul)," November 2014, https://seoul solution.kr/content/recycling-smart-waste-management-seoul?language=en.

13. John Craig, "'Seattle Stomp' Blamed in Garbage Rate Increase," *The Spokesman-Review* (Spokane, WA), January 26, 1995.

14. Amy Brittain and Steven Rich, "Is D.C.'s 5-cent Fee for Plastic Bags Actually Serving Its Purpose?" *Washington Post*, May 9, 2015.

15. Women in Informal Employment: Globalizing and Organizing (WIEGO), *Urban Informal Workers and the Green Economy* (Cambridge, MA: undated).

16. Daniel Hoornweg, Laura Thomas, and Lambert Otten, *Composting and Its Applicability in Developing Countries* (Washington, DC: World Bank, 2000).

17. Nikita Naik, Ekaterina Tkachenko, and Roy Wung, *The Anaerobic Digestion of Municipal Solid Waste in California* (Berkeley, CA: University of California at Berkeley, 2013); Thomas DiStefano and Lucas Belenky, "Life-Cycle Analysis of Energy and Greenhouse Gas Emissions from Anaerobic Biodegradation of Municipal Solid Waste," *Journal of Environmental Engineering* (November 2009): 1,097–1,105.

18. CalRecycle, *Municipal Solid Waste Thermal Technologies* (Sacramento: September 17, 2013).

19. Table 14–3 from Ibid., 5.

20. Waste-to-Energy Research and Technology Council, "Answers to Frequently Asked Questions Regarding Waste-to-Energy," www.seas.columbia.edu/earth/wtert/faq.html.

21. World Business Council for Sustainable Development, "Cement Sustainability Initiative (CSI)" (Geneva, Washington, DC, and New Delhi: 2014), 1.

22. Youngchul Byun et al., "Thermal Plasma Gasification of Municipal Solid Waste (MSW)," in Yongseung Yun, ed., *Gasification for Practical Applications* (Rijeka, Croatia: InTech, 2012).

23. Ibid.

24. U.S. Environmental Protection Agency, *Reducing Greenhouse Gases Through Recycling and Composting* (Washington, DC: May 2011); Global Methane Initiative, "Global Methane Emissions and Mitigation Opportunities" (Washington, DC: undated).

City View: Barcelona, Spain

1. Data from Barcelona City Statistics Department website, www.bcn.cat/estadistica/. The authors would like to thank the following individuals for their contributions to this City View: Sito Alarcón, Teresa Casasayas, Antoni Farrero, Teresa Franquesa, Carles Llop, Jaume Marlès, Joan Nogué, Margarita Parés, Montse Rivero, Salvador Rueda, Coloma Rull, Jaume Terradas, and Anna Zahonero.

2. Martí Boada and Laia Capdevila, *Barcelona: Biodiversitat Urbana* (Barcelona: Ajuntament de Barcelona, 2000).

3. N. M. Rubió i Tudurí, *El problema de los espacios libres. Divulgación de su teoría y notas para su solución práctica* (Barcelona: Institut d'Estudis Metropolitans de Barcelona. Ed. Alta Fulla, 1926).

4. Boada and Capdevila, *Barcelona: Biodiversitat Urbana.*

5. Salvador Rueda, *Las supermanzanas: reinventando el espacio público, Ciudades (im)propias: la tensión entre lo global y lo local* (Universidad Politécnica de Valencia: Centro de investigación arte y entorno, 2010).

6. Xavier Argimon, *Estudi de la biodiversitat vegetal dels parcs i jardins de Barcelona* (Barcelona: Fundació de l'enginyeria agrícola catalana. Parcs i jardins de Barcelona, Institut Municipal, 2009); J. A. Burriel Moreno, Juan José Ibàñez Martí, and Jaume Terradas, "El mapa ecológico de Barcelona: Los cambios de la ciudad en las últimas tres décadas," *Cuadernos Geográficos* 39 (2006): 167–84; Jaume Puig, Daniel Renalías, and David Valero, *Biodiversitat florística a Collserola. El cas dels prats d'albellatge, Diagnosi ambiental al Parc de Collserola* (Barcelona: Barcelona Provincial Government, 2008), 113–21.

7. Martí Boada et al., *Diagnosi ambiental del Parc de Montjuïc. Biodiversitat* (Cerdanyola del Vallès: Universitat Autònoma de Barcelona, 2004, unpublished); Guillem Boix et al., *La pressió urbanística en l'àmbit del Parc de Collserola: estudi ambiental Diagnosi ambiental al Parc de Collserola* (Barcelona: Barcelona Provincial Government, 2008), 147–54.

8. Barcelona City Council, *Green and Biodiversity Plan Barcelona 2020* (Barcelona: 2013).

9. Ibid.; Barcelona City Council, *Annex 2. El Verd: planejament i diagnosi* (Barcelona: 2010).

10. Barcelona City Council, *Green and Biodiversity Plan Barcelona 2020*.

11. Barcelona City Council, *El Compromís ciutadà per a la sostenibilitat. Agenda 21 BCN* (Barcelona: 2002); Barcelona City Council, *Cap a l'Agenda 21 de Barcelona: document per al debat* (Barcelona: 2001); ICLEI–Local Governments for Sustainability website, http://iclei.org.

12. Barcelona City Council, *Green and Biodiversity Plan Barcelona 2020*.

13. Patuxent Wildlife Research Center website, www.pwrc.usgs.gov; BioBlitz website, http://bioblitzbcn.museu ciencies.cat/.

14. Pan-European Common Bird Monitoring Scheme website, www.ebcc.info/pecbm.html; Catalonian Ornithology Institute (ICO), "SOCC," www.ornitologia.org/ca/quefem/monitoratge/seguiment/socc.

15. Bosc Turull website, www.boscturull.cat.

16. Amics del Jardí Botànic de Barcelona website, www.amicsjbb.org.

17. Falcons de Barcelona website, www.falconsbarcelona.net.

18. Fàbrica del Sol website, http://lafabricadelsol.bcn.cat.

Chapter 15. Rural-Urban Migration, Lifestyles, and Deforestation

1. Union of Concerned Scientists, "Deforestation and Global Warming," 2013, citing work by Winrock International and Woods Hole Research Center, www.ucsusa.org/global_warming/solutions/stop-deforestation/deforestation-global-warming-carbon-emissions.html; Ruth DeFries et al., "Deforestation Driven by Urban Population Growth and Agricultural Trade in the Twenty-first Century," *Nature GeoScience* 3 (2010): 178–81.

2. Eric Lambin and Patrick Meyfroidt, "Global Land Use Change, Economic Globalization, and the Looming Land Scarcity," *Proceedings of the National Academy of Sciences* 108, no. 9 (2011): 3,465–72; DeFries et al., "Deforestation Driven by Urban Population Growth and Agricultural Trade in the Twenty-first Century." Figure 14–1 from the following sources: per capita gross national income, in constant 2005 U.S. dollars, from World Bank Databank, http://data.worldbank.org/indicator/NY.GNP.PCAP.KD; per capita meat availability from United Nations Food and Agriculture Organization (FAO), extract prepared by Tomasz Filipczuk from http://faostat3.fao.org/download/FB/CL/.

3. Doug Boucher et al., *The Root of the Problem: What's Driving Tropical Deforestation Today* (Cambridge, MA: Union of Concerned Scientists, 2011).

4. City population growth from Andy Gouldson et al., *Accelerating Low-Carbon Development in the World's Cities,* New Climate Economy Working Paper (Washington, DC: Global Commission on the Economy and Climate, 2015); city population estimate in 2050 from Karen Seto et al., "Urban Land Teleconnections and Sustainability," *Proceedings of the National Academy of Sciences* 109, no. 20 (2012): 7,687–92; urban land area expansion from Karen Seto, Burak Güneralp, and Lucy R. Hutyra, "Global Forecasts of Urban Expansion to 2030 and Direct Impacts on Biodiversity and Carbon Pools," *Proceedings of the National Academy of Sciences* 109, no. 40 (2012): 16,083–88; farmland loss from Lambin and Meyfroidt, "Global Land Use Change, Economic Globalization, and the Looming Land Scarcity."

5. Jim Robbins, "Deforestation and Drought, *New York Times,* October 11, 2015, SR7; Antonio Donato Nobre, *The Future Climate of Amazonia,* Scientific Assessment Report sponsored by CCST-INPE, INPA and ARA (São Paulo: 2014).

6. FAO, *Global Forest Resources Assessment 2015* (Rome: 2015), 14; Will Steffen et al., "Planetary Boundaries: Guiding Human Development on a Changing Planet," *Science* 47, no. 6223 (2015).

7. Seto et al., "Urban Land Teleconnections and Sustainability," 7,687.

8. Lambin and Meyfroidt, "Global Land Use Change, Economic Globalization, and the Looming Land Scarcity."

9. Ibid.; Wanqing Zhou, *The Triangle: The Evolution and Future of Industrial Animal Agriculture in the U.S., China and Brazil* (New York: Brighter Green, 2015).

10. Lambin and Meyfroidt, "Global Land Use Change, Economic Globalization, and the Looming Land Scarcity"; C. Nellemann et al., *The Environmental Food Crisis: The Environment's Role in Averting Future Food Crises* (Arendal, Norway: United Nations Environment Programme/GRID-Arendal, 2009).

11. Lambin and Meyfroidt, "Global Land Use Change, Economic Globalization, and the Looming Land Scarcity."

12. Ibid.

13. Ibid., 3,471.

14. Edward Glaeser and Joshua Gottlieb, *The Wealth of Cities: Agglomeration Economies and Spatial Equilibrium in the United States,* National Bureau of Economic Research Working Paper 14806 (Cambridge, MA: November 2011); Michael Hermann and David Svarin, *Environmental Pressures and Rural-Urban Migration: The Case of Bangladesh* (Munich: UNCTAD/Munich Personal RePEc Archive paper, January 2009; Simon Fairlie, "A Short History of Enclosure in Britain," *The Land* 7 (Summer 2009).

15. World Bank, "Remittances Growth to Slow Sharply in 2015, as Europe and Russia Stay Weak; Pick-up Expected Next Year," press release (Washington, DC: April 13, 2015).

16. See, for example, Walden Bello, *The Food Wars* (London: Verso, 2009), 11 and ff. Box 15–1 based on Chris Smaje, "Three Urban Myths," Small Farm Future blog, March 30, 2014, and on Chris Smaje, "The Ungreen City— Or the Polluting Countryside?" *Significance* (Royal Statistical Society) 8, no. 2 (2011): 61–64.

17. Felix Creutzig, personal communication with author, July 26, 2015; Glaeser and Gottlieb, *The Wealth of Cities,* 3, 50–51.

18. FAO, *Global Food Losses and Food Waste: Extent, Losses, and Prevention* (Rome: 2011); Michael Hamm, "City Region Food Systems – Part 1 – Conceptualization," Food Climate Research Network blog, May 20, 2015; Zhou, *The Triangle.*

19. Brian Machovina et al., "Biodiversity Conservation: The Key Is Reducing Meat Consumption," *Science of the Total Environment* 536 (December 1, 2015): 419–31.

Chapter 16. Remunicipalization, the Low-Carbon Transition, and Energy Democracy

1. Martin Pigeon et al., eds. *Putting Water Back in Public Hands* (Amsterdam: Transnational Institute, 2012); Susanne Halmer and Barbara Hauenschild, *Remunicipalisation of Public Services in the EU* (Vienna: Austrian Association for Political Consulting and Development (OGPP), 2014); Satoko Kishimoto, Emanuele Lobina, and Olivier Petitjean, *Our Public Water Future: The Global Experience with Remunicipalisation* (Amsterdam: Transnational Institute, 2015); Oliver Wagner and Kurt Berlo, *The Wave of Remunicipalisation of Energy Networks and Supply in Germany: The Establishment of 72 New Municipal Power Utilities* (Stockholm: European Council for an Energy Efficient Economy, 2015); David Hall, *Re-municipalising Municipal Services in Europe* (Greenwich, U.K.: Public Services International Research Unit (PSIRU), 2012).

2. John Vickers and George Yarrow, *Privatization: An Economic Analysis* (Cambridge, MA: MIT Press, 1988).

3. Andrew Cumbers, *Reclaiming Public Ownership: Making Space for Economic Democracy* (London: Zed Books, 2012); Kishimoto, Lobina, and Petitjean, *Our Public Water Future*; Hall, *Re-municipalising Municipal Services in Europe*.

4. Emanuele Lobina, "Introduction: Calling for Progressive Water Policies," in Kishimoto, Lobina, and Petitjean, *Our Public Water Future*, 7.

5. Lobina, "Introduction: Calling for Progressive Water Policies," 17; Cumbers, *Reclaiming Public Ownership*.

6. Lobina, "Introduction: Calling for Progressive Water Policies," 17; Cumbers, *Reclaiming Public Ownership*. Box 16–1 from David Hall, Stephen Thomas, and Violeta Corral, *Global Experience with Electricity Liberalisation* (Greenwich, U.K.: PSIRU, 2009), and from Kishimoto, Lobina, and Petitjean, *Our Public Water Future*, 109.

7. Halmer and Hauenschild, *Remunicipalisation of Public Services in the EU*; Hall, *Re-municipalising Municipal Services in Europe*. Table 16–1 from the following sources: Hall, idem; Charleen Fei and Ian Rinehart, *Taking Back the Grid: Municipalization Efforts in Hamburg, Germany and Boulder, Colorado* (Washington, DC: Heinrich Böll Stiftung, 2014); Thomas Blanchet, "Struggle Over Energy Transition in Berlin: How Do Grassroots Initiatives Affect Local Energy Policy-making? *Energy Policy* 78 (March 2015): 246–54.

8. Vattenfall from Fei and Rinehart, *Taking Back the Grid*; London and Kiel from Halmer and Hauenschild, *Remunicipalisation of Public Services in the EU*; U.K. from Hall, *Re-municipalising Municipal Services in Europe*.

9. Hall, *Re-municipalising Municipal Services in Europe*.

10. Andrew Cumbers et al., *Repossessing the Future: A Common Weal Strategy for Community and Democratic Ownership of Scotland's Energy Resources* (Biggar, Scotland: Jimmy Reid Foundation, 2013); Andrew Bowman et al., *The Great Train Robbery: Rail Privatisation and After* (Manchester, U.K.: Centre for Research on Socio-Cultural Change, 2013).

11. John Farrell, *Beyond Utility 2.0 to Energy Democracy* (Minneapolis, MN: Institute for Local Self-Reliance, 2014).

12. Halmer and Hauenschild, *Remunicipalisation of Public Services in the EU*.

13. Community Power Network, "Introduction to Municipalization," http://communitypowernetwork.com /node/990, viewed November 2015; Center for Social Inclusion, *Community-Scale Energy: Models, Strategies and Racial Equity* (New York: 2013).

14. EcoDistricts, "Vision + Values," http://ecodistricts.org/about/vision-values/#, viewed November 2015.

15. Hall, *Re-municipalising Municipal Services in Europe*; Conrad Kunze and Sören Becker, "Collective Ownership in Renewable Energy and Sustainable Degrowth," *Sustainability Science* 10, no. 3 (2015), 425–37.

16. Author's own assessment.

17. Halmer and Hauenschild, *Remunicipalisation of Public Services in the EU*.

18. Ibid.

19. Wagner and Berlo, *The Wave of Remunicipalisation of Energy Networks and Supply in Germany*. Box 16–2 from Halmer and Hauenschild, *Remunicipalisation of Public Services in the EU*.

20. Halmer and Hauenschild, *Remunicipalisation of Public Services in the EU*; Kunze and Becker, "Collective Ownership in Renewable Energy and Sustainable Degrowth," 425–37.

21. Halmer and Hauenschild, *Remunicipalisation of Public Services in the EU*.

22. Wagner and Berlo, *The Wave of Remunicipalisation of Energy Networks and Supply in Germany*.

23. Former Green Party official in Hamburg government responsible for energy policy, personal communication with author, February 2015.

24. Timothy Moss, Sören Becker, and Matthias Naumann, "Whose Energy Transition Is It, Anyway? Organisation and Ownership of the *Energiewende* in Villages, Cities and Regions," *Local Environment: The International Journal of Justice and Sustainability* 20, no. 13 (2015): 1,547–63; activist from personal communication with author, July 2014.

25. Philipp Terhorst and David Hall, *Remunicipalisation of the Germany Energy Sector* (London: PSIRU, 2011); official in Hessen Energy Ministry, personal communication with author, July 2015.

26. Stephen Hall, Timothy Foxon, and Ronan Bolton, "The New 'Civic' Energy Sector: Implications for Ownership, Governance and Financing of Low Carbon Energy Infrastructure," *Energy Research & Social Science* (forthcoming 2016); Caroline Julian, *Creating Local Energy Economies: Lessons from Germany* (London: Respublica, 2014). Figure 16–1 from German Renewable Energies Agency (AEE), "Renewable Energy in the Hands of the People," April 2013, www.unendlich-viel-energie.de/media-library/charts-and-data/renewable-energy-in-the-hands-of-the-people.

27. Hall, Foxon, and Bolton, "The New 'Civic' Energy Sector"; official in Hessen Energy Ministry, personal communication with author, July 2015.

28. Assistant director of local municipal utility, personal communication with author, July 2014; see also Stadtwerke München, "SWM Renewable Energies Expansion Campaign," https://www.swm.de/english/company/energy-generation/renewable-energies.html.

29. Figure 16–2 from U.S. Energy Information Administration, "Total Primary Coal Production (Thousand Short Tons)," International Energy Statistics, www.eia.gov/cfapps/ipdbproject/iedindex3.cfm?tid=1&pid=7&aid=1.

30. "Coming to a standstill" from energy official in regional government, personal communication with author, September 2015.

31. Andrew Cumbers et al., *Repossessing the Future*.

32. Caroline Kuzemko, "Energy Depoliticisation in the UK: Destroying Political Capacity," *British Journal of Politics & International Relations* (April 16, 2015): 1.

33. Cumbers, *Reclaiming Public Ownership*.

34. Ibid.

City View: Portland, Oregon, United States

1. Data are adapted from Mike Steinhoff et al., *Measuring Up 2015* (Washington, DC: WWF-US and ICLEI USA, 2013). Population data are for 2013.

2. City of Portland and Multnomah County, *Climate Action Plan 2009* (Portland, OR: 2009).

3. City of Portland staff, personal communication, August 2014.

4. Ibid.

5. Ibid.

6. City of Portland Bureau of Transportation, "Bicycles in Portland Factsheet," https://www.portlandoregon.gov/transportation/article/407660.

7. City of Portland staff, personal communication, August 2014.

8. Buildings from City of Portland and Multnomah County, *Climate Action Plan 2009: Year Two Progress Report* (Portland, OR: April 2012); solar energy systems based on City of Portland staff, personal communication, August 2014.

9. American Council for an Energy-Efficient Economy, *Clean Energy Works Portland* (Washington, DC: March 2011).

10. City of Portland staff, personal communication, December 2014.

11. City of Portland staff, personal communication, August 2014.

12. Ibid.

13. Ibid.

14. Ibid.

15. Ibid.

16. Emissions reductions from Ibid.; C40 Cities, "Portland – Healthy Connected City Network," www.c40.org/awards/1/profiles/16.

17. City of Portland, "2015 Sustainable City Principles, 2030 Objectives," www.portlandoregon.gov/bps/article/527220.

18. City of Portland staff, personal communication, August 2014.

19. Ibid.

Chapter 17. The Vital Role of Biodiversity in Urban Sustainability

1. Martí Boada and Francisco Javier Gómez, *Biodiversidad. Cuadernos de Medio Ambiente* (Barcelona: Rubes, 2008).

2. Jaume Terradas, *Ecologia Urbana* (Barcelona: Monografies de medi ambient, Generalitat de Catalunya, 2001).

3. Martí Boada and Roser Maneja Zaragoza, *The Socio-environmental Heritage of the UAB Campus* (Bellaterra: Universitat Autònoma de Barcelona, 2005). Box 17–1 from the following sources: Secretariat of the Convention on Biological Diversity, *Cities and Biodiversity Outlook* (Montreal: 2012), 8; stormwater management and return on investment from European Environment Agency (EEA), *Exploring Nature-based Solutions: The Role of Green Infrastructure in Mitigating the Impacts of Weather- and Climate Change-related Natural Hazards* (Brussels: 2015); lawn from Svenskt Vatten, *Hållbar dag – och dränvattenhantering. Råd vid planering och utformning* (Broma, Sweden: 2011); stress levels from Omid Kardan et al., "Neighborhood Greenspace and Health in a Large Urban Center, *Scientific Reports* 5, no. 11610 (2015), from Catharine Ward Thompson et al., "More Green Space Is Linked to Less Stress in Deprived Communities: Evidence from Salivary Cortisol Patterns," *Landscape and Urban Planning* 105, no. 3 (2012): 221–29, and from Patrik Grahn and Ulrika A. Stigsdotter, "Landscape Planning and Stress," *Urban Forestry & Urban Greening* 2, no. 1 (2003): 1–18; natural characteristics from Matilda Annerstedt, *Nature and Public Health: Aspects of Promotion, Prevention, and Intervention*, doctoral thesis (Alnarp: Swedish University of

Agricultural Sciences, 2010); jaybirds from Cajsa Houghner, Johan Colding, and Tore Söderqvist, "Economic Valuation of a Seed Dispersal Service in the Stockholm National Urban Park, Sweden," *Ecological Economics* 59, no. 3 (2006): 364–74; pigeons from Anders Eriksson and Tommy Eriksson, *Duvhök i norra Stockholm. Accipiter gentilis* (Stockholm: City of Stockholm, 2007); canopy cover from S. Thorsson, "The Urban Climate – Measures to Reduce the Temperature in Urban Areas," FOI-R-3415-SE, 2012; noise barriers from Mats Nilsson et al., "Novel Solutions for Quieter and Greener Cities" (Stockholm: 2013); erosion from EEA, *Exploring Nature-based Solutions*, and from Secretariat of the Convention on Biological Diversity, *Cities and Biodiversity Outlook*.

4. Ernst Haeckel, *Generelle Morphologie der Organismen* (Berlin: Druck und Verlag von Georg Reimer, 1866). Figure 17–1 from M. Boada and L. Capdevila, *Barcelona, Biodiversitat Urbana* (Barcelona: Barcelona City Council, 2000).

5. Virginio Bettini, *Elementos de ecología urbana* (Torino: Editorial Trotta, 1996); Jaume Terradas et al., *Ecologia Urbana* (Barcelona: Revista investigación i tecnologia, 2011).

6. Salvador Rueda, *Barcelona, ciutat mediterrània, compacta i complexa. Una visió de futur més sostenible.* (Barcelona: Barcelona City Council, 2002).

7. Salit Kark et al., "Living in the City: Can Anyone Become an 'Urban Exploiter'?" *Journal of Biogeography* 34 (2007): 638–51; Barcelona City Council, "El falcó a Barcelona," www.bcn.cat/agenda21/falco/.

8. Terradas et al., *Ecologia Urbana*.

9. Salvador Rueda, *Green Roofs and Walls in Barcelona. A Study on Existing and Potential Implementation Strategies* (Barcelona: Urban Ecology Agency of Barcelona, 2010).

10. Box 17–2 from the following sources: MedPAN website, www.medpan.org; IUCN website, www.iucn.org; Fernando. Valladares, "El hábitat mediterráneo continental: un sistema humanizado, cambiante y vulnerable," in Mariano Paracuellos, *Ambientes mediterráneos. Funcionamiento, biodiversidad y conservación de los ecosistemas mediterráneos* (Almeria: Intituto de Estudios Almerienses, 2007); Salvador Rueda, *Un modelo medioambientlal de ordenación, Plan estrategico metropolitano* (Barcelona: Barcelona City Council, 2004), 109–13.

11. Richard T. T. Forman, *Land Mosaics: The Ecology of Landscapes and Regions* (Cambridge, U.K.: Cambridge University Press, 1995).

12. Esteban Fernández-Juricic, "Avifaunal Use of Wooded Streets in an Urban Landscape," *Conservation Biology* 14, no. 2 (2000): 513–21; Martí Boada and Sònia Sànchez, *Naturaleza y cultura, biodiversidad urbana. Ecoinovação para a Melhoria Ambiental de Produtos e Serviços: Experiências Espanholas e Brasileiras nos Setores Industrial, Urbano e Agrícola* (São Carlos, Brazil: Diagrama Editorial, 2012), 131–42; Alexis A. Alvey, "Promoting and Preserving Biodiversity in the Urban Forest," *Urban Forestry & Urban Greening* 5, no. 4 (2006): 195–201; E. Gregory McPherson and Charles Nilon, "A Habitat Suitability Index Model for Gray Squirrel in an Urban Cemetery," *Landscape Journal* 6, no. 1 (1987): 21–30; F. Munyenyembe et al., "Determinants of Bird Populations in an Urban Area," *Australian Journal of Ecology* 14, no. 4 (1989): 549–57; P. Clergeau et al., "Bird Abundance and Diversity Along an Urban–Rural Gradient: A Comparative Study Between Two Cities on Different Continents, *The Condor* 100, no. 3 (1998): 413–25; Michael A. Steele and John L. Koprowski, *North American Tree Squirrels* (Washington, DC and London: Smithsonian Institution Press, 2001); David Palomino and Luis M. Carrascal, "Urban Influence on Birds at a Regional Scale: A Case Study with the Avifauna of Northern Madrid Province," *Landscapes Urban Planning* 77, no. 3 (2006): 276–90; Gang Yang et al., "Evaluation of Microhabitats for Wild Birds in a Shanghai Urban Area Park," *Urban Forestry & Urban Greening* 14, no. 2 (2015): 246–54.

13. Boada and Capdevila, *Barcelona, Biodiversitat Urbana*; Boada and Gómez, *Biodiversidad. Cuadernos de Medio Ambiente*; Tommy S. Parker and Charles H. Nilon, "Gray Squirrel (*Sciurus carolinensis*) Density, Habitat Suitability, and Behavior in Urban Parks," *Urban Ecosystems* 11 (2008): 243–55; Boada and Sànchez, *Naturaleza y cultura, biodiversidad urbana*; John Marzluff and Amanda Rodewald, "Conserving Biodiversity in Urbanizing Areas:

Nontraditional Views from a Bird's Perspective," *Cities and the Environment* 1, no. 2 (2008); McPherson and Nilon, "A Habitat Suitability Index Model for Grey Squirrel in an Urban Cemetery; Esteban Fernández-Juricic et al., "Alert Distance as an Alternative Measure of Bird Tolerance to Human Disturbance: Implications for Park Design, *Environmental Conservation* 28, no. 3 (2001): 263–69; Tommy S. Parker and Charles H. Nilon, "Urban Landscape Characteristics Correlated with the Synurbization of Wildlife," *Landscape and Urban Planning* 106, no. 4 (2012): 316–25; Anders Pape Møller, "Flight Distance of Urban Birds, Predation, and Selection for Urban Life," *Behavioral Ecology and Sociobiology* 63 (2008): 63–75; Boada and Sànchez, *Naturaleza y cultura, biodiversidad urbana.*

14. Boada and Capdevila, *Barcelona, Biodiversitat Urbana*; Boada and Gómez, *Biodiversidad. Cuadernos de Medio Ambiente.*

15. Boada and Capdevila, *Barcelona, Biodiversitat Urbana*; Boada and Gómez, *Biodiversidad. Cuadernos de Medio Ambiente.*

16. Montserrat Pallarès-Barbera et al., "Bienestar, planificación urbana y biodiversidad. El caso de Barcelona," presented at XXXVIII Reunión de estudios regionales, 2012; Irene van Kamp et al., "Urban Environmental Quality and Human Well-being: Towards a Conceptual Framework and Demarcation of Concepts: A Literature Study," *Landscape and Urban Planning* 65, no. 1–2 (2003): 5–18; Francisco J. Goerlich Gisbert and Isidro Cantarino Martí, *Zonas de morfología urbana: coberturas del suelo y demografía* (Madrid: Fundación BBVA, 2013); Rachel Kaplan, "Urban Forestry and the Workplace," in Paul H. Gobster, ed., *Managing Urban and High Use Recreation Settings* (St. Paul, MN: U.S. Forest Service, North Central Forest Experiment Station, 1993), 41–45; Rachel Kaplan and Stephen Kaplan, *The Experience of Nature: A Psychological Perspective* (New York: Cambridge University Press, 1989); Roger S. Ulrich, "View Through a Window May Influence Recovery from Surgery," *Science* 224 (April 27, 1984): 420–21; R. B. Hull IV, "Brief Encounters with Urban Forests Produce Moods That Matter," *Journal of Arboriculture* 18, no. 6 (1992): 322–24.

17. Harvard University Center for Health and the Global Environment, *Sustaining Life: How Human Health Depends on Biodiversity* (Oxford, U.K.: Oxford University Press, 2008). Table 17–1 from Marie Svensson and Ingegärd Eliasson, *Grönstrukturens betydelse för stadens ventilation* (Stockholm: Swedish Environmental Protection Agency, 1997), and from Per Bolund and Sven Hunhammar, "Ecosystem Services in Urban Areas," *Ecological Economics* 29 (1999): 293–30.

18. Erik Gómez-Baggethun and David N. Barton, "Classifying and Valuing Ecosystem Services for Urban Planning," *Ecological Economics* 86 (February 2013): 235–45; Leonie Pearson, ed., "Sustainable Urbanisation: A Resilient Future," Special Issue, *Ecological Economics* 86 (February 2013): 1–300; Jari Lyytimäki and Maija Sipilä, "Hopping on One Leg—The Challenge of Ecosystem Disservices for Urban Green Management," *Urban Forestry & Urban Greening* 8, no. 4 (2009): 309–15.

19. Table 17–2 from D. C. Dearborn and S. Kark, "Motivations for Conserving Urban Biodiversity," *Conservation Biology* 24, no. 2 (2010): 432–40; Assaf Shwartz et al., "Outstanding Challenges for Urban Conservation Research and Action," *Global Environmental Change* 28 (September 2014): 39–49.

20. City of Melbourne, *Urban Forest Strategy: Making a Great City Greener 2012–2032* (Melbourne: 2012); Barcelona City Council, *Green and Biodiversity Plan Barcelona 2020* (Barcelona: 2013).

21. Millennium Ecosystem Assessment, *Ecosystems and Human Well-being: Biodiversity Synthesis* (Washington, DC: World Resources Institute, 2005); Ryo Kohsaka et al., "Indicators for Management of Urban Biodiversity and Ecosystem Services: City Biodiversity Index," in Thomas Elmqvist et al., eds., *Urbanization, Biodiversity and Ecosystem Services: Challenges and Opportunities: A Global Assessment* (Springer, 2013), 699–718.

22. Pallarès-Barbera et al., "Bienestar, planificación urbana y biodiversidad"; Boada and Sànchez, *Naturaleza y cultura, biodiversidad urbana.*

City View: Jerusalem, Israel

1.   Data from Jerusalem Institute for Israel Studies (JIIS), "Chapter 1: Area" and "Chapter 3: Population," in *The Statistical Yearbook of Jerusalem* (Jerusalem: 2015). The authors would like to thank the following individuals for their contributions to this City View: Amir Balaban, Carles Barriocanal, Yoav Farago, Yaara Israeli, Salit Kark, Noam Levin, Amiram Rotem, Helene Roumani, and Assaf Shartz.

2.   JIIS, "Chapter 1: Area."

3.   JIIS, "Chapter 3: Population."

4.   Menahem Marcus, *Geomorphology, The New Israel Guide* (Jerusalem: Keter Publishing, 2001); Isaac Schattner, "Surroundings," in Michael Avi-Yonah, *Sepher Yerushalayim: Jerusalem, Its Natural Conditions, History and Development from the Origins to the Present Day* (Jerusalem: The Bialik Institute and the Dvir Publishing House, 1956).

5.   Dov Ashbel, "Climate," in Avi-Yonah, *Sepher Yerushalayim.*

6.   Helene Roumani, *City of Jerusalem Biodiversity Report* (Jerusalem: Local Action for Biodiversity, 2013).

7.   Martí Boada and Laia Capdevila, *Barcelona: Biodiversitat Urbana* (Barcelona: Ajuntament de Barcelona, 2000); Ron Frumkin, Berry Pinshow, and Shani Kleinhaus, "Review of Bird Migration Over Israel," *Journal of Ornithology* 136, no. 2 (1995): 127–47.

8.   Martin Sicker, *Between Rome and Jerusalem: 300 Years of Roman-Judean Relations* (Westport, CT: Greenwood Publishing Group, 2001); Roumani, *City of Jerusalem Biodiversity Report.*

9.   Mauro Bernabei, "The Age of the Olive Trees in the Gardens of Gethsemane," *Journal of Archaeological Science* 53 (January 2015): 43–48; Society for the Protection of Nature in Israel (SPNI), *Survey of Mature and Unique Trees in the City of Jerusalem* (Jerusalem: 2013).

10. SPNI, *Summary Jerusalem Urban Nature Infrastructure Survey. Goals, Methods Findings and Conclusions* (Jerusalem: 2010).

11. SPNI, *Jerusalem Urban Nature Infrastructure Survey* (Jerusalem: 2010).

12. Figure based on Jerusalem Green Map, "Partners and Stakeholders – Jerusalem Bioregion Center," www.greenmap.org.il/content?lang=en&pageid=70.

13. SPNI, "Jerusalem Bird Observatory," http://natureisrael.org/JBO.

14. Roumani, *City of Jerusalem Biodiversity Report.*

15. Jerusalem Green Map website, www.greenmap.org.il/?lang=en.

16. Naomi Tsur, "The Story of Jerusalem's Railway Park: Getting the City Back on Track, Economically, Environmentally and Socially," TheNatureofCities.com, August 18, 2014.

17. Green Pilgrimage Network, "Jerusalem," http://greenpilgrimage.net/other-pilgrim-sites/jerusalem/.

18. Gazelle Valley website, www.zvaiim.jerusalem.muni.il.

Chapter 18. The Inclusive City: Urban Planning for Diversity and Social Cohesion

1.   UN-Habitat, *State of the World's Cities 2008/2009: Harmonious Cities* (Nairobi: 2009); United Nations Department of Economic and Social Affairs (DESA), Population Division, *World Urbanization Prospects: The 2014 Revision, Highlights* (New York: 2014); International Organization for Migration (IOM), *World Migration Report 2015. Migrants and Cities: New Partnerships to Manage Mobility* (Geneva: 2015).

2. Mary J. Hickman and Nicola Mai, "Migration and Social Cohesion. Appraising the Resilience of Place in London," *Population, Space and Place* 21, no. 5 (2015): 431.

3. UN-Habitat, *State of the World's Cities 2012/2013: Prosperity of Cities* (London: Routledge, 2013), 150.

4. U.K. Department for Communities and Local Government, *English Housing Survey 2010 to 2011: Headline Report* (London: 2012); Patrick Butler, "'Inadequate, Unaffordable, Insecure': UK Housing's Decline and Fall," *The Guardian* (U.K.), September 11, 2013.

5. Karin Peters, Birgit Elands, and Arjen Buijs, "Social Interactions in Urban Parks. Stimulating Social Cohesion?" *Urban Forestry & Urban Greening* 9, no. 2 (2010): 93–100.

6. UN-Habitat, *Urban Planning and Design for Social Cohesion. Concept Note World Urban Forum* (Medellín, Colombia: April 2014), 2; IOM, *World Migration Report 2015*, 4.

7. Gerard Boucher and Yunas Samad, "Introduction. Social Cohesion and Social Change in Europe," *Pattern of Prejudice* 47, no. 3 (2013): 197; UN DESA, United Nations Development Programme (UNDP), and Office of the United Nations High Commissioner for Human Rights (OHCHR), *Habitat III Issue Papers – 1 Inclusive Cities* (New York: 2015). Figure 18–1 from UN-Habitat, *State of the World's Cities 2010/2011: Cities for All – Bridging the Urban Divide* (New York: 2011), 73.

8. UN DESA, UNDP, and OHCHR, *Habitat III Issue Papers*; Tiit Tammaru et al., eds., *Socio-Economic Segregation in European Capital Cities. East Meets West* (London: Routledge, 2015); Richard Fry and Paul Taylor, *The Rise of Residential Segregation by Income* (Washington, DC: Pew Research Center, August 1, 2012).

9. Tammaru et al., eds., *Socio-Economic Segregation in European Capital Cities.*

10. Hartmut Häussermann, "Wohnen und Quartier: Ursachen sozialräumlicher Segregation," in Ernst-Ulrich Huster, Jürgen Boeckh, and Hildegard Mogge-Grothjahn, *Handbuch Armut und soziale Ausgrenzung* (Wiesbaden: VS, 2008), 335–49.

11. Jane Parry, *Issue Paper on Secure Tenure for Urban Slums. From Slums to Sustainable Communities: The Transformative Power of Secure Tenure* (Atlanta and Brussels: Habitat for Humanity and Cities Alliance, 2015); UN-Habitat, *Urban Planning and Design for Social Cohesion*, 2.

12. Boucher and Samad, "Introduction. Social Cohesion and Social Change in Europe"; Peters, Elands, and Buijs, "Social Interactions in Urban Parks"; Talja Blokland, Carlotta Giustozzi, and Franziska Schreiber, "The Social Dimensions of Urban Transformation: Contemporary Diversity in Global North Cities and the Challenges for Urban Cohesion," in Harald A. Mieg and Klaus Töpfer, *Institutional and Social Innovation for Sustainable Urban Development* (Oxon and New York: Routledge, 2013), 125.

13. UN-Habitat, *Urban Planning and Design for Social Cohesion*, 1.

14. Box 18–1 from the following sources: German Federal Ministry for the Environment, Nature Conservation, Building and Nuclear Safety, *Social City Program* (Berlin: 2015); Alexandra Galeshewe et al., *National Urban Renewal Programme. Implementation Framework* (Pretoria: Department of Provincial and Local Government, Republic of South Africa, undated); Michael E. Leary and John McCarthy, *The Routledge Companion to Urban Regeneration* (London and New York: Routledge, 2013), 402; Hans Skifter Andersen and Louise Kielgast, *Area-based Initiatives in Denmark – "Kvarterløft": Addressing Increasing Social Problems and Concentration of Immigrants and Refugees in Seven Neighborhoods* (Copenhagen: Danish Building Research Institute, June 2003).

15. German Institute of Urban Affairs, *Status Report. The Programme "Social City" (Soziale Stadt) – Summary* (Berlin: Federal Ministry of Transport, Building and Urban Affairs, 2008).

16. Galeshewe et al., *National Urban Renewal Programme*; Thomas Franke and Wolf-Christian Strauss, *Management*

gebietsbezogener integrativer Stadtteilentwicklung. Ansätze in Kopenhagen und Wien im Vergleich zur Programmumsetzung "Soziale Stadt" in deutschen Städten (Berlin: German Institute of Urban Affairs, 2005).

17. Franke and Strauss, *Management gebietsbezogener integrativer Stadtteilentwicklung*; Ivan Turok, *The Evolution of National Urban Policies: A Global Overview* (Nairobi: UN-Habitat and Cities Alliance, 2014).

18. Senatsverwaltung für Stadtentwicklung, "The Neighborhood Council Within the Neighborhood Management Process," handout at the 3rd Congress of Berlin's Neighborhood Councils (Berlin: March 20, 2010).

19. Franke and Strauss, *Management gebietsbezogener integrativer Stadtteilentwicklung*.

20. Ellen Højgaard Jensen and Asger Munk, *Kvaterløft. 10 Years of Urban Regeneration. Ministry of Refugees, Immigration and Integration Affairs* (Copenhagen: 2007).

21. Jan Gehl, *Cities for People* (Washington, DC: Island Press, 2010).

22. Chris Firth, Damian Maye, and David Pearson, "Developing 'Community' in Community Gardens," *Local Environment: The International Journal of Justice and Sustainability* 16, no. 6 (2011): 555–68. Box 18–2 based on the following sources: International Network for Economic, Social & Cultural Rights, "Report and Recommendation on Request for Inspection, Re: Argentina – Special Structural Adjustment Loan 4405-AR (Pro-Huerta Case)," 2012, https://www.escr-net.org/node/364789; Ana Bell, "Community Gardens Boost Self-sufficiency in Argentina," Panos London, August 31, 2012; Ministry of Foreign Affairs and Worship of Argentina, "Desarrollo sustentable: Haiti - autoproduccíon de alimentos frescos Pro Huerta," http://cooperacionarg.gob.ar/en/haiti/autopro duccion-de-alimentos-frescos-pro-huerta; Instituto Nacional de Tecnología Agropecuaria (INTA), "Pro Huerta," http://prohuerta.inta.gov.ar/; Walter Alberto Pengue, "Aún nos quedan las manos y la tierra," *El Diplo* 38 (August 2002); Municipality of Rosario, "Indicadores Demograficos," November 23, 2015, www.rosario.gov.ar/sitio/carac teristicas/indicadores.jsp; United Nations Food and Agriculture Organization, "Rosario," in *Growing Greener Cities in Latin America and the Caribbean* (Rome: 2013); Ferne Edwards, "Sustainable City & Model – Urban Agriculture in Argentina," Sustainable Cities Network, July 13, 2007, www.sustainablecitiesnet.com/models/model-urban-ag riculture-in-rosario-argentina/; Ministry of Social Development of Argentina, "Pro Huerta," 2013, www.desarrollo social.gob.ar/wp-content/uploads/2015/08/1.-M--s-sobre-PRO-HUERTA.pdf; Canadian International Development Agency and Inter-American Institute for Cooperation on Agriculture (IICA), "Argentina, Canada and Haiti Join Efforts to Improve Food Security. Project for Self-sufficiency in the Production of Fresh Foods in Haiti Is Expanded," press release (Haiti: June 2008); IICA, "Program for Fresh Food Self-sufficiency in Haiti: Pro-Huerta 2005-2008," *Comuniica*, January–April 2008; Pan-American Health Organization and Ministerio de Relaciones Exteriores, Comercio Internacional y Culto de la República Argentina, *South-South Cooperation: Triangular Cooperation Experience Between the Government of the Argentine Republic and the Pan-American Health Organization/World Health Organization* (Buenos Aires: October 2009); "Lessons Learned in Argentina Helping Haiti Cope with Cholera," *New Agriculturalist*, December 2010; "Haiti Agriculture: True Success of Pro Huerta Program in Haiti," *Haiti Libre*, March 23, 2015; "Haiti – Agriculture: The Argentinean Program Pro Huerta Extended Until 2016," *Haiti Libre*, January 17, 2014.

23. Jacqueline Groth and Eric Corjin, "Reclaiming Urbanity: Intermediate Spaces, Informal Actors and Urban Agenda Setting," *Urban Studies* 42, no. 3 (2005): 503–26; David Harvey, *Rebel Cities: From the Right to the City to the Urban Revolution* (London and New York: Verso, 2012); Franziska Schreiber, "Viele viele Frei(t)räume: The Prinzessinnengarten and Contemporary Land Use Conflicts in Berlin," anstiftung.de/downloads/send/15-forschungsar beiten-urbane-gaerten/173-the-prinzessinnengarten-and-contemporary-land-use-conflicts-in-berlin.

24. The Queensland Government, *Transit Oriented Development: Guide to Community Diversity* (Brisbane: Queensland Department of Infrastructure and Planning, 2010); Gehl, *Cities for People*; Xuemei Zhu et al., "A Retrospective Study on Changes in Residents' Physical Activities, Social Interactions, and Neighborhood Cohesion After Moving to a Walkable Community," *Preventive Medicine* 69, no. 1 (2014): 93–97.

25. Gehl, *Cities for People*, 7.

26. Institute for Transportation and Development Policy – China, *Best Practices in Urban Development in the Pearl River Delta* (Guangzhou: December 2012), 81–88.

27. UN-Habitat, *Streets as Tools for Urban Transformation in Slums. A Street-led Approach to Citywide Slum Upgrading* (Nairobi: 2012), 15.

28. Julio D. Dávila and Diana Daste, "Aerial Cable-Cars in Medellín, Colombia: Social Inclusion and Reduced Emissions," in Mark Swilling et al., *City-Level Decoupling: Urban Resource Flows and the Governance of Infrastructure Transitions. Case Studies from Selected Cities.* A Report of the Working Group on Cities of the International Resource Panel (Paris: United Nations Environment Programme, 2013), 47–48.

29. Ibid.

30. Gehl, *Cities for People*, xii.

31. Link Arkitektur, "Stranden – Aker Brygge," http://linkarkitektur.com/en/Projects/Stranden-Aker-Brygge.

32. Justus Uitermark, "'Social Mixing' and the Management of Disadvantaged Neighbourhoods: The Dutch Policy of Urban Restructuring Revisited, *Urban Studies* 40, no. 3 (2003): 531–49.

City View: Durban, South Africa

1. Population data from eThekwini Municipality, *eThekwini Municipality Annual Report 2009–2010, Chapter 1: Mayor's Foreword and Executive Summary* (eThekwini Municipality, Durban: 2010).

2. eThekwini Municipality, *eThekwini Quality of Life Household Survey 2010–2011, A Survey of Municipal Services and Living Conditions* (eThekwini Municipality, Durban: 2011). Hotspots are areas with a high number of endemic species (i.e., more than 1,500 species of endemic vascular plants) and where at least 70 percent of the original habitat has been lost; see Conservation International, "Hotspots," www.conservation.org/How/Pages/Hotspots.aspx. Population from eThekwini Municipality, *eThekwini Municipality Annual Report 2009–2010*; poverty and unemployment from eThekwini Municipality, *Draft Economic Development and Job Creation Strategy 2012* (eThekwini Municipality, Durban: Economic Development and Investment Promotion Unit, 2012).

3. David Satterthwaite et al., "Adapting to Climate Change in Urban Areas: The Possibilities and Constraints in Low- and Middle-income Nations," in Jane Bicknell, David Dodman, and David Satterthwaite, eds., *Adapting Cities to Climate Change: Understanding and Addressing the Development Challenges* (London: Earthscan, 2009); Stanley W. Burgiel and Adrianna A. Muir, *Invasive Species, Climate Change and Ecosystem-based Adaptation: Addressing Multiple Drivers of Global Change* (Washington, DC and Nairobi: Global Invasive Species Programme, 2010).

4. Temperature projections from Golder Associates, *Community-based Adaptation to Climate Change in Durban*, report prepared for eThekwini Municipality (Durban: 2011); rainfall variations from eThekwini Municipality, *eThekwini Municipality Integrated Development Plan, Five-Year Plan: 2011–2016: 2011–2012 Plan* (eThekwini Municipality, Durban: 2011); disaster impacts from Golder Associates, *eThekwini Municipality Integrated Assessment Tool for Climate Change*, prepared for eThekwini Municipality (Durban: 2010).

5. Debra Roberts, "Thinking Globally, Acting Locally – Institutionalizing Climate Change at the Local Government Level in Durban, South Africa," *Environment & Urbanization* 20, no. 2 (2008): 521–37; Debra Roberts, "Prioritizing Climate Change Adaptation and Local Level Resiliency in Durban, South Africa," *Environment & Urbanization* 22, no. 2 (2010): 397–413; JoAnn Carmin, Debra Roberts, and Isabelle Anguelovski, *Planning Climate Resilient Cities: Early Lessons from Early Adapters*, prepared for the World Bank 5th Urban Research Symposium: Cities and Climate Change: Responding to an Urgent Agenda, Marseille, France, June 28–30, 2009.

6. Andrew A. Mather, Debra Roberts, and Geoffrey Tooley, "Adaptation in Practise: Durban, South Africa," in Konrad Otto-Zimmermann, ed., *Resilient Cities: Cities and Adaptation to Climate Change. Proceedings of the Global Forum 2010* (Dordrecht, The Netherlands: Springer, 2011), 543–63.

7. eThekwini Municipality, "Durban Metropolitan Open Space System FAQ," www.durban.gov.za/City_Services /development_planning_management/environmental_planning_climate_protection/Durban_Open_Space /Pages/MOSS_FAQ.aspx; 2013 area from Debra Roberts and Sean O'Donoghue, "Urban Environmental Challenges and Climate Change Action in Durban, South Africa," *Environment & Urbanization* 25, no. 2 (2013): 299–319.

8. Nicci Diederichs Mander and Debra Roberts, *Greening Durban 2010: Summary Review of the eThekwini Municipality's 2010 FIFA World Cup Event Greening Programme* (eThekwini Municipality, Durban: Environmental Planning and Climate Protection Department, 2010).

9. Greater Capital, *Social Assessment of the Buffelsdraai Landfill Site Community Reforestation Project*, prepared for the Wildlands Conservation Trust (Hilton, South Africa: 2011).

10. Durban CEBA website, www.durbanceba.org.

11. Josh Foster, Ashley Lowe, and Steve Winkelman, *The Value of Green Infrastructure for Urban Climate Adaptation* (Washington, DC: Center for Clean Air Policy, 2011).

12. Durban Adaptation Charter, "About the Charter," www.durbanadaptationcharter.org/about-the-charter, and "Signatories," www.durbanadaptationcharter.org/signatories.

Chapter 19. Urbanization, Inclusion, and Social Justice

1. United Nations Population Division, *World Urbanization Prospects: The 2014 Revision, Highlights* (New York: 2014); U.K. Government Office for Science, *Foresight, Migration and Global Environmental Change. Final Project Report* (London: 2011).

2. "Cyclone Nargis," Wikipedia, https://en.wikipedia.org/wiki/cyclone_nargis; Airah Cadiogan, "Two Years After Typhoon Haiyan, Leaders Have a Duty to Act on Climate Change," *The Guardian* (U.K.), November 8, 2015; "Super Typhoon Chan-hom Batters Chinese Coast, *Associated Press*, July 13, 2015.

3. Richard Stren, "Urban Service Delivery in Africa and the Role of International Assistance," *Development Policy Review* 32, no. 1 (2014): 19–37.

4. World Health Organization (WHO)/UNICEF Joint Monitoring Programme for Water Supply and Sanitation, *Progress on Drinking Water and Sanitation 2012* (Geneva and New York: 2012); Tanvi Misra, "How Urban Planning Failed Kathmandu," CityLab.com, April 26, 2015; Elisa Muzzini and Gabriela Aparicio, *Urban Growth and Spatial Transition in Nepal. An Initial Assessment* (Washington, DC: World Bank, 2013).

5. Gavin Shatkin, "The City and the Bottom Line: Urban Megaprojects and the Privatization of Planning in Southeast Asia," *Environment and Planning A* 40, no. 2 (2008): 383; World Bank, "Improved Sanitation Facilities, Urban (% of Urban Population with Access)," http://data.worldbank.org/indicator/SH.STA.ACSN.UR; Karen Bakker et al., *Disconnected: Poverty, Water Supply and Development in Jakarta, Indonesia*, Human Development Report Occasional Paper (Jakarta: United Nations Development Programme (UNDP), 2006); "Indonesia: Jakarta's Slums Struggle with Sanitation," *IRIN*, April 16, 2010.

6. David Satterthwaite, *Adapting to Climate Change in Urban Areas: The Possibilities and Constraints in Low- and Middle-Income Nations* (London: International Institute for Environment and Development (IIED), 2007); Richard Friend et al., "Mainstreaming Urban Climate Resilience into Policy and Planning; Reflections from Asia," *Urban Climate* 7 (March 2014): 6–19; Justus Kithiia, "Climate Change Risk Responses in East African Cities: Need, Barriers and Opportunities," *Current Opinion in Environmental Sustainability* 3, no. 3 (2011): 176–80; Helen Pidd, "India Blackouts Leave 700 Million Without Power," *The Guardian* (U.K.), July 31, 2012.

7. Robert McDonald et al., "Urban Growth, Climate Change, and Freshwater Availability," *Proceedings of the National Academy of Sciences* 108, no. 15 (2011): 6,312–17; WHO, "Chronic Respiratory Diseases, Deaths per

100 000. Data by Country," Global Health Observatory Data Repository, http://apps.who.int/gho/data/node.main .A866?lang=en; WHO, "WHO's Ambient Air Pollution Database – Update 2014" (Geneva: 2014); Castrol Magnatec, "Castrol Magnatec Stop-Start Index," http://interone2.azurewebsites.net/campaigns/stop-start-index.html.

8. David Harvey, "The Right to the City," *New Left Review* (September–October 2008): 23–40; A. Kim, "Changes Talking Back: The Role of Narrative in Vietnam's Recent Land Compensation," *Urban Studies* 48, no. 3 (2011): 493; Krisztina Kis-Katos and Günther G. Schulze, "Corruption in Southeast Asia: A Survey of Recent Research," *Asian-Pacific Economic Literature* 27, no. 1 (2013): 79–109; Thomas Tanner et al., *Urban Governance for Adaptation: Assessing Climate Change Resilience in Ten Asian Cities, Report to Rockefeller Foundation* (Sussex, U.K.: University of Sussex Institute of Development Studies, 2008).

9. Kis-Katos and Schulze, "Corruption in Southeast Asia"; Tanner et al., *Urban Governance for Adaptation*; Marcus Mietzner, "Dysfunction by Design: Political Finance and Corruption in Indonesia," *Critical Asian Studies* 47, no. 4 (2015): 587–610.

10. David Harvey, "From Managerialism to Entrepreneurialism: The Transformation in Urban Governance in Late Capitalism," *Geografiska Annaler. Series B, Human Geography* 71, no. 1 (2006): 3–17.

11. Shatkin, "The City and the Bottom Line," 383.

12. Ibid., 383; Harvey, "The Right to the City"; Willem Paling, "Planning a Future for Phnom Penh: Mega Projects, Aid Dependence and Disjointed Governance," *Urban Studies* 49, no. 13 (2012): 2,889–2,912.

13. Z/Yen Group Ltd., *Financing the Transition: Sustainable Infrastructure in Cities* (London: Long Finance and WWF, 2015); Siemens, PwC, and Berwin Leighton Paisner, *Investor Ready Cities: How Cities Can Create and Deliver Infrastructure Value* (London: 2014); Mike Douglass and Gavin Jones, "The Morphology of Mega-Urban Regions Expansion," in Gavin W. Jones and Mike Douglass, eds., *Mega-Urban Regions in Pacific Asia* (Singapore: NUS Press, 2008), 19–40; Friend et al., "Mainstreaming Urban Climate Resilience into Policy and Planning," 6–19.

14. Gordon McGranahan and David Satterthwaite, *Urbanisation Concepts and Trends* (London: IIED, 2014); Friend et al., "Mainstreaming Urban Climate Resilience into Policy and Planning"; James Jarvie et al., "Lessons for Africa from Urban Climate Change Resilience Building in Indonesia," *Current Opinion in Environmental Sustainability* 13 (April 2015): 19–24; UN-Habitat and United Nations Economic and Social Commission for Asia and the Pacific (UN ESCAP), *The State of Asian and Pacific Cities 2015* (Bangkok: 2015); Richard Stren, "Urban Service Delivery in Africa and the Role of International Assistance."

15. UN-Habitat and UN ESCAP, *The State of Asian and Pacific Cities 2015*; Sixth Asia-Pacific Urban Forum, "Chair's Summary" (Bangkok: 2015); UNDP, "Goal 11: Sustainable Cities and Communities," www.undp.org/content/undp/en/home/mdgoverview/post-2015-development-agenda/goal-11.html.

16. Habitat International Coalition website, www.hic-gs.org/index.php; Henri Lefebvre, "Le droit à la ville," *L'Homme et la Société* 6, no. 1 (1967): 29–35; UN, "Outcomes on Sustainable Development," www.un.org/en/development/devagenda/sustainable.shtml; Habitat III, "The New Urban Agenda. The Transformative Power of Urbanization," https://www.habitat3.org/the-new-urban-agenda; Michael Kane, "Statement of Habitat International Coalition Before the Second Session of the First Meeting of the Preparatory Committee for Habitat III," Habitat International Coalition, September 17, 2014; Global Platform for the Right to the City, "Call for the Inclusion of the Right to the City in the Habitat III Agenda" (São Paolo: Global Platform for the Right to the City, July 31, 2015).

17. Roanne Van Voorst and Rita Padawangi, "Floods and Forced Evictions in Jakarta," *New Mandala*, August 21, 2015; Owen Gibson and Jonathon Watts, "World Cup: Rio Favelas Being 'Socially Cleansed' in Runup to Sporting Events," *The Guardian* (U.K.), December 5, 2013; Owen Gibson and Pete Pattisson, "Death Toll Among Qatar's 2022 World Cup Workers Revealed," *The Guardian* (U.K.), December 23, 2014; Human Rights Watch, "Cambodia: Drop Restrictive Organizations Law. No Public Draft for Consultation Signals Government's Bad Faith," April 26, 2015; Seema Guha, "Crushing Dissent; NGOs Under Threat in India," OpenDemocracy.net, July 15, 2015;

UN-Habitat, *UN-Habitat Global Activities Report 2013. Our Presence and Partnerships* (Nairobi: 2013).

18. David Satterthwaite, "The Millennium Development Goals and Urban Poverty Reduction: Great Expectations and Nonsense Statistics," *Environment and Urbanization* 15, no. 2 (2003): 179–90; Diana Mitlin and David Satterthwaite, "How the Scale and Nature of Urban Poverty Are Under-estimated – The Limitations of the US $1 a Day Poverty Line," 2002, www.ucl.ac.uk/dpu-projects/drivers_urb_change/urb_society/pdf_liveli_vulnera /IIED_Mitlin_David_urban_poverty_under_estimated.pdf; Sabina Alkire and James Foster, "Understandings and Misunderstandings of Multidimensional Poverty Measurement," *Journal of Economic Inequality* 9, no. 2 (2011): 289–314; Gordon McGranahan and David Satterthwaite, *Urbanisation Concepts and Trends* (London: IIED, 2014); Robert Marshall, Nguyen Bui Linh, and Sarah Reed, "The Poor by Any Other Name," UNDP in Asia and the Pacific blog, July 24, 2015; UN-Habitat, "Habitat III," *Urban Visions,* no. 3 (2013); Joseph Schechla, "Fractured Continuity: Moving from Habitat II to Habitat III," Cityscope.org, August 7, 2015; Diana Mitlin and David Satterthwaite, *Urban Poverty in the Global South: Scale and Nature* (London and New York: Routledge, 2013).

19. Satterthwaite, "The Millennium Development Goals and Urban Poverty Reduction"; Mitlin and Satterthwaite, "How the Scale and Nature of Urban Poverty Are Under-estimated"; Alkire and Foster, "Understandings and Misunderstandings of Multidimensional Poverty Measurement"; McGranahan and Satterthwaite, *Urbanisation Concepts and Trends*; Marshall, Bui Linh, and Reed, "The Poor by Any Other Name"; UN-Habitat, "Habitat III"; Schechla, "Fractured Continuity"; Mitlin and Satterthwaite, *Urban Poverty in the Global South*; Richard Friend and Marcus Moench, "Rights to Urban Climate Resilience: Moving Beyond Poverty and Vulnerability," *Wiley Interdisciplinary Reviews: Climate Change* 6, no. 6 (2015), 643–51.

20. Shruti Ravindran, "Is India's 100 Smart Cities Project a Recipe for Social Apartheid," *The Guardian* (U.K.), May 7, 2015.

21. "The Great Divide," *The Economist*, September 15, 2012.

22. Asian Cities Climate Change Resilience Network website, http://acccrn.net.

23. Richard Friend and Marcus Moench, "What Is the Purpose of Urban Climate Resilience? Implications for Addressing Poverty and Vulnerability?" *Urban Climate* 6 (December 2013): 98–113.

Index